Metal Oxide Semiconductors

Metal Oxide Semiconductors

Synthesis, Properties, and Devices

Zhigang Zang
Wensi Cai
Yong Zhou

WILEY-VCH

The Authors

Prof. Zhigang Zang
Chongqing University
174 Shazheng Street
1206 Main Building, Campus A
Chongqing
China
400044

Dr. Wensi Cai
Chongqing University
174 Shazheng Street
1410 Main Building, Campus A
Chongqing
China
400044

Prof. Yong Zhou
Chongqing University
174 Shazheng Street
1412 Main Building, Campus A
Chongqing
China
400044

Cover Image: © Yuichiro Chino/Getty Images

Library of Congress Card No.: applied for

British Library Cataloguing-in-Publication Data
A catalogue record for this book is available from the British Library.

Bibliographic information published by the Deutsche Nationalbibliothek
The Deutsche Nationalbibliothek lists this publication in the Deutsche Nationalbibliografie; detailed bibliographic data are available on the Internet at <http://dnb.d-nb.de>.

Print ISBN: 978-3-527-35225-8
ePDF ISBN: 978-3-527-84254-4
ePub ISBN: 978-3-527-84256-8
oBook ISBN: 978-3-527-84255-1

Typesetting Straive, Chennai, India
Printing and Binding CPI Group (UK) Ltd, Croydon, CR0 4YY

C9783527352258_301123

Contents

Preface

Electronics are undergoing a disruptive evolution, transitioning from heavy, rigid, and bulky devices to lightweight, soft, and flexible ones. This shift has given rise to emerging applications such as epidermal devices, artificial skins, and transparent displays, which have the potential to revolutionize our daily lives. However, traditional semiconductors like silicon (Si) face challenges in meeting the requirements of flexibility, lightweightness, transparency, and biocompatibility for these new applications.

In recent years, metal oxide semiconductors have emerged as a novel class of semiconductors and have garnered significant attention. Compared to traditional Si-based technologies, oxide semiconductors offer unique advantages such as high transparency and flexibility, making them highly promising for next-generation wearable and transparent electronics. Currently, thin-film transistors (TFTs) with an InGaZnO channel layer have already been commercialized for driving high-definition and energy-saving LCDs in smartphones, tablets, PC monitors, and OLED TVs. Additionally, oxide semiconductors are being studied for applications in diodes, gas sensors, solar cells, photodetectors, and memory devices, challenging conventional semiconductor technologies.

Despite rapid advancements in the materials science and device physics of oxide semiconductors over the past decade, the understanding of science and technology in this field remains incomplete due to its relatively short research history and the differences in chemical bonding between oxides and covalent-type semiconductors. Nevertheless, oxide semiconductors hold significant promise, particularly in the backplane of advanced displays and gas sensors, with applications in solar cells, photodetectors, and memory devices on the horizon. Regrettably, no comprehensive monograph has been published to date that covers all aspects of oxide semiconductors. This book aims to bridge the gap between the rapidly advancing research progress in this field and the demand for relevant materials and devices by researchers, engineers, and students.

The monograph provides a current understanding of oxide semiconductors and their applications in electronics and optoelectronics. It covers fundamentals, synthesizing methods, and applications in diodes, TFTs, gas sensors, solar cells, UV detectors, and memories. We present state-of-the-art information along with fundamental prerequisites for understanding, discuss the current challenges in pursuing commercialization, and provide our insights to guide the future directions of this field.

We express our gratitude to our colleagues at Chongqing University for their assistance in organizing the seminar and editing this book. Special thanks go to Dr. Huaxin Wang for editing Chapter 6 and Dr. Shuangyi Zhao for editing Chapter 7. We also extend our appreciation to the editor of Wiley for inviting us to write this book, and we are delighted to have the opportunity to publish this book on the novel class of metal oxide semiconductors.

Zhigang Zang
Wensi Cai
Yong Zhou

1

Metal Oxide Semiconductors: State-of-the-Art and New Challenges

1.1 Introduction

Metal oxide semiconductors (MOS) are abundant materials found in the Earth's crust and are commonly used in traditional ceramics. However, they differ significantly from conventional inorganic counterparts like silicon and III–V compounds in various aspects. These distinctions encompass materials design concepts, electronic structure, charge transport mechanisms, defect states, thin-film processing, and optoelectronic properties. As a result, oxide semiconductors enable the realization of both established and innovative functionalities [1].

In comparison to inorganic semiconductors, oxide semiconductors possess unique characteristics. These include exceptional carrier mobilities even in the amorphous state, resilience against mechanical stress, compatibility with organic dielectric and photoactive materials, and high optical transparency. These properties make oxide semiconductors particularly appealing for various applications.

In recent decades, MOS have garnered significant attention in various research fields, including optoelectronics, thin-film transistors (TFTs), photocatalysts, gas sensors, solar cells, and memristors [1–14], as shown in Figure 1.1. MOS can be categorized into two types based on their conductivity: *n*-type, where electrons are the majority carriers, and *p*-type, where holes are the majority carriers. These semiconducting properties arise from factors such as doped aliovalent cations or oxygen nonstoichiometry [15, 16]. Among the *n*-type MOS, ZnO, SnO_2, TiO_2, In_2O_3, and Ga_2O_3 are widely studied concerning synthesis, characterization, and applications [17–21]. As for the *p*-type MOS, research efforts primarily focus on Cu_xO (CuO and Cu_2O), SnO, and NiO_x [22–24].

1.2 *n*-Type Metal Oxide Semiconductors

1.2.1 ZnO

Zinc oxide (ZnO), as an *n*-type semiconductor, has sparked significant research interest due to its distinctive physical and chemical properties [25–27]. In the field of materials science, ZnO is classified as a II–VI compound semiconductor, possessing

Metal Oxide Semiconductors: Synthesis, Properties, and Devices,
First Edition. Zhigang Zang, Wensi Cai, and Yong Zhou.

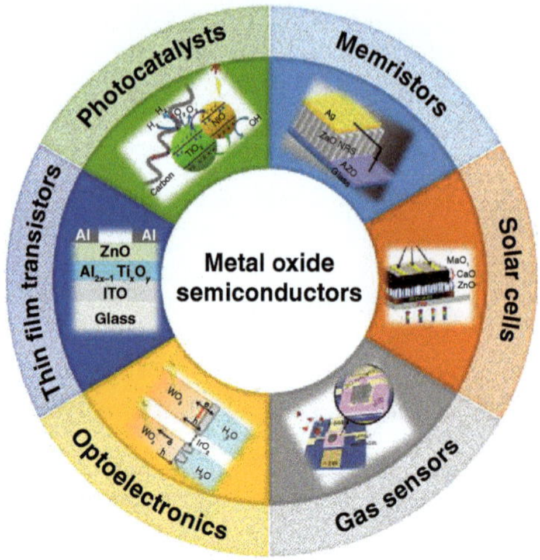

Figure 1.1 Applications of metal oxide semiconductors. Source: Reprinted with permission from Refs. [9–14]. Copyright 2015 American Chemical Society, 2016 American Chemical Society, 2020 American Chemical Society, 2020 The Royal Society of Chemistry.

a covalence level between ionic and covalent semiconductors. Its attributes, such as a direct wide bandgap ($E_g \sim 3.3$ eV at 300 K), large free exciton binding energy (60 meV), and high thermal and mechanical stability at room temperature, make ZnO a promising candidate for various applications in electronic devices, optoelectronics, gas sensors, and laser technology [28, 29]. Additionally, ZnO can be utilized as an energy collector due to its piezo- and pyroelectric properties, as well as a photocatalyst for hydrogen production [30–32].

The crystal structure of ZnO can be classified into three types: wurtzite (B4), zinc blende (B3), and rocksalt (B1), as shown in Figure 1.2. Thereinto, wurtzite is the most stable thermodynamic phase under ambient conditions. Wurtzite ZnO possesses a hexagonal structure (space group *C6mc*) with lattice parameters $a = 0.3296$ nm and $c = 0.52065$ nm. The structure of ZnO can be simply described as a number of alternating planes composed of tetrahedrally coordinated O^{2-} and Zn^{2+} ions, stacked alternately along the c-axis. The tetrahedral coordination in ZnO results in noncentral symmetric structures and consequently piezoelectricity and pyroelectricity [33].

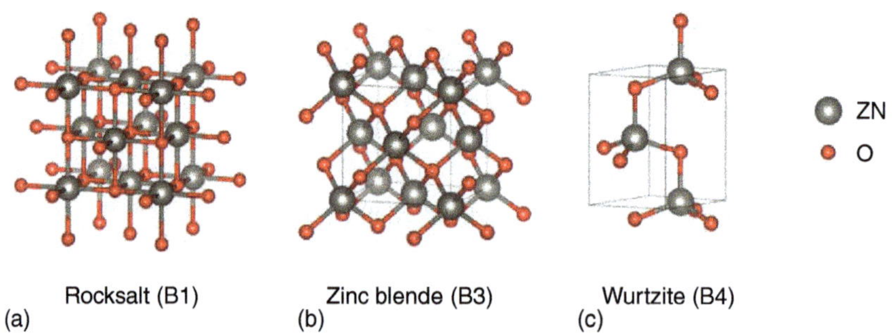

Figure 1.2 Stick and ball representation of ZnO crystal structures: (a) cubic rocksalt (B1), (b) cubic zinc blende (B3), and (c) hexagonal wurzite (B4). The gray and red spheres denote Zn and O atoms, respectively.

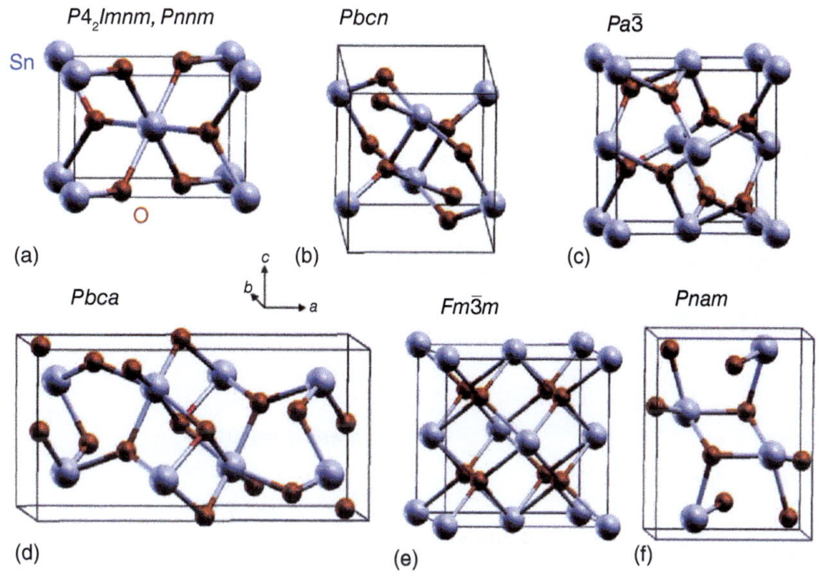

Figure 1.3 Bulk structures of SnO_2 polymorphs (gray and red colors represent Sn and O atoms, respectively). (a) Rutile (*P4$_2$/mnm*) and $CaCl_2$-type (*Pnnm*), (b) R-PbO_2-type (*Pbcn*), (c) pyrite-type (*Pa$\bar{3}$*), (d) ZrO_2-type (*Pbca*), (e) fluorite-type (*Fm$\bar{3}$hm*), and (f) cotunnite-type (*Pnam*). Source: Reproduced from Gracia et al. [40]/with permission of American Chemical Society.

1.2.2 SnO_2

SnO_2 is also a wide-bandgap (~3.6 eV) *n*-type semiconductor, belongs to the group-IV compounds, and exhibits remarkable transparency and conductivity simultaneously [20, 34, 35]. The unique chemical, electronic, and optical properties of SnO_2 have led to extensive research on its applications in various devices, including solar cells [36], catalytic materials [37], and gas sensors [38]. Additionally, SnO_2 serves as a distinctive transparent metal oxide and finds wide-ranging applications as transparent conducting oxide electrodes in optoelectronic devices. This is due to its excellent chemical and thermal stabilities in atmospheric environments, as well as its high optical transmission properties [39].

SnO_2 possesses several polymorphs including rutile-type (*P4$_2$/mnm*), $CaCl_2$-type (*Pnnm*), a-PbO_2-type (*Pbcn*), pyrite-type (*Pa$\bar{3}$*), ZrO_2-type orthorhombic phase I (*Pbca*), fluorite-type (*Fm$\bar{3}$m*), and cotunnite-type orthorhombic phase II (*Pnam*) with ninefold coordination, as shown in Figure 1.3. All these structures are sequentially obtained when the most commonly available and stable rutile phase is subjected to a high mechanical pressure [40].

1.2.3 In_2O_3

Another widely studied *n*-type semiconductor is indium oxide (In_2O_3), which has a bandgap ranging from 3.5 to 3.7 eV. It finds numerous applications in electronic

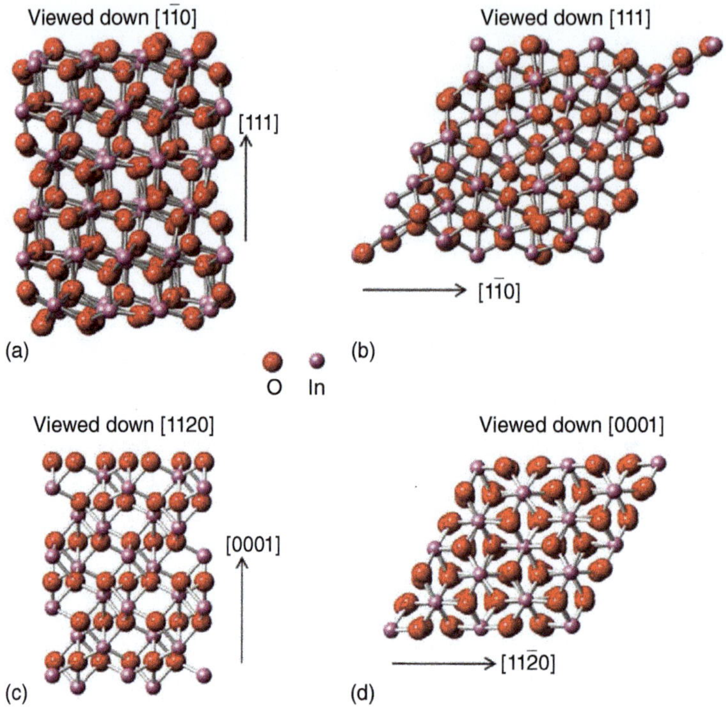

Figure 1.4 Ball and stick representations of crystal structures of *bcc*-In_2O_3 (a, b) and *rh*-In_2O_3 (c, d). In atoms are pale pink, and O atoms are dark red. The viewing directions and the in-plane orientation are indicated in the figure. Source: Reproduced from Zhang et al. [42]/with permission of American Chemical Society.

and optoelectronic fields such as solar cells, gas sensors based on TFTs, and Schottky contacts and diodes [21, 41]. In_2O_3 can exist in two well-established crystal structures: body-centered cubic (*bcc*) and rhombohedral (*rh*), as depicted in Figure 1.4 [42]. The phase of In_2O_3 thermodynamically stable under ambient conditions adopts the body-centered cubic bixbyite structure, with the space group $Ia\bar{3}$ (#206) and a lattice constant of 10.118 Å. The rhombohedral structure is stabilized under high-pressure conditions. The rhombohedral cell belongs to the $R\bar{3}c$ space group with lattice constants $a = 5.478$ Å and $c = 14.51$ Å. There are six formula units per hexagonal cell, and the volume per formula of 62.85 $Å^3$ for the *rh* phase is much smaller than the value of 64.72 $Å^3$ for the ambient *bcc* phase.

1.2.4 TiO_2

As a member of transition metal oxides, TiO_2 is a well-known *n*-type semiconductor [43–46]. It exhibits excellent electronic and optical properties, making it highly suitable for various applications in the fields of gas sensors [47], solar cells [48], and photocatalysis [49]. In its natural form, TiO_2 exists in three different

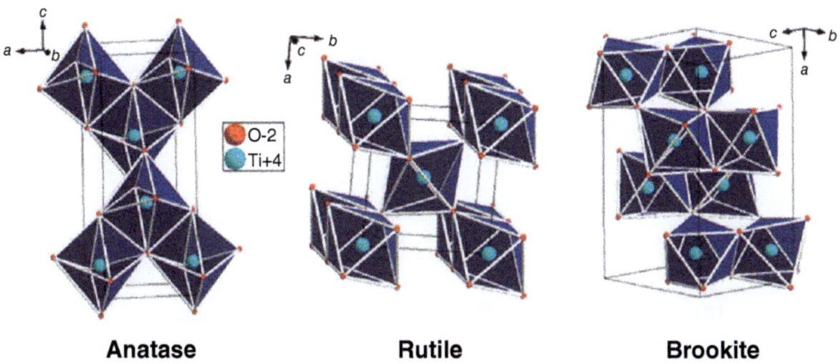

Figure 1.5 Structure of anatase, rutile, and brookie TiO_2. Source: Reproduced from Macwan et al. [51]/with permission of American Chemical Society.

phase structures, namely anatase (tetragonal), brookite (orthorhombic), and rutile (tetragonal), as illustrated in Figure 1.5. These phases have energy bandgaps of 3.2 eV (anatase), 3.02 eV (brookite), and 2.96 eV (rutile) [50, 51]. Among these phases, anatase and rutile are widely applied due to their superior stability, while rutile TiO_2, with a tetragonal structure containing six atoms per unit cell, exhibits a slight distortion in the TiO_6 octahedron [52]. Anatase TiO_2 also possesses a tetragonal structure, but with a slightly larger distortion of the TiO_6 octahedron. In general, rutile TiO_2 is more thermodynamically stable than anatase under typical temperature and pressure conditions.

1.2.5 Ga_2O_3

Ga_2O_3 is an emerging ultra-wide bandgap semiconductor with a bandgap of 4.8 eV. This unique property enables Ga_2O_3 to simultaneously achieve high breakdown voltage and low on-resistance, making it highly promising for both direct current (DC) and radiofrequency (RF) applications [53, 54]. As a result, Ga_2O_3 offers exciting prospects in various fields such as electronics (high-power devices, field-effect transistors) [55], optoelectronics (solar cells, solar-blind ultraviolet photodetectors) [56–58], and sensors (gas sensors, radiation detectors) [59]. Ga_2O_3 exists in five polymorphs, including corundum (α), monoclinic (β), defective spinel (γ), orthorhombic (δ), and hexagonal (ε), as shown in Figure 1.6 [53]. Among these, β-Ga_2O_3 is the most stable phase under ambient conditions, while the other phases can transform into β-Ga_2O_3 through heating treatment [60].

1.3 *p*-Type Metal Oxide Semiconductors

1.3.1 Copper Oxides (CuO/Cu_2O)

Tenorite (CuO) and cuprite (Cu_2O) are two stable phases in copper oxides at an ambient environment. As *p*-type semiconductors, CuO and Cu_2O separately possess a band gap of 1.9–2.1 and 2.1–2.6 eV [61]. Both binary copper oxides are

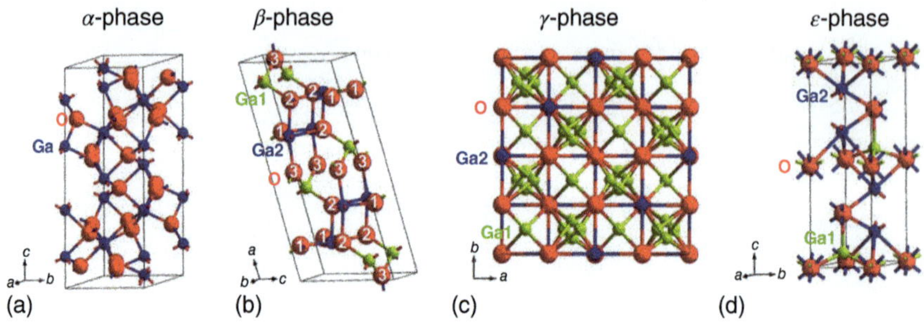

Figure 1.6 Crystal structure Ga_2O_3 polymorphs. (a) Rhombohedral crystal structure of corundum-like α-Ga_2O_3. (b) Monoclinic β-Ga_2O_3 crystal structure. (c) Cubic defective spinel lattice structure of γ-phase Ga_2O_3. (d) Orthorhombic ε-Ga_2O_3 structure. Source: Reproduced from Zhang et al. [53]/with permission of AIP Publishing.

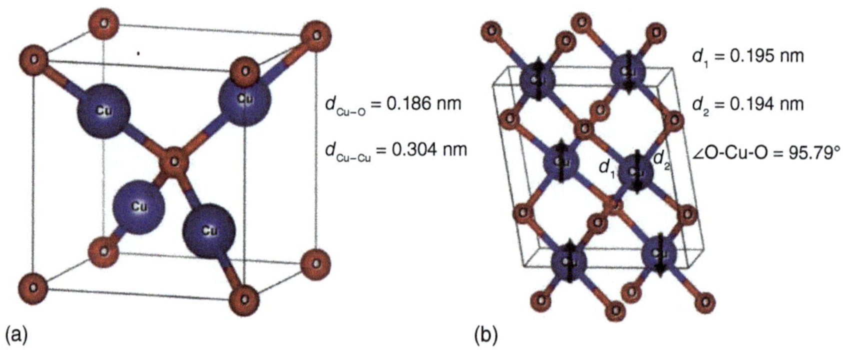

Figure 1.7 Crystal structure of binary copper oxides. (a) Cubic Cu_2O and (b) Monoclinic CuO. Source: Reproduced from Gupta et al. [61]/with permission of Elsevier.

attractive for versatile applications such as solar cells, TFTs, gas sensors, and photoelectric detectors [62–67]. When annealed at a high temperature, Cu_2O could transform to CuO by the oxidation reaction [63]. In addition to these two stable phases, paramelaconite (Cu_4O_3) is another metastable mixed-valence intermediate compound between Cu_2O and CuO and could decompose into Cu_2O and CuO beyond the stability limit under vacuum environment at 670–800 K. Figure 1.7 illustrates these crystal structures of Cu_2O and CuO [61].

1.3.2 SnO

Recently, tin monoxide (SnO) has garnered significant attention due to its inherent *p*-type conductivity, structural stability, and electric properties in ambient environment [68]. The bandgap of SnO ranges from 2.2 to 3.0 eV at 300 K [69]. Figure 1.8 illustrates the crystal structure of SnO and the hybridization of valence-band maximum (VBM) [62]. It is worth noting that the energy level of Sn 5*s* is in close proximity to that of O 2*p* near the VBM, as well as Sn 5*p* near the conduction band. This arrangement effectively reduces hole localization and results in high hole

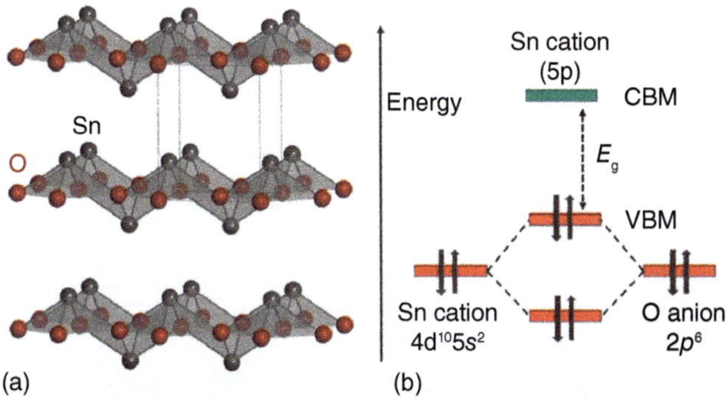

Figure 1.8 (a) Crystal structure of SnO. (b) Schematic illustration of the valence-band maximum (VBM) hybridization in SnO. Source: Reproduced from Wang et al. [62]/with permission of Elsevier.

mobility. The presence of Sn vacancies and oxygen interstitials contributes to the *p*-type conductivity observed in SnO [70]. The native *p*-type behavior of SnO makes it suitable for various applications in the fields of TFTs [71], gas sensors [72], and optoelectronics [69].

1.3.3 NiO_x

NiO_x is a widely studied *p*-type semiconductor that exhibits a green color in ambient environments and possesses a wide bandgap ranging from 3.6 to 4.0 eV [73]. Due to its exceptional thermodynamic stability and unique optical/chemical properties [74, 75], NiO_x has attracted significant research interest in various fields, including solar cells [76], supercapacitors [77], catalysis [78], electrodes for lithium-ion batteries [74], and gas sensors [79].

Figure 1.9a illustrates the crystal structure of NiO. Nalage et al. proposed a sol–gel synthesis method for NiO thin films [80]. In this typical procedure, nickel acetate serves as the source of Ni, and the flow diagram for the production of NiO films is depicted in Figure 1.9b. The NiO films prepared through this process and subsequently sintered at 700 °C exhibited an electron carrier concentration and mobility of 3.75×10^{19} cm^{-3} and 4.2×10^{-5} cm^2 V^{-1}s^{-1}, respectively, along with a bandgap of 3.47 eV.

In various application scenarios like TFTs, photocatalysts, gas sensors, solar cells, and memristors, the morphological and structural characteristics of metal oxides play a crucial role in determining their performance. Factors such as crystal phases, terminal atom species, crystal size, shape, defects, vacancies, and crystal planes significantly influence the ultimate performance of these materials. Therefore, depending on the desired performance requirements, it is possible to design and improve the synthesis parameters while adhering to a fixed preparation strategy. This approach allows us to obtain the desired metal oxide with specific properties, albeit with some compromises between performance merits such as sensitivity and stability.

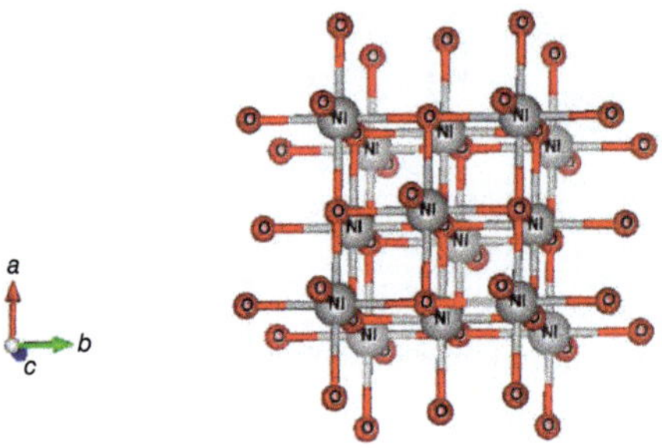

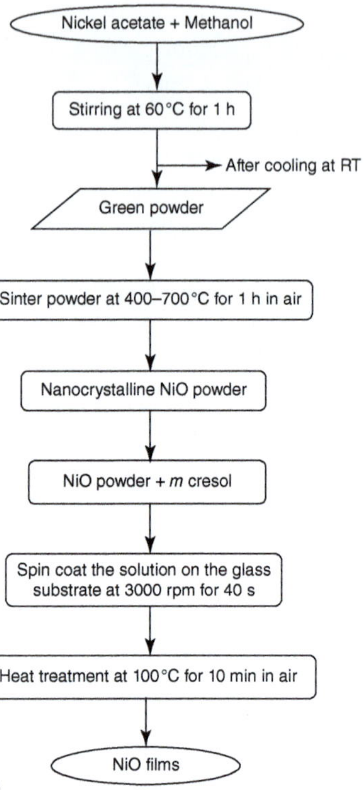

Figure 1.9 (a) Crystal structure of NiO_x. Source: Reproduced from Goel et al. [75]/with permission of Elsevier. Copyright 2020, Elsevier. (b) Flow diagram for NiO films synthesis by sol–gel method. Source: Reproduced from Nalage et al. [80]/with permission of Elsevier.

References

1 Yu, X., Marks, T., and Facchetti, A. (2016). Metal oxides for optoelectronic applications. *Nat. Mater.* 15 (4): 383–396.

2 Karthikeyan, C., Arunachalam, P., Ramachandran, K. et al. (2020). Recent advances in semiconductor metal oxides with enhanced methods for solar photocatalytic applications. *J. Alloys Compd.* 828: 154281.

3 Valdes, A., Brillet, J., Gratzel, M. et al. (2012). Solar hydrogen production with semiconductor metal oxides: new directions in experiment and theory. *Phys. Chem. Chem. Phys.* 14 (1): 49–70.

4 Khan, M., Adil, S., and Al-Mayouf, A. (2015). Metal oxides as photocatalysts. *J. Saudi Chem. Soc.* 19 (5): 462–464.

5 Jose, R., Thavasi, V., and Ramakrishna, S. (2009). Metal oxides for dye-sensitized solar cells. *J. Am. Ceram. Soc.* 92 (2): 289–301.

6 Park, J., Maeng, W., Kim, H. et al. (2012). Review of recent developments in amorphous oxide semiconductor thin-film transistor devices. *Thin Solid Films* 520 (6): 1679–1693.

7 Mohammad, B., Jaoude, M., Kumar, V. et al. (2016). State of the art of metal oxide memristor devices. *Nanotechnol. Rev.* 5 (3): 311–329.

8 Barsan, N., Koziej, D., and Weimar, U. (2007). Metal oxide-based gas sensor research: how to? *Sens. Actuators, B* 121 (1): 18–35.

9 Ping, Y., Goddard, W., and Galli, G. (2015). Energetics and solvation effects at the photoanode/catalyst interface: ohmic contact versus schottky barrier. *J. Am. Chem. Soc.* 137 (16): 5264–5267.

10 Sun, Y., Yan, X., Zheng, X. et al. (2015). High on–off ratio improvement of ZnO-based forming-free memristor by surface hydrogen annealing. *ACS Appl. Mater. Interfaces* 7 (13): 7382–7388.

11 Afouxenidis, D., Mazzocco, R., Vourlias, G. et al. (2015). ZnO-based thin film transistors employing aluminum titanate gate dielectrics deposited by spray pyrolysis at ambient air. *ACS Appl. Mater. Interfaces* 7 (13): 7334–7341.

12 Wang, M., Han, J., Hu, Y. et al. (2016). Carbon-incorporated NiO/TiO_2 mesoporous shells with p-n heterojunctions for efficient visible light photocatalysis. *ACS Appl. Mater. Interfaces* 8 (43): 29511–29521.

13 Cho, I., Sim, Y., Cho, M. et al. (2020). Monolithic micro light-emitting diode/metal oxide nanowire gas sensor with microwatt-level power consumption. *ACS Sens.* 5 (2): 563–570.

14 Kaphle, A., Echeverria, E., McLlroy, D. et al. (2020). Enhancement in the performance of nanostructured CuO-ZnO solar cells by band alignment. *RSC Adv.* 10 (13): 7839–7854.

15 Szczuko, D., Werner, J., Oswald, S. et al. (2001). XPS investigations of surface segregation of doping elements in SnO_2. *Appl. Surf. Sci.* 179 (1–4): 301–306.

16 Kılıç, Ç. and Zunger, A. (2002). Origins of coexistence of conductivity and transparency in SnO_2. *Phys. Rev. Lett.* 88 (9): 095501.

17 Guo, D., Guo, Q., Chen, Z. et al. (2019). Review of Ga_2O_3-based optoelectronic devices. *Mater. Today Phys.* 11: 100157.

18 Hashimoto, K., Irie, H., and Fujishima, A. (2005). TiO_2 photocatalysis: a historical overview and future prospects. *Jpn. J. Appl. Phys.* 44 (12R): 8269–8285.

19 Klingshirn, C. (2007). ZnO: material, physics and applications. *ChemPhysChem* 8 (6): 782–803.

20 Wang, H. and Rogach, A. (2014). Hierarchical SnO_2 nanostructures: recent advances in design, synthesis, and applications. *Chem. Mater.* 26 (1): 123–133.

21 Lim, S., Hwang, S., Chang, D. et al. (2010). Preparation of mesoporous In_2O_3 nanofibers by electrospinning and their application as a CO gas sensor. *Sens. Actuators, B* 149 (1): 28–33.

22 Kastner, M., Birgeneau, R., Shirane, G. et al. (1998). Magnetic, transport, and optical properties of monolayer copper oxides. *Rev. Mod. Phys.* 70 (3): 897–928.

23 Saji, K., Venkata, S.Y., Tian, K. et al. (2016). P-type SnO thin films and SnO/ZnO heterostructures for all-oxide electronic and optoelectronic device applications. *Thin Solid Films* 605: 193–201.

24 Silva, V., Simões, T., Grilo, J. et al. (2020). Impact of the NiO nanostructure morphology on the oxygen evolution reaction catalysis. *J. Mater. Sci.* 55: 6648–6659.

25 Huang, F., Lin, Z., Lin, W. et al. (2014). Research progress in ZnO single-crystal: growth, scientific understanding, and device applications. *Chin. Sci. Bull.* 59: 1235–1250.

26 Wang, Z. (2004). Zinc oxide nanostructures: growth, properties and applications. *J. Phys. Condens. Matter* 16 (25): 829–858.

27 Janotti, A. and Van de Walle, C. (2009). Fundamentals of zinc oxide as a semiconductor. *Rep. Prog. Phys.* 72 (12): 126501.

28 Wang, Z. (2007). Novel nanostructures of ZnO for nanoscale photonics, optoelectronics, piezoelectricity, and sensing. *Appl. Phys. A* 88: 7–15.

29 Harun, K., Hussain, F., Purwanto, A. et al. (2017). Sol–gel synthesized ZnO for optoelectronics applications: a characterization review. *Mater. Res. Express* 4 (12): 122001.

30 Tian, C., Zhang, Q., Wu, A. et al. (2012). Cost-effective large-scale synthesis of ZnO photocatalyst with excellent performance for dye photodegradation. *Chem. Commun.* 48 (23): 2858–2860.

31 Hsiao, C., Hu, Y., Chang, R. et al. (2009). Residual stresses and mechanical properties of a ZnO pyroelectric sensor. *Theor. Appl. Fract. Mech.* 52 (1): 1–6.

32 Dhiman, P., Rana, G., Kumar, A. et al. (2022). ZnO-based heterostructures as photocatalysts for hydrogen generation and depollution: a review. *Environ. Chem. Lett.* 20: 1047–1081.

33 Yi, G., Wang, C., and Park, W. (2005). ZnO nanorods: synthesis, characterization and applications. *Semicond. Sci. Technol.* 20 (4): 22.

34 Akgul, F., Gumus, C., Er, A. et al. (2013). Structural and electronic properties of SnO_2. *J. Alloys Compd.* 579: 50–56.

35 Das, S. and Jayaraman, V. (2014). SnO_2: a comprehensive review on structures and gas sensors. *Prog. Mater Sci.* 66: 112–255.

36 El-Etre, A. and Reda, S. (2010). Characterization of nanocrystalline SnO_2 thin film fabricated by electrodeposition method for dye-sensitized solar cell application. *Appl. Surf. Sci.* 256 (22): 6601–6606.

37 Kar, A., Sain, S., Kundu, S. et al. (2015). Influence of size and shape on the photocatalytic properties of SnO_2 nanocrystals. *ChemPhysChem* 16 (5): 1017–1025.

38 Srivastava, A. (2003). Detection of volatile organic compounds (VOCs) using SnO_2 gas-sensor array and artificial neural network. *Sens. Actuators, B* 96 (1–2): 24–37.

39 Yadava, Y., Denicoló, G., Arias, A. et al. (1997). Preparation and characterization of transparent conducting tin oxide thin film electrodes by chemical vapour deposition from reactive thermal evaporation of $SnCl_2$. *Mater. Chem. Phys.* 48 (3): 263–267.

40 Gracia, L., Beltrán, A., and Andrés, J. (2007). Characterization of the high-pressure structures and phase transformations in SnO_2. A density functional theory study. *J. Phys. Chem. B* 111 (23): 6479–6485.

41 Spencer, J., Mock, A., Jacobs, A. et al. (2022). A review of band structure and material properties of transparent conducting and semiconducting oxides: Ga_2O_3, Al_2O_3, In_2O_3, ZnO, SnO_2, CdO, NiO, CuO, and Sc_2O_3. *Appl. Phys. Rev.* 9 (1): 011315.

42 Zhang, K., Lazarov, V., Galindo, P. et al. (2012). Domain matching epitaxial growth of In_2O_3 thin films on α-Al_2O_3(0001). *Cryst. Growth Des.* 12 (2): 1000–1007.

43 Roy, P., Berger, S., and Schmuki, P. (2011). TiO_2 nanotubes: synthesis and applications. *Angew. Chem. Int. Ed. Engl.* 50 (13): 2904–2939.

44 Chen, X. and Selloni, A. (2014). Introduction: titanium dioxide (TiO_2) nanomaterials. *Chem. Rev.* 114 (19): 9281–9282.

45 Siegel, R., Ramasamy, S., Hahn, H. et al. (2011). Synthesis, characterization, and properties of nanophase TiO_2. *J. Mater. Res.* 3 (6): 1367–1372.

46 Kusior, A., Banas, J., Trenczek-Zajac, A. et al. (2018). Structural properties of TiO_2 nanomaterials. *J. Mol. Struct.* 1157: 327–336.

47 Garzella, C., Comini, E., Tempesti, E. et al. (2000). TiO_2 thin films by a novel sol–gel processing for gas sensor applications. *Sens. Actuators, B* 68 (1–3): 189–196.

48 Wang, Z., Yamaguchi, T., Sugihara, H. et al. (2005). Significant efficiency improvement of the black dye-sensitized solar cell through protonation of TiO_2 films. *Langmuir* 21 (10): 4272–4276.

49 Nakata, K. and Fujishima, A. (2012). TiO_2 photocatalysis: design and applications. *J. Photochem. Photobiol. C* 13 (3): 169–189.

50 Bai, J. and Zhou, B. (2014). Titanium dioxide nanomaterials for sensor applications. *Chem. Rev.* 114 (19): 10131–10176.

51 Macwan, D., Dave, P., and Chaturvedi, S. (2011). A review on nano-TiO_2 sol–gel type syntheses and its applications. *J. Mater. Sci.* 46: 3669–3686.

52 Linsebigler, A., Lu, G., and Yates, J. (1995). Photocatalysis on TiO_2 surfaces: principles, mechanisms, and selected results. *Chem. Rev.* 95 (3): 735–758.

53 Zhang, J., Shi, J., Qi, D. et al. (2020). Recent progress on the electronic structure, defect, and doping properties of Ga_2O_3. *APL Mater.* 8 (2): 020906.

54 Pearton, S., Yang, J., Cary, P. et al. (2018). A review of Ga_2O_3 materials, processing, and devices. *Appl. Phys. Rev.* 5 (1): 011301.

55 Higashiwaki, M., Sasaki, K., Murakami, H. et al. (2016). Recent progress in Ga_2O_3 power devices. *Semicond. Sci. Technol.* 31 (3): 034001.

56 Zhang, M., Kang, S., Wang, L. et al. (2021). Facile synthesis of β–Ga_2O_3 nanowires network for solar-blind ultraviolet photodetector. *J. Phys. D: Appl. Phys.* 54 (17): 175106.

57 Ping, L., Berhanuddin, D., Mondal, A. et al. (2021). Properties and perspectives of ultrawide bandgap Ga_2O_3 in optoelectronic applications. *Chin. J. Phys.* 73: 195–212.

58 Galazka, Z. (2018). β-Ga_2O_3 for wide-bandgap electronics and optoelectronics. *Semicond. Sci. Technol.* 33 (11): 113001.

59 Liang, H., Cui, S., Su, R. et al. (2018). Flexible X-ray detectors based on amorphous Ga_2O_3 thin films. *ACS Photonics* 6 (2): 351–359.

60 Wang, C., Zhang, J., Xu, S. et al. (2021). Progress in state-of-the-art technologies of Ga_2O_3 devices. *J. Phys. D: Appl. Phys.* 54 (24): 243001.

61 Gupta, D., Meher, S., Illyaskutty, N. et al. (2018). Facile synthesis of Cu_2O and CuO nanoparticles and study of their structural, optical and electronic properties. *J. Alloys Compd.* 743: 737–745.

62 Wang, Z., Nayak, P., Caraveo-Frescas, J. et al. (2016). Recent developments in p-type oxide semiconductor materials and devices. *Adv. Mater.* 28 (20): 3831–3892.

63 Al-Jawhari, H. (2015). A review of recent advances in transparent p-type Cu_2O-based thin film transistors. *Mater. Sci. Semicond. Process.* 40: 241–252.

64 Teng, F., Hu, K., Ouyang, W. et al. (2018). Photoelectric detectors based on inorganic p-type semiconductor materials. *Adv. Mater.* 30 (35): 1706262.

65 Kim, H. and Lee, J. (2014). Highly sensitive and selective gas sensors using p-type oxide semiconductors: overview. *Sens. Actuators, B* 192: 607–627.

66 Zhang, Q., Zhang, K., Xu, D. et al. (2014). CuO nanostructures: synthesis, characterization, growth mechanisms, fundamental properties, and applications. *Prog. Mater Sci.* 60: 208–337.

67 Grigore, M., Biscu, E., Holban, A. et al. (2016). Methods of synthesis, properties and biomedical applications of CuO nanoparticles. *Pharmaceuticals* 9 (4): 75.

68 Allen, J., Scanlon, D., Piper, L. et al. (2013). Understanding the defect chemistry of tin monoxide. *J. Mater. Chem. C.* 1 (48): 8194–8208.

69 Singh, M., Gaspera, E., Ahmed, T. et al. (2017). Soft exfoliation of 2D SnO with size-dependent optical properties. *2D Mater.* 4 (2): 025110.

70 Eqbal, E. and Anila, E. (2018). Properties of transparent conducting tin monoxide(SnO) thin films prepared by chemical spray pyrolysis method. *Physica B* 528: 60–65.

71 Yabuta, H., Kaji, N., Hayashi, R. et al. (2010). Sputtering formation of p-type SnO thin-film transistors on glass toward oxide complimentary circuits. *Appl. Phys. Lett.* 97 (7): 072111.

72 Suman, P., Felix, A., Tuller, H. et al. (2015). Comparative gas sensor response of SnO_2, SnO and Sn_3O_4 nanobelts to NO_2 and potential interferents. *Sens. Actuators, B* 208: 122–127.

73 Mokoena, T., Swart, H., and Motaung, D. (2019). A review on recent progress of p-type nickel oxide based gas sensors: future perspectives. *J. Alloys Compd.* 805: 267–294.

74 Varghese, B., Reddy, M., Yanwu, Z. et al. (2008). Fabrication of NiO nanowall electrodes for high performance lithium ion battery. *Chem. Mater.* 20 (10): 3360–3367.

75 Goel, R., Jha, R., and Ravikant, C. (2020). Investigating the structural, electrochemical, and optical properties of p-type spherical nickel oxide (NiO) nanoparticles. *J. Phys. Chem. Solids* 144: 109488.

76 Kaviyarasu, K., Manikandan, E., Kennedy, J. et al. (2016). Synthesis and characterization studies of NiO nanorods for enhancing solar cell efficiency using photon upconversion materials. *Ceram. Int.* 42 (7): 8385–8394.

77 Kate, R., Khalate, S., and Deokate, R. (2018). Overview of nanostructured metal oxides and pure nickel oxide (NiO) electrodes for supercapacitors: a review. *J. Alloys Compd.* 734: 89–111.

78 Wu, J., Huang, Y., Xia, Q. et al. (2013). Decomposition of toluene in a plasma catalysis system with NiO, MnO_2, CeO_2, Fe_2O_3, and CuO catalysts. *Plasma Chem. Plasma Process.* 33: 1073–1082.

79 Zhou, Q., Lu, Z., Wei, Z. et al. (2018). Hydrothermal synthesis of hierarchical ultrathin NiO nanoflakes for high-performance CH_4 sensing. *Front. Chem.* 6: 194.

80 Nalage, S., Chougule, M., Sen, S. et al. (2012). Sol–gel synthesis of nickel oxide thin films and their characterization. *Thin Solid Films* 520 (15): 4835–4840.

2

Fabrication Techniques and Principles

2.1 Introduction

Owing to their vast application potential in the over-increasingly developed Internet of Things (IoT) and artificial intelligence (AI), oxide semiconductors, encompassing both *n*-type and *p*-type, have gained increasing attention in recent years. To attain devices with exceptional performance, the implementation of high-quality oxide layers is generally required.

The quality of prepared oxide semiconductors heavily relies on the employed deposition methods. Currently, both vacuum- and solution-based methods are utilized for depositing oxide semiconductors. Vacuum-based methods, such as sputtering, atomic layer deposition (ALD), and evaporation, have been reported to exhibit excellent uniformity over large areas and precise thickness control, making them ideal for applications in devices like thin-film transistors (TFTs) and diodes. On the other hand, solution-based methods have demonstrated significant potential in depositing oxide semiconductors with reduced dimensions, thus being highly promising for sensing applications. Additionally, there is a growing interest in solution-processed oxide thin films due to their compatibility with existing printing techniques, which aligns with the increasing demand for printable electronics. This chapter primarily focuses on discussing the common fabrication techniques and principles of oxide semiconductors.

2.2 Vacuum-Based Methods

The deposition techniques of oxide semiconductors can be mainly separated into two categories: vacuum-based methods and solution-based methods. Vacuum-based techniques, when compared to their solution-based counterparts, provide distinct advantages such as precise thickness control and the ability to achieve highly conformal films at low deposition temperatures. Sputtering, ALD, and evaporation are among the most commonly utilized vacuum-based methods for depositing oxide semiconductors. Here, in this subchapter, oxide semiconductors are reviewed based on their respective deposition techniques.

Metal Oxide Semiconductors: Synthesis, Properties, and Devices,
First Edition. Zhigang Zang, Wensi Cai, and Yong Zhou.

2.2.1 Sputtering

As a type of physical vapor deposition, sputtering can be separated as direct current (DC), radio frequency (RF), reactive and magnetron sputtering. Sputtering refers to the physical removal of atoms from the surface of the target material through collision with energetic particles (i.e. an ionized gas known as plasma) on the surface of materials. The atoms that are removed from the target surface due to this energy bombardment are then transferred to the adjacent surface (i.e. substrate), forming a film with the desired thickness. A typical sputtering growth process is shown in Figure 2.1.

Compared with other vacuum-based methods, sputtering offers several advantages, including (i) suitability for various materials, including metals, insulators, alloys, and composites, (ii) high film quality and good step coverage, (iii) good uniformity over large areas, and (iv) potential low-temperature or even room-temperature deposition. These advantages have positioned sputtering as one of the most used vacuum techniques for depositing high-quality oxide semiconductor films, particularly in thin-film transistors.

Initially, research on sputtered oxide semiconductors primarily focused on binary compounds, i.e. ZnO and SnO_2. For example, Baud and coworkers conducted a study on the structural and optical properties of ZnO films sputtered using a ZnO target, showing that the composition of Zn and O exhibited a narrow dependence on the sputtering parameters [1]. The O/Zn atomic ratio ranged from 0.95 to 1.06, with an increase observed in correlation with the total pressure and oxygen partial pressure, while a decrease was observed with an increase of sputtering power.

Notably, to achieve high mobility, crystallization of ZnO is necessary, which poses challenges in achieving good uniformity and reproducibility over a large area.

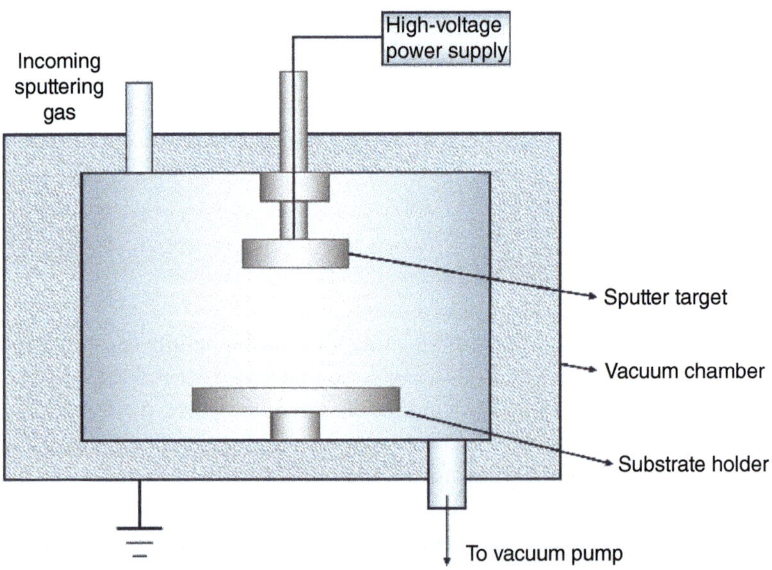

Figure 2.1 A schematic diagram showing the sputtering process.

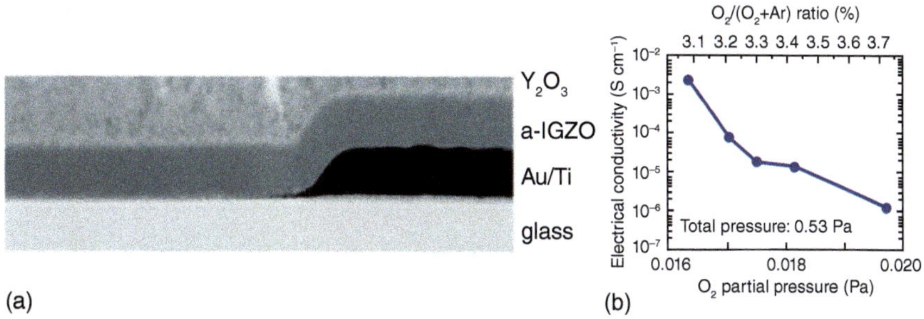

Figure 2.2 (a) A cross-sectional TEM image of a-IGZO TFTs. Source: Reproduced with permission from Yabuta et al. [2]. © 2006, AIP Publishing. (b) Electrical conductivity as a function of O_2 partial pressure. Source: Reproduced from [2]/with permission of AIP Publishing.

Using amorphous oxides with good uniformity is an alternative. An adorable achievement in sputtering of oxide semiconductors was reported by Hosono and coworkers in depositing amorphous InGaZnO (IGZO) TFTs [2]. As shown in the cross-sectional TEM images (Figure 2.2a), the deposited a-IGZO layer was uniform and dense, showing excellent step coverage without any observable grain boundaries. By varying the O_2 partial pressure during sputtering (i.e. from 3.1% to 3.7% for $O_2/(O_2 + Ar)$), the conductivity of the a-IGZO films was controlled from 10^{-3} to 10^{-6} S cm^{-1} (Figure 2.2b). The prepared IGZO TFTs exhibited a remarkable current on/off ratio of ~10^8, a mobility of 12 cm^2 V^{-1}s^{-1} and a subthreshold swing of 0.2 V dec^{-1}, showcasing the potential of sputtering for depositing high-performance TFTs on flexible substrates.

The requirement of complementary circuits necessitates the preparation of both *n*-type and *p*-type oxide semiconductors with the same methods. Currently, the sputtering of *n*-type oxide semiconductors, especially IGZO, has been successfully commercialized. However, the deposition of *p*-type oxide semiconductors with comparable performance to *n*-type ones remains limited. Such an issue has been partially addressed with the demonstration of sputtered *p*-type SnO, as reported by Fortunato et al. [3]. The sputtering window for achieving high-performance *p*-type SnO films is relatively narrow, with parameters including post-annealing conditions (temperature and atmosphere), oxygen partial pressure, and sputtering power affecting the overall film quality. For example, Alshareef and coworkers reported a comprehensive study on the deposition of high-performance SnO film using DC reactive magnetron sputtering [4]. Oxygen partial pressure and the whole deposition pressure were carefully controlled by them, with the former varying from 3 to 25% and the latter from 1.2 to 2.2 mTorr. XRD patterns indicated that the phase formation strongly depended on both parameters, resulting in polycrystalline SnO with or without metallic phases only within a narrow window between oxygen partial pressure of 7% and 15%, and pressures of 1.5 and 2 mTorr. The deposition phase map in Figure 2.3a suggested that for each process pressure between 1.5 and 2.0 mTorr, the deposited film composition exhibited a specific trend. A higher strain value was found in mixed-phase (mp-) SnO, indicating the presence of more

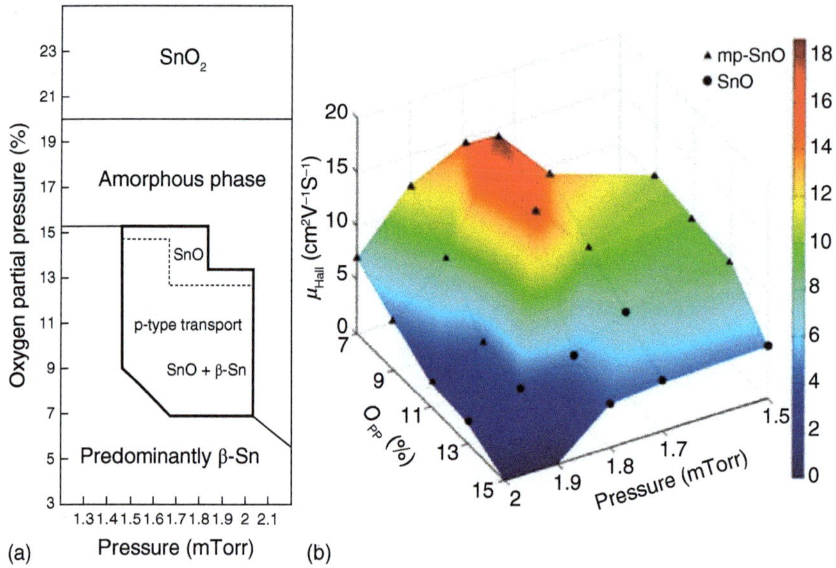

Figure 2.3 (a) Phase stability map showing the narrow window of obtaining p-type SnO. (b) Room temperature Hall mobility of the pure SnO and mp-SnO films. Source: Reproduced from Caraveo-Frescas et al. [4]/with permission of American Chemical Society.

lattice defects, which might cause additional carrier scatterings. Compared with pure-phase SnO, mp-SnO exhibited improved Hall mobility, reaching a peak hole mobility of 18.7 $cm^2\ V^{-1}s^{-1}$ for films consisting of c. 3 at% β-Sn second phase (Figure 2.3b). DFT simulations suggested that defects (i.e. Sn interstitial and oxygen vacancy) generated under Sn-rich growth conditions gave a more metallic character to the valence band of the SnO, contributing to the higher mobility. The high hole mobility might attribute to the balance between the modulation of valence band at low concentrations of the β-Sn second phase (<3 at%) and the increased charge-carrier scattering at higher concentrations of the β-Sn second phase. Using mp-SnO with the highest mobility, TFTs were fabricated on both rigid glass and flexible polyimide substrates, showing a linear field-effect mobility of 6.75 and 5.87 $cm^2\ V^{-1}s^{-1}$, respectively.

2.2.2 Atomic Layer Deposition (ALD)

Atomic layer deposition (ALD) is a chemical gas phase method capable of depositing thin films over a large area with good uniformity. This method typical involves four essential steps in each cycle, (i) precursor exposure to the substrate, (ii) evacuation or purging of the precursors as well as byproducts from the chamber, (iii) reactant species exposure, typically oxidants or other reagents, and (iv) evacuation or purging of the reactants and byproduct molecules from the chamber. A typical growth process of ALD is shown in Figure 2.4. ALD offers advantages such as excellent conformality, precise thickness control, and relatively low growth temperature, all of which are beneficial for nanofabrication.

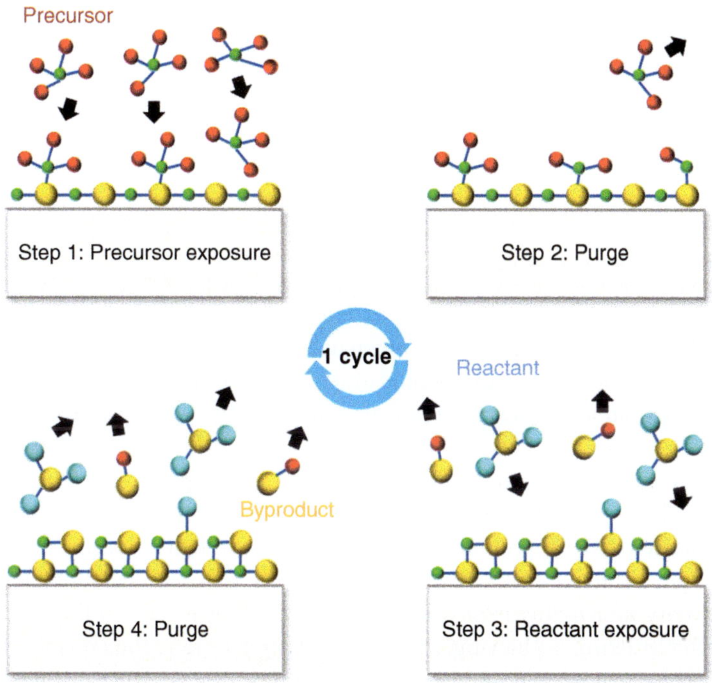

Figure 2.4 Schematic diagrams showing the ALD process.

The initial reports of using ALD in depositing oxides mainly focused on oxide insulators, i.e. hafnium oxide, aluminum oxide, titanium oxide and their mixtures. This emphasis was due to the highly conformal nature of ALD methods, enabling the deposition of ultra-thin films with high density. These materials have potential applications not only as dielectric layers in TFTs to achieve low operating voltage but also as insulator layers in MIM diodes and potentially in RRAMs. More recently, ALD has also been reported in depositing high-performance oxide semiconductors, demonstrating performance comparable to those deposited by sputtering. For example, Tsai et al. demonstrated ALD-deposited flexible ZnO TFTs with a mobility of over $20\,cm^2\,V^{-1}s^{-1}$ [5]. IGZO TFTs with a mobility of $\sim 70\,cm^2\,V^{-1}s^{-1}$ and a current on/off ratio exceeding 10^8 were reported by Park and coworkers [6].

As a binary oxide with a bandgap of 3.37 eV, ZnO has shown great potential in solar cells and TFTs. The polycrystalline structure of ZnO itself contributes to the high reliability of film quality during deposition. Synthesizing ZnO with ALD typically involves the use of diethyl zinc (DEZ) as the precursor, while the reaction temperature might vary, although keeping at a relatively low value. Deposition of ZnO at a high temperature could result in the generation of a great number of defects, i.e. oxygen vacancies, thereby increasing carrier mobility [7]. However, the deposition with a low temperature could effectively passivate the film defects, showing a reduced mobility but an increase in current on/off ratio [8].

Oxygen source is another important factor that affect the performance of ALD-ZnO films. Liu's work suggested that compared with H_2O, strong oxidants

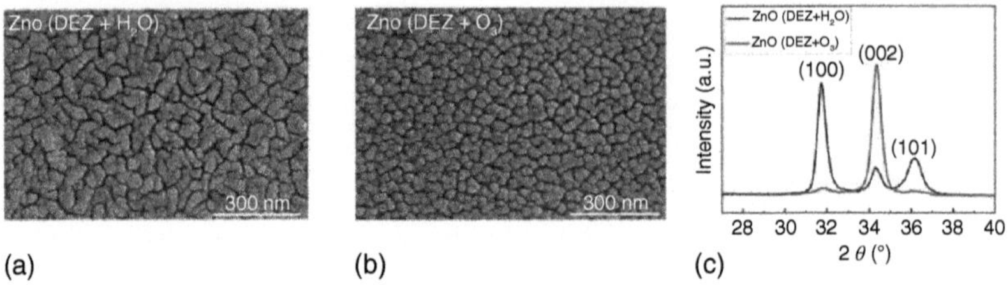

Figure 2.5 SEM images of ZnO using (a) H_2O and (b) O_3 as the oxidant. (c) XRD patterns of the ZnO films. Source: Reproduced with permission from Chen et al. [9]. © 2020, ELSEVIER.

such as O_3 yield smaller average grain sizes when used with DEZ (Figure 2.5a, b) [9]. The XRD patterns in Figure 2.5c also demonstrated the additional changes in the preferred grain orientation of ZnO film, indicating the tunability of growth direction through oxidant selection.

Another highly used binary oxide is In_2O_3, which is also capable of depositing using ALD with high mobility and good transparency [10]. Similar to ZnO, the growth temperature of In_2O_3 also plays a crucial role, since temperatures higher than 200 °C might result in In_2O_3 films with metal-like behavior owing to low electrical resistivity and high carrier concentration, leading to crystallinity and molecular orbital ordering. Achieving a semiconducting nature is thus doable by reducing the growth temperature. Additionally, methods such as post-annealing in an oxygen-rich environment, N_2O plasma treatment, and doping strategies have been previously reported to reduce the overall carrier concentration [11–13].

Deposition of high-quality SnO_2 is also doable using ALD, with a wide range of tin precursors including TDMASn, dimethylamino-2-methyl-2-propoxy-tin(II) [Sn(dmamp)$_2$], tetrakis(dimethylamino)propyl tin(IV) [Sn(DMP)$_4$] being reported previously [14–16]. However, the commercialization of ALD-SnO_2 is still limited, possibly due to the lack of commercially available Sn precursors. In 2019, Kim and coworkers used commercially available tetraethyltin (TET) as the precursor and H_2O_2 as the reactant to deposit SnO_2 films [17]. No significant changes in properties were found within a wide range of growth temperature from 250 to 400 °C, showing a high density of ~6.2 g cm^{-3} and an optical bandgap of 3.7–3.9 eV, comparable to that of bulk SnO_2. Negligible impurities remained in the films grown over the entire temperature range. Although the microstructure and crystallinity of SnO_2 films varied with the growth temperature and number of cycles, the films exhibited negligible changes of electrical properties with the growth temperature and had low resistivity values around 10^{20} cm^{-3}, suitable for a wide range of applications.

Apart from the great success made in ALD-deposited *n*-type oxide semiconductors, *p*-type oxide semiconductors grown by ALD have also attracted considerable attention. For example, high-quality single-phase SnO thin films with negligible impurity levels were reported by Chung and coworkers deposited using ALD at low temperatures of 90–210 °C using bis(1-dimethyamino-2-methyl-2propoxy)tin(II) and H_2O as precursors [18]. Despite *p*-type behaviors appearing in a wide temperature range, the crystallinity of the ALD-grown SnO films was determined by the growth temperature, as suggested by their work. While (001) textured crystalline

SnO was deposited at temperature over 150 °C, amorphous SnO was achieved at a temperature below 120 °C. Using the optimized SnO deposited using ALD, the same group also reported SnO TFTs with a high I_{ON}/I_{OFF} ratio of 2×10^6, and a mobility of ~1 $cm^2 V^{-1}s^{-1}$ within a thermal limit of 250 °C [19].

Apparently, several ALD processes have also been reported for copper oxides. Cu_2O is a promising *p*-type transparent conductive oxide that has demonstrated a Hall mobility of over 100 $cm^2 V^{-1}s^{-1}$ [20]. ALD Cu_2O films were reported by Horwat and coworkers using $Cu(HfAc)_2$ as the Cu precursor and DI water as reactant [21]. The deposited films showed a relatively high carrier concentration of ~10^{16} cm^{-3}, a low resistivity of 9–150 Ω cm, and a high mobility of 19 $cm^2 V^{-1}s^{-1}$. The combination of Cu_2O with sputtered ZnO also enabled a semi-transparent Cu_2O/ZnO heterojunction capable of exhibiting photoconductive behaviors with a stable and fast self-powered photoresponse under 1 Sun illumination.

Based on the growth mechanism of ALD, high-quality oxide semiconductor films can be grown at relatively low temperatures, making it promising for plastic substrates. The low deposition temperature is also beneficial for avoiding native vacancies and interfacial defects, thus resulting in a higher electrical and optical properties of the deposited films. However, the deposition speed of ALD is quite slow, limiting the potential in applications where a thicker film is required. The research on oxide semiconductors, especially *p*-type ones, deposited by ALD is still insufficient to meet the requirement in uses such as photovoltaics and gas sensors. Nevertheless, with the ever-decreasing scale of many novel applications, ALD with a high degree of conformality should increasingly gain importance in oxide semiconductor technologies in the coming years.

2.2.3 Evaporation

Apart from the well-documented techniques of sputtering and ALD for depositing oxide semiconductors, evaporation has also been reported as a potential vacuum-based method for depositing high-performance oxide semiconductor films. As early as 1964, Klasens and Koelmans proposed the use of evaporated SnO_2 as the channel layer for TFTs [22].

As shown in Figure 2.6, evaporation consists of three main steps: the evaporation of the oxide source, the transfer of vapor atoms/molecules, and the condensation on the surface of the substrate to form the thin film. Depending on the heat source employed, evaporation can be categorized as thermal evaporation or e-beam evaporation. In thermal evaporation, the source material is heated by heating the crucible through the Joule effect. On the other hand, e-beam evaporation involves the generation of an electron "gun" through thermionic emission. By applying a high voltage, the electron beam is accelerated toward the source of the material, causing it to evaporate.

Oxide semiconductors such as CuO, NiO_x, and Cu_2O have been reported to be capable of depositing by thermal evaporation. A conventional method for producing oxide thin films involves evaporating metal targets under vacuum conditions, followed by heat treatment in oxide-rich environments to achieve

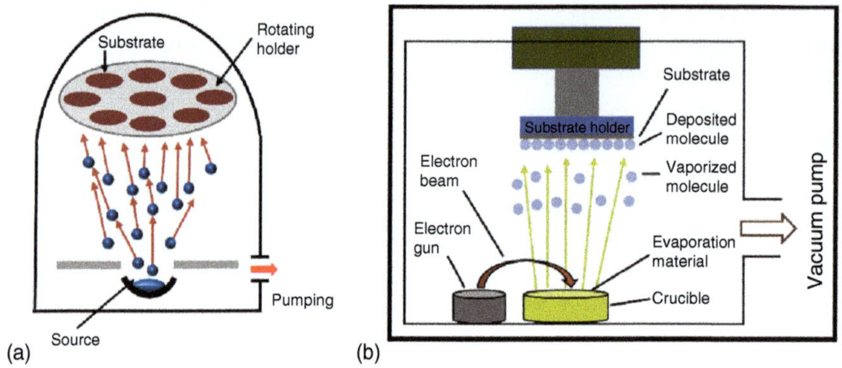

Figure 2.6 Schematic diagrams showing (a) thermal evaporation and (b) e-beam evaporation.

oxidation. For instance, CuO thin films were grown by evaporating Cu powders, followed by heat treatment at different temperatures, as reported by Mohapatra and coworkers [23]. The size of the CuO particles was found to be closely related to the annealing temperature, showing an increase of particle size from 23 to 293 nm with the increase of annealing temperature to 600 °C.

By direct thermal evaporating of Cu_2O powders, Al-Kuhaili reported and characterized Cu_2O thin films [24]. The films were deposited on substrates at various temperatures and the as-deposited films were subjected to post-deposition annealing in air. Chemical analysis indicated that only the as-deposited films on unheated substrates had a dominant Cu_2O composition, while those deposited on heated substrates and annealed films showed a predominant CuO composition, suggesting that the temperature was a crucial parameter in avoiding phase mixture. The as-deposited films were amorphous with poor transmittance in the visible and infrared regions. However, the annealed films exhibited greater crystallinity and higher transparency, with a bandgap ranging from 1.76 to 2.16 eV.

Alternatively, e-beam evaporation has also been utilized for depositing oxide semiconductors, as reported by Pereira et al. in depositing NiO_x thin films at room temperature [25], Mahmood and coworkers in synthesizing SnO_x [26], Chen and coworkers in preparing SnO films [27], and others. Furthermore, tin-doped indium oxide thin films could also be deposited by such a method, as reported by Fallah et al., demonstrating high electrical conductivity of 7×10^{-4} Ω cm and high visible transmittance exceeding 93% [28]. The annealing temperature was found to play a crucial role in controlling the electrical properties and optical transmittance of ITO films. XRD patterns revealed that the crystallinity of the ITO thin films improved with annealing, exhibiting a grain size of 37 nm at 550 °C. The DC conductivity of the ITO thin films exhibited a well-defined behavior in the range of 350–550 °C, with conductance increasing as the annealing temperature increased.

It has been observed that the performance of evaporated oxide films is highly dependent on the temperature. Higher temperatures can lead to larger particle sizes, but they may not be suitable for integrating with flexible electronics, thereby limiting practical applications. Besides, unlike sputtering and ALD, evaporation is reported less in n-type oxide semiconductors but more on p-type ones. We speculate that the possible reason for this disparity may be the considerably high evaporation temperature required for source materials in n-type oxide semiconductors. If the electrical properties can be further improved, evaporation could emerge as a promising method for depositing p-type oxide semiconductors.

2.3 Solution-Based Methods

Until now, the deposition of high-performance oxide semiconductors has predominantly relied on vacuum-based methods, particularly for applications in displays and flexible electronics. However, as the demand for printable electronics continues to rise, there has been a growing research focus on depositing oxide semiconductors using solution-based methods. Additionally, it has been observed that synthesizing oxide semiconductors with reduced dimensions using vacuum-based techniques is challenging, at least compared to solution-based techniques. In applications such as gas sensing, achieving a large surface-to-volume ratio for the active material is crucial, highlighting the significance of solution deposition for oxide semiconductors. In this subsection, we will introduce the most commonly reported solution-based methods for depositing oxide semiconductors based on the material dimensions.

2.3.1 0D Oxide Semiconductors

Zero-dimensional (0D) nanostructures encompass materials whose dimensions are entirely within the nanoscale range, including nanoparticles (NPs), quantum dots (QDs), and nanocrystals (NCs). These structures exhibit high surface-to-volume ratios and quantum confinement effects, making them well-suited for various applications such as sensing and lighting. For example, Hu et al. demonstrated the utilization of ZnO QDs@SnO_2 hollow nanospheres in sensors [29]. Our group reported Cu_2O NPs/reduced graphene oxide nanocomposites with photoconductive behaviors [30].

Until now, a typical approach for preparing colloidal oxide QDs has been through sol–gel techniques involving hydrolysis. This process entails dissolving metal salts and stabilizers in proper solvents, such as ethanol. For example, Fu et al. mixed zinc acetate solutions with LiOH solutions (both in ethanol) and successfully obtained a transparent and colorless solution containing ZnO QDs [31]. The reactions involved in this process were

$$Zn(CH_3CO_2)_2 + 2\,LiOH \rightarrow ZnO + H_2O + 2\,LiCH_3CO_2 \tag{2.1}$$

The ZnO QDs obtained after washing and centrifugating had a bandgap of 3.8 eV and an average grain size of 4.8 nm. Besides, the peaks observed in the XRD patterns closely matched those of tetragonal rutile ZnO, confirming the successful formation of ZnO QDs.

However, colloidal QDs synthesized using traditional sol–gel methods often suffer from issues such as aggregation and growth, primarily due to their high

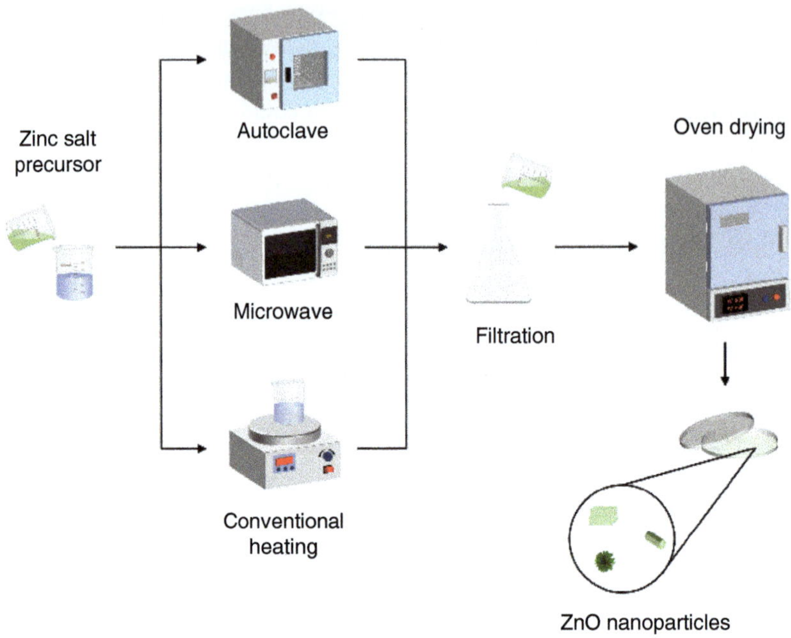

Figure 2.7 Schematic diagrams showing the hydrothermal synthesis.

surface energy. This renders them unstable, particularly when exposed to ambient air. To address this challenge and improve stability, as well as enhance the photoluminescence (PL) emission, silane surface modification can be employed. For example, Shi et al. reported a sol-gel strategy for fabricating ZnO QDs with strong and tunable fluorescence [32]. (3-(2,3-epoxypropoxy)-propyl)trimethoxysilane was chosen as the stabilizer for silanization, resulting in ZnO colloids with stable luminescence that persisted for over nine weeks.

Alternatively, the hydrothermal/solvothermal process has been reported as a viable method for depositing 0D oxide semiconductors, offering advantages of fine particle production, catalyst-free growth, simplicity of equipment, and cost-effectiveness. While in hydrothermal method the solution is aqueous (i.e. water), solvothermal method uses nonaqueous solutions. A typical hydrothermal process is shown in Figure 2.7.

Generally, the control of particle size and morphology in the hydrothermal/solvothermal process is achieved by adjusting the reaction temperature, time, and precursor concentration. In such a route, the metal ions in oxides are expected to be solvated by a shell of water molecules, where the raising of PH in the reaction is necessary. Jayaraj and coworkers synthesized ZnO nanoparticles through hydrothermal synthesis using precursor solutions containing $Zn(CH_3COO)_2{\cdot}2H_2O$ and NaOH in methanol, with a pH value between 8 and 11 [33]. These solutions were then transferred into Teflon-lined sealed stainless-steel autoclaves and maintained at various temperatures between 100 and 200 °C for 6 to 12 hours under autogenous pressure. In the precursor solutions, the source of Zn was $Zn(OH)_2$ and

$Zn(OH)_4{}^{2-}$, and under hydrothermal conditions, $Zn(OH)_2$ would dissolve to form Zn^{2+} and OH^- ions. Once the product of $[Zn^{2+}]$ and $[OH^-]$ exceeded the critical value necessary for the formation of ZnO crystals, ZnO crystals would precipitate from the solution. Since the solubility of ZnO was significantly smaller than that of $Zn(OH)_2$ under the hydrothermal conditions, $Zn(OH)_2$ precipitates tended to be transformed into ZnO crystals during the process, according to equations below:

$$Zn(OH)_2 = Zn^{2+} + 2OH^- \tag{2.2}$$

$$Zn^{2+} + 2OH^- = ZnO + H_2O \tag{2.3}$$

The XRD patterns demonstrated the hexagonal wurtzite structure of the prepared nanoparticles, with the particle size increasing with the growth temperature and decreasing with the concentration of the precursors. Additionally, ZnO NPs synthesized at 150 °C for 6 hours with 0.5 M NaOH had an average size of about 25 nm, as confirmed by the TEM image.

Other oxide nanoparticles have also been previously reported using hydrothermal synthesis. For example, our group synthesized cubic-like Cu_2O nanoparticles at a relatively low temperature of 50 °C, showing diameters ranging from 100 to 400 nm [30]. Combining the hydrothermal route with an ice-water bath stirring method, Wang et al. reported the synthesis of SnO_2 NPs [34]. Zheng and coworkers hydrolyzed aqueous solutions of $TiCl_4$ hydrothermally at temperatures ranging from 100 to 300 °C for varying durations of 6 to 44 hours, resulting in controlled size, morphology, and phase of TiO_2 particles [35]. Similar to other oxide nanoparticles prepared through hydrothermal synthesis, the amount of rutile increased with time, temperature, and the initial $TiCl_4$ concentration. Small particle sizes were observed with low initial $TiCl_4$ concentrations and short reaction times.

As the multi-functionality of oxide nanomaterials continues to expand, there has been exploration into the development of new synthesis methods. One such method that has recently gained attention is the microwave hydrothermal method for preparing nanocrystals. As a combination of hydrothermal and microwave method, such a method fully exerts the advantages of microwaves and water heat. In comparison to traditional hydrothermal synthesis, microwave hydrothermal methods enable the use of microwaves to heat the precursors, resulting in significantly improved reaction speeds and the growth of nanoparticles with a narrow particle size distribution and uniform morphology. For example, as shown in Figure 2.8, Will's work demonstrated that TiO_2 prepared by microwave hydrothermal method had higher crystallinity than those prepared using convection hydrothermal method but requiring less time and energy consumption [36].

Owing to their high surface energy, 0D oxides are typically highly active and thermodynamically unstable, making them be easily aggregated and grown into larger sizes. Consequently, the key factors in synthesizing 0D oxide nanomaterials lie in controlling the growth of nanocrystals and preventing aggregation. Currently, the synthesis of 0D nanomaterials with well-controlled sizes and shapes necessitates complex processes, stringent synthesis parameters, and expensive raw materials. Moreover, these methods often suffer from low production yields and involve the

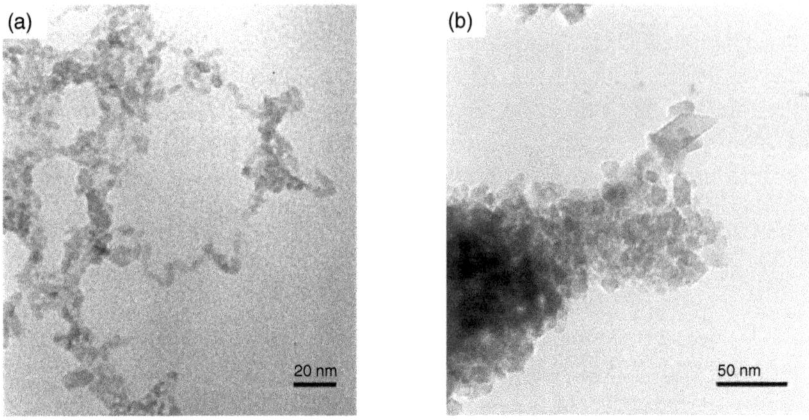

Figure 2.8 TEM images of TiO_2 colloids prepared by (a) microwave and (b) convection hydrothermal methods. Source: Reproduced from Wilson et al. [36]/with permission of Royal Society of Chemistry.

use of non-eco-friendly chemicals and processes. Further advancements in this field should prioritize the precise control of preparation processes, as well as improving reproducibility and stability. Additionally, the incorporation of 0D oxide nanomaterials into micro- and nanoscale devices and their assembly in mobile electronics is highly desirable, especially considering the involvement of heterojunctions between 0D–1D and 0D–2D nanocomposites.

2.3.2 1D Oxide Semiconductors

One-dimensional (1D) nanostructures encompass various forms such as nanorods, nanowires, nanobelts, nanofibers, and nanotubes. These structures exhibit significant potential, particularly as sensing materials, owing to their large surface-to-volume ratio, high porosity, and unique electronic properties. Previous reports have highlighted methods like electrospinning and hydrothermal synthesis as capable of producing high-performance 1D oxide semiconductors [37]. Here, we discuss the advantages and limitations of each synthetic approach.

Electrospinning is a widely utilized technique for depositing nanofibers, yielding fibers with diameters ranging from a few nanometers to a few micrometers, depending on the composition of precursors, electrospinning parameters, and environmental conditions [38]. A schematic showing the electrospinning process is given as Figure 2.9. During electrospinning, a charged polymer-based precursor solution is ejected through a small orifice of the needle under an applied high electric field, where the ejected solution either solidifies or coalesces into a filamentous morphology. The polymer-based precursor solution in the syringe becomes charged through the conductive metal needle, which is electrically connected to a high-voltage power supply. As the precursor solution is gradually pushed out of the needle, the high-voltage electric field between the needle and the collector provides an electrostatic repulsive force that overcomes the surface tension of the

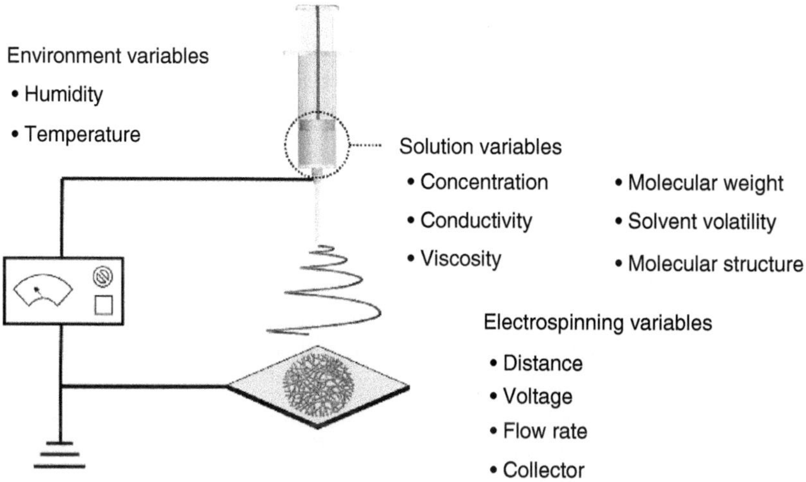

Figure 2.9 A schematic diagram showing the electrospinning process and possible variables.

droplet formed at the needle orifice to eject the charged precursor solution toward the electrically grounded collector. As the ejected solution reaches the grounded collector, it undergoes solidification due to solvent evaporation, resulting in the deposition of solid nanofibers.

Several parameters might affect the morphology of nanofibers, including solution properties, environmental conditions, and electrospinning conditions. Among them, the molecular weight of polymers and the purity of metal salts are the most affecting parameters especially for the nanofiber diameter, since they can affect the solution viscosity. A lower solution viscosity, which occurs with lower polymer content, leads to increased polymer chain fragmentation during the electrospinning process and consequently results in thinner diameters. Conversely, a precursor solution with higher polymer content and viscosity produces nanofibers with larger diameters because of the higher degree of polymer entanglement along the axis of the fiber [39, 40]. Additional parameters that influence nanofiber morphologies include absolute humidity and temperature of the environmental conditions, feed rate, applied voltage, and working distance during electrospinning.

Until now, various 1D oxide semiconductors including *n*-type ZnO, SnO_2, and In_2O_3 nanofibers [41–46], as well as *p*-type CuO and NiO_x [47, 48], have been synthesized using electrospinning, showing great potentials in applications such as gas sensors, glucose detectors, and ion batteries. For example, Shao et al. reported the synthesis of NiO nanofibers with a diameter of 50–150 nm through electrospinning of poly(vinyl acetate) (PVA)/nickel acetate composites followed by calcination treatment [47]. Similarly, Wu et al. prepared ultra-thin ZnO nanofibers using sol–gel processing and electrospinning techniques with PVA and zinc acetate as precursors [42]. In this work, they found that the diameter of the fibers was closely related to the concentration of zinc acetate, while the morphology was affected by the calculation time.

Clearly, electrospinning is a facile and versatile process for 1D nanostructure synthesis through the ease of tuning composition and morphology, as well as large-scale synthesis of nanomaterials. Advantages like simple processing and scalable fabrication suggest that electrospinning is one of the most compatible methods to synthesize 1D oxide semiconductors with diverse morphologies. However, 1D nanostructures prepared by electrospinning predominantly exhibit a planar structure, limiting their application in devices requiring a vertical structure.

An alternative choice for depositing 1D nanostructures is hydrothermal synthesis. For example, our group reported Cu_2O nanowires grown by a hydrothermal synthesis, which had a wire-shaped morphology with a diameter size of around 180–200 nm and an average length exceeding 13 μm [49]. It is found that with the use of a seed layer, such a process can grow nanostructures in a hierarchical structure. Various parameters, including temperature, reaction times, PH values, solvent, salt concentration, and additives, might control the morphology and structure of the resulting nanomaterials.

Wang and coworkers demonstrated the hydrothermal growth mechanism of ZnO nanowires, as shown in Figure 2.10 [50]. They conducted optimization of several parameters, including the pH values of the aqueous solution, growth time, and solution temperature, to achieve well-oriented and homogeneous arrays of ZnO nanowires. The study revealed that the two different growth approaches depicted in Figure 2.10, namely way 1 and way 2, led to distinct diameter size distributions. Way 1 resulted in a narrow distribution, while way 2 exhibited a wider distribution of nanowire diameters. Additionally, the research indicated that at lower temperature ranges, both lateral and axial growth of the nanowires were promoted, whereas

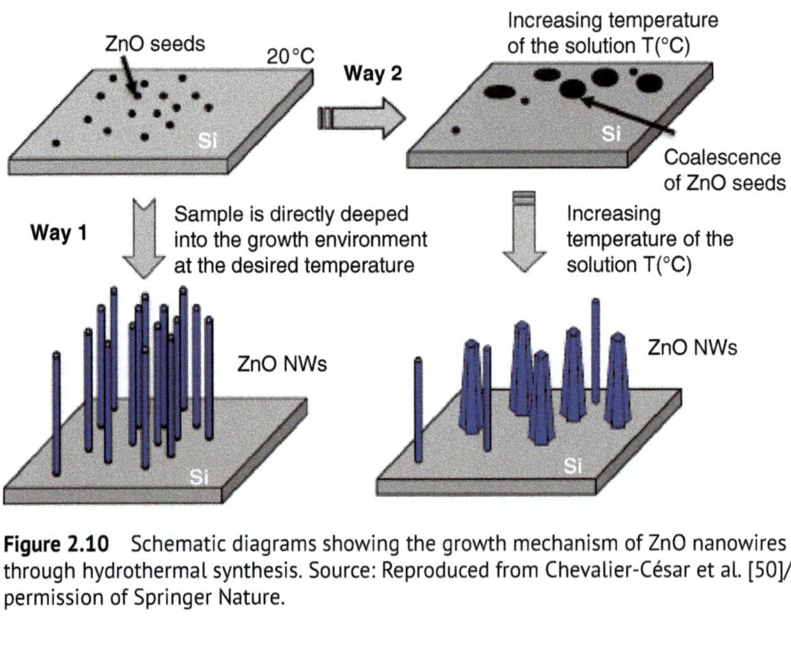

Figure 2.10 Schematic diagrams showing the growth mechanism of ZnO nanowires through hydrothermal synthesis. Source: Reproduced from Chevalier-César et al. [50]/with permission of Springer Nature.

at higher temperatures, axial growth dominated. This observation emphasized the significant influence of the starting temperature of the seed layer on the nanowire morphology.

Apparently, one can also deposit the seed layer using physical methods, as reported by our group in depositing ZnO nanorods [51], Gao and coworkers in depositing Ga_2O_3 nanorods [52], and Torres-Martinez and coworkers in depositing TiO_2 nanorods [53], and so on. Nanorods grown on seed layers have their growth direction being limited, beneficial for uses in photodiodes. However, hydrothermal synthesis generally requires a lengthy reaction time and imposes restrictions on substrate choice, limiting their mass production capabilities, especially in printable electronics. Further advancements in the field of 1D oxide nanomaterials may focus on combining hydrothermal synthesis with electrospinning techniques to enable the synthesis of nanocomposites and the formation of hierarchical structures.

2.3.3 2D Oxide Semiconductors

Two-dimensional (2D) nanomaterials encompass various structures such as nanosheets, nanoplates, and nanodisks. Currently, the most commonly reported methods for depositing 2D materials involve exfoliation and vacuum-based techniques, including chemical vapor deposition and pulsed layer deposition. However, solution-phase synthesis has also gained attention, especially in sensing-related fields, due to its low cost and ease of deposition. While the number of reported 2D oxide semiconductors is relatively fewer compared to other dimensional materials, their unique 2D structure makes them highly suitable for crystal facet engineering and related applications.

In 2014, Kim and coworker reported the first ZnO nanowall structure on both sides of a flexible substrate using a low-temperature hydrothermal method [54]. The process involved immersing the seed-layer-coated substrate in a solution containing equimolar amounts of zinc nitrate ($Zn(NO_3)_2{\cdot}6H_2O$) and hexamethylenetetramine (HMTA,$(CH_2)_6N_4$), and polyethylenimine (PEI) in DI water, and maintaining it at 90 °C for 8 h. After rinsing the substrate with DI water to remove any residual substances on the surface and then drying in air at 100 °C overnight, the wanted ZnO nanowall structure was resulted, as shown in Figure 2.11. The prepared ZnO nanowalls exhibited vertical alignment on the substrate, with a (002) orientation, a length of 2–3 μm, and a wall thickness of 60–80 nm. This structure showed great potential for applications in smart and wearable electronic applications.

In a separate study, Sun and coworkers reported the synthesis of ultrathin anatase/rutile 2D TiO_2 nanosheets using a wet-chemistry approach (Figure 2.12a) [55]. The prepared TiO_2 nanosheets had sizes ranging from 200 to 500 nm and displayed clear single-crystalline characteristics without sub-grains. As shown in Figure 2.12b, the nanosheets were found to be stacked together with 4–5 monolayers under an interlayer spacing of 0.62 nm. This arrangement led to a calculated bandgap of 3.65 eV, as determined from the UV-Vis absorption spectrum.

In another approach, Li et al. involved a solvothermal method to deposit TiO_2 nanosheets, which showed a thickness down to ~2.5 nm and a side length of

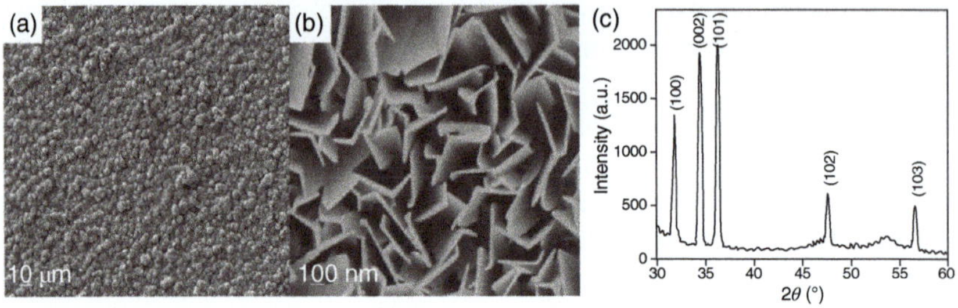

Figure 2.11 FE-SEM images in (a) low and (b) high resolutions, and (c) XRD patterns of ZnO nanowall. Source: Reprinted with permission from Saravanakumar and Kim [54]/with permission of American Chemical Society.

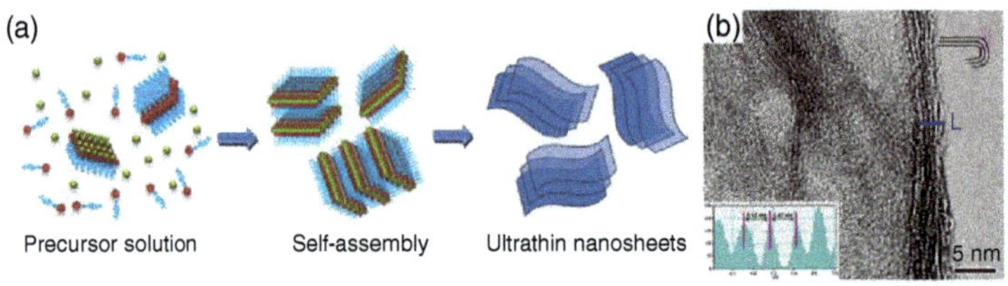

Figure 2.12 (a) Schematic diagrams showing the synthesis of TiO_2 nanosheets. (b) HRTEM images of TiO_2. Source: Reproduced from Sheng et al. [55]/with permission of Elsevier.

~200 nm [56]. The dominant facet of such TiO_2 was (001). This was due to the use of ethanol solvent, which significantly enhanced the content of surface chemisorbed F^- and lowered the surface energy of (001) facets. By loading the nanosheets with 1 wt% Pt, they demonstrated an H_2 evolution rate of 17.86 mmol hg^{-1} and an apparent quantum efficiency of 34.2%.

However, 2D nanomaterials often face challenges related to aggregation caused by van der Waals interactions during the synthesis process, resulting in a reduced surface area and porosity. As an alternative, 2D oxide semiconductors can be engineered into ultrathin nanomaterials or incorporated into composite films with other materials. For example, Li et al. reported the synthesis of photocatalytic composite thin films consisting of 2D ZnO nanosheets/regenerated cellulose (ZNSRC) using a two-step method at room temperature [57]. The study revealed that the 2D ZnO nanosheets were dispersed both on the surface and within the ZnO thin films. These films exhibited excellent photocatalytic activity and organic dye degradation, with the best performance observed using 80-nm-thick and 350–450 nm-sized 2D ZNSRC, obtained with a reaction time of 13 hours.

2D oxide semiconductors beyond graphene and other 2D nanomaterials have become a promising class of nanomaterials because of their desirable properties for various applications. Currently, characterization techniques such as TEM, XPS, and Raman spectroscopy are enough for studying the growth mechanism, but synthesizing methods for high-performance 2D oxide materials are still insufficient. The future advancement of this field will likely focus on the exploration of novel synthesis approaches and their potential integration into commercial technologies.

2.3.4 3D Oxide Semiconductors

Three-dimensional (3D) oxide semiconductors typically refer to oxide thin films that are synthesized using metal precursors dissolved in a solvent and deposited onto a substrate using various coating methods such as spin-coating, spray pyrolysis, and inkjet printing. The process involves coating the substrate with the oxide precursor solution and subsequently annealing the film to remove the solvent and develop an active layer. The process can be divided into three main stages: decomposition, hydrolysis, and dihydroxylation, as well as alloying if applicable. These stages help in the transformation of the precursor into the desired oxide semiconductor film with the desired properties.

Spin-coating is a commonly used method in depositing oxide thin films in applications such as thin-film transistors, solar cells, and memories, owing to their easy processing, low cost, and compatibility with various substrates. For example, our group used such a method to deposit ZnO films as the electron transporting layer in solar cells [58], as well as InZnO and InZnO:La films as the channel layer in transistors [59, 60]. The spin-coating process involves placing the substrate on a rotating stage and carefully dispensing the solution onto the surface of the substrate. The rotation creates a centrifugal force that helps to remove excess solution from the substrate surface. Subsequently, the coated films undergo drying and annealing processes to achieve the desired properties. The thickness of the spin-coated films is influenced by various parameters, including spin speed, acceleration, steps, time, repetition number of the spinning process, as well as the viscosity and molar concentrations of the solutions. However, spin-coating does have its limitations. It can result in the waste of precursors, and it is not suitable for large-area or curved substrates, limiting its applicability in certain scenarios. The uniformity of spin-coated films over a large area is often poor, and controlling the thickness can be challenging. Although techniques such as slit-die coating have been developed to improve uniformity, they are still insufficient for meeting the requirements of commercial-scale production. Therefore, further advancements and alternative methods are needed to address these limitations and enable the commercialization of spin-coated oxide films.

A potential alternative is spray coating, which is readily suitable for continuous large-area mass production of oxide films. In spray coating, a chamber containing carrier gases is utilized, and an atomizer or nebulizer is used to generate fine droplets from low-viscosity solutions. These droplets are directed onto the substrate surface by the combined force of the carrier gas and gravity. Additionally, applying an electrostatic field to the substrate helps aggregate the droplets into a thin film. The thickness, roughness, and quality of the films are influenced by factors such as droplet size, solution viscosity, number of coating cycles, and substrate temperature during deposition.

A pioneer of this method is Anthopoulos's group, who reported oxide TFTs based on ZnO, ZnO:Li, In_2O_3, Cu_2O, and so on [61–64]. In 2017, they reported an in-depth study on the growth mechanism of In_2O_3 layers processed by spray coating [62]. Depending on the substrate and processing temperature, a strong influence of the Leidenfrost effect on the quality of the deposited In_2O_3 layer was observed. Based on this, they showed that compared with the actual temperature, the Leidenfrost temperature played a more crucial role in the film performance. Benefiting from the Leidenfrost effect, In_2O_3 TFTs exhibited superior electron transport characteristics with an electron mobility up to 40 and 26 $cm^2\ V^{-1}s^{-1}$ at a processing temperature of 252 and 240 °C, respectively, on conventional Si/SiO_2 substrates.

However, films deposited by spray coating often exhibit high surface roughness and have limitations in terms of available solutions with low viscosity. Among various coating methods for solution-processed 3D oxides, inkjet printing stands out as one of the most promising alternatives to vacuum and photolithographic processes. In inkjet printing, the key component is the printer head, which contains micrometer-scale nozzles for ejecting droplets. There are two main types of mechanisms for nozzle ejection: thermal inkjet and piezoelectric inkjet. In a thermal inkjet, a small amount of the ink solvent is heated and evaporated with the evaporated gas forcing ink out of the nozzle. On the other hand, piezoelectric inkjet printing utilizes a nozzle tip made of piezoelectric material. Applying an electrical signal, such as a voltage, to the nozzle tip causes it to change shape, either opening or closing to expel ink. The diameter, temperature, and type of nozzles employed in inkjet printing directly influence the morphology of the printed films.

As early as 2007, Chang and coworkers reported oxide TFTs prepared by ink-jet printing at a high post-annealing temperature of 600 °C, as shown in Figure 2.13 [65]. The HRTEM image in Figure 2.13b suggests no clear crystalline patterns of the prepared films with the electron diffraction pattern further confirming the amorphous structure of the IZO film. Such IZO thin films with a thickness of 50 nm showed an amorphous structure with a bandgap of 3.4 eV (Figure 2.13c). By adopting the

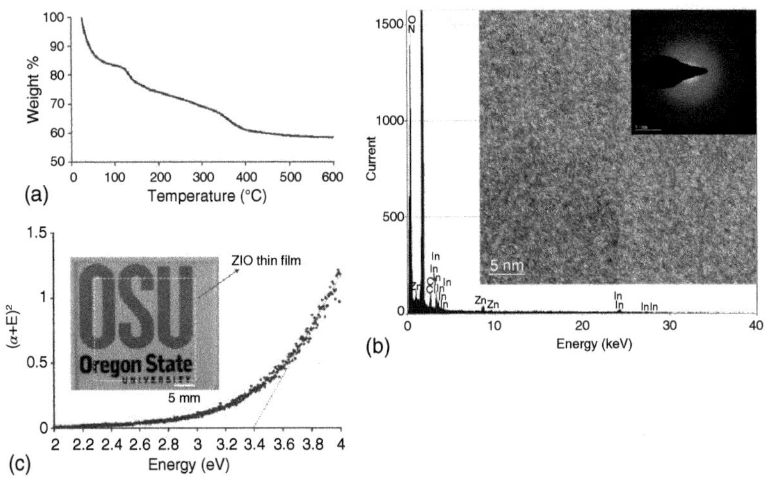

Figure 2.13 IZO film characterizations with (a) TGA of precursors, (b) a TEM image, and (c) the UV-Vis absorption measurement. Source: Reproduced from Lee et al. [65]/with permission of John Wiley & Sons.

IZO films as channel layer, the authors also prepared TFTs, showing a mobility of 16.1 $cm^2 V^{-1}s^{-1}$ and a current on/off ratio of ~10^4.

Inkjet printing offers several advantages, including precise control over droplet placement, compatibility with various substrates, and the ability to print on non-flat or curved surfaces. The technique allows for the deposition of materials with high spatial resolution, making it particularly suitable for patterning and device fabrication. However, the throughput of ink-jet printing is found still to be quite low, especially when compared with roller-based printing methods such as gravure and flexography. Besides, the viscosity of ink limits the range of potential ink choices. The low uniformity of printed film is also a drawback of ink-jet printing, resulting from the coffee-ring effect caused by the difference in evaporation rate between the edge and the center region of the printed droplet.

Recently, a method so-called electrohydrodynamic (EHD) jet printing has been developed as an alternative to ink-jet printing, offering advantages of high printing resolution in the range of sub-micrometer scale, the ability to change ejection form, the possibility to use high-viscosity solutions, and controllable nozzle shape and electric field intensity. In contrast to traditional inkjet printing, where thermal or piezoelectric techniques are the main ejection methods, EHD jet printing utilizes the electric field as the core driving force. This is achieved by using a conductive nozzle and substrate for electric field actuation, which stabilizes the jetting characteristics. For example, Rogers and Paik prepared IZO films with a linewidth resolution down to ~1.5 μm by EHD jet printing [66]. Park et al. reported In_2O_3 TFTs prepared by EHD jet printing at temperature of less than 250 °C, showing a high mobility of ~230 $cm^2 V^{-1}s^{-1}$ [67].

It should be pointed out that EHD jet printing also has its limitations. It may encounter difficulties when printing on insulating, rugged, or large-area substrates, as the applied electric field may not be uniformly sustained, resulting in the ink not being ejected or the droplets having uneven shapes. Additionally, EHD jet printing can suffer from low throughput due to the slow movement of stages and the use of fine-sized nozzles.

Until now, various methods have been developed for synthesizing oxide thin films, but most of them still require high annealing temperatures to cure the film and achieve a reasonably high performance. Lowering the temperature of fabrication process and improving the performance and reliability are of great importance for pursuing applications in flexible, bendable, and stretchable devices. Oxide thin films prepared by a solution process at low temperatures show great potential for uses in emerging applications regarding logic circuits, memory devices, and active matrices for sensors (i.e. gas sensors, photodetectors, and biosensors). So far, methods including microwave-assisted annealing process [68, 69], photoactivation process [70–72], and high-pressure annealing process [73] have been reported in lowering the deposition temperature, showing good progress despite still far from commercialization.

Furthermore, printing technologies should be extensively investigated, considering that current reports on such methods are relatively rare and indicate low film quality and poor device performance. Reducing the coffee-ring effects and improving the film uniformity are the main paths, which require more development, not only from the printers themselves but also the ink formulations. By addressing the aforementioned issues, solution processing, including printing techniques, has the potential to become the mainstream method in the fourth industrial revolution for oxide-based next-generation IoT and large-area electronics.

References

1 Bachari, M., Baud, G., Ben, S. et al. (1999). Structural and optical properties of sputtered ZnO films. *Thin Solid Films* 348 (1): 165–172.

2 Yabuta, H., Sano, M., Abe, K. et al. (2006). High-mobility thin-film transistor with amorphous $InGanO_4$ channel fabricated by room temperature rf-magnetron sputtering. *Appl. Phys. Lett.* 89 (11): 112123.

3 Fortunato, E., Barros, R., Barquinha, P. et al. (2010). Transparent p-type snox thin film transistors produced by reactive rf magnetron sputtering followed by low temperature annealing. *Appl. Phys. Lett.* 97 (5): 052105.

4 Caraveo-Frescas, A., Nayak, K., Al-Jawhari, A. et al. (2013). Record mobility in transparent p-type tin monoxide films and devices by phase engineering. *ACS Nano* 7 (6): 5160–5167.

5 Lin, Y., Hsu, C., Tseng, M. et al. (2015). Stable and high-performance flexible ZnO thin-film transistors by atomic layer deposition. *ACS Appl. Mater. Interfaces* 7 (40): 22610–22617.

6 Sheng, J., Hong, T., Lee, H. et al. (2019). Amorphous IGZO TFT with high mobility of ~70 $cm^2/(vs)$ via vertical dimension control using PEALD. *ACS Appl. Mater. Interfaces* 11 (43): 40300–40309.

7 Cho, S., Ahn, C., Yun, M. et al. (2014). Effects of growth temperature on performance and stability of zinc oxide thin film transistors fabricated by thermal atomic layer deposition. *Thin Solid Films* 562: 597–602.

8 Klingshirn, C. (2007). ZnO: from basics towards applications. *Phys. Status Solidi B* 244 (9): 3027–3073.

9 Chen, X., Wan, J., Wu, H. et al. (2020). ZnO bilayer thin film transistors using H_2O and O_3 as oxidants by atomic layer deposition. *Acta Mater.* 185: 204–210.

10 Sheng, J., Choi, D., Lee, S. et al. (2016). Performance modulation of transparent ald indium oxide films on flexible substrates: transition between metal-like conductor and high performance semiconductor states. *J. Mater. Chem. C* 4 (32): 7571–7576.

11 Yeom, H., Ko, J., Mun, G. et al. (2016). High mobility polycrystalline indium oxide thin-film transistors by means of plasma-enhanced atomic layer deposition. *J. Mater. Chem. C* 4 (28): 6873–6880.

12 Sheng, J., Park, J., Choi, D. et al. (2016). A study on the electrical properties of atomic layer deposition grown InO_x on flexible substrates with respect to N_2O plasma treatment and the associated thin-film transistor behavior under repetitive mechanical stress. *ACS Appl. Mater. Interfaces* 8 (45): 31136–31143.

13 Sheng, J., Park, E., Shong, B. et al. (2017). Atomic layer deposition of an indium gallium oxide thin film for thin-film transistor applications. *ACS Appl. Mater. Interfaces* 9 (28): 23934–23940.

14 Alshammari, F., Hota, M., Wang, Z. et al. (2017). Atomic-layer-deposited SnO_2 as gate electrode for indium-free transparent electronics. *Adv. Electron. Mater.* 3 (9): 1700155.

15 Lee, B., Jung, E., Kim, S. et al. (2012). Physical/chemical properties of tin oxide thin film transistors prepared using plasma-enhanced atomic layer deposition. *Mater. Res. Bull.* 47 (10): 3052–3055.

16 Mai, L., Zanders, D., Subaşı, E. et al. (2019). Low-temperature plasma-enhanced atomic layer deposition of tin(iv) oxide from a functionalized alkyl precursor: fabrication and evaluation of SnO_2-based thin-film transistor devices. *ACS Appl. Mater. Interfaces* 11 (3): 3169–3180.

17 Lim, S., Baek, I., Kim, K. et al. (2019). Atomic layer deposition of SnO_2 thin films using tetraethyltin and H_2O_2. *Cera. Intern.* 45 (16): 20600–20605.

18 Han, J., Chung, Y., Park, B. et al. (2014). Growth of p-type tin(ii) monoxide thin films by atomic layer deposition from bis(1-dimethylamino-2-methyl-2propoxy)tin and H_2O. *Chem. Mater.* 26 (21): 6088–6091.

19 Kim, S., Baek, I., Kim, D. et al. (2017). Fabrication of high-performance p-type thin film transistors using atomic-layer-deposited SnO films. *J. Mater. Chem. C* 5 (12): 3139–3145.

20 Li, B., Akimoto, K., and Shen, A. (2009). Growth of Cu_2O thin films with high hole mobility by introducing a low-temperature buffer layer. *J. Cryst. Growth* 311 (4): 1102–1105.

21 Melo, C., Jullien, M., Battie, Y. et al. (2019). Semi-transparent p-Cu_2O/n-ZnO nanoscale-film heterojunctions for photodetection and photovoltaic applications. *ACS Appl. Nano Mater.* 2 (7): 4358–4366.

22 Klasens, H. and Koelmans, H. (1964). A tin oxide field-effect transistor. *Sol. St. Electron.* 7 (9): 701–702.

23 Sahu, K., Choudhary, S., Khan, S. et al. (2019). Thermal evolution of morphological, structural, optical and photocatalytic properties of CuO thin films. *Nano-Struct. Nano-Objects* 17: 92–102.

24 Al-Kuhaili, M. (2008). Characterization of copper oxide thin films deposited by the thermal evaporation of cuprous oxide (Cu_2O). *Vacuum* 82 (6): 623–629.

25 Pereira, S., Gonçalves, A., Correia, N. et al. (2014). Electrochromic behavior of nio thin films deposited by e-beam evaporation at room temperature. *Sol. Energy Mater. Sol. Cells* 120: 109–115.

26 Alnuwaiser, M., Javaid, K., Jacob, J. et al. (2022). Annealing induced morphology evolution and phase transition in SnO_x thin films grown by e-beam evaporation method. *Inorg. Chem. Commun.* 140: 109473.

27 Pan, L., Li, W., Yang, S. et al. (2019). Effects of annealing conditions on the properties of SnO films deposited by e-beam evaporation process. *Mater. Lett.* 257: 126737.

28 Fallah, H., Ghasemi, M., Hassanzadeh, A. et al. (2007). The effect of annealing on structural, electrical and optical properties of nanostructured ITO films prepared by e-beam evaporation. *Mater. Res. Bull.* 42 (3): 487–496.

29 Sun, Y., Yang, H., Zhao, Z. et al. (2020). Fabrication of ZnO quantum dots@SnO_2 hollow nanospheres hybrid hierarchical structures for effectively detecting formaldehyde. *Sens. Actuators, B* 318: 128222.

30 Wei, J., Zang, Z., Zhang, Y. et al. (2017). Enhanced performance of light-controlled conductive switching in hybrid cuprous oxide/reduced graphene oxide (Cu_2O/rGO) nanocomposites. *Opt. Lett.* 42 (5): 911–914.

31 Deng, J., Fu, Q., Luo, W. et al. (2016). Enhanced H_2S gas sensing properties of undoped ZnO nanocrystalline films from QDs by low-temperature processing. *Sens. Actuators, B* 224: 153–158.

32 Shi, H., Li, W., Sun, L. et al. (2011). Synthesis of silane surface modified ZnO quantum dots with ultrastable, strong and tunable luminescence. *Chem. Commun.* 47 (43): 11921–11923.

33 Aneesh, N., Vanaja, K., and Jayaraj, M. (2007). Synthesis of ZnO nanoparticles by hydrothermal method. *Proc. SPIE* 6639: 66390J.

34 Meng, X., Yao, M., Mu, S. et al. (2019). Oxygen vacancies enhance triethylamine sensing properties of SnO_2 nanoparticles. *ChemistrySelect* 4 (38): 11268–11274.

35 Yanqing, Z., Erwei, S., Zhizhan, C. et al. (2001). Influence of solution concentration on the hydrothermal preparation of titania crystallites. *J. Mater. Chem.* 11 (5): 1547–1551.

36 Wilson, G., Will, G., Frost, R. et al. (2002). Efficient microwave hydrothermal preparation of nanocrystalline anatase TiO_2 colloids. *J. Mater. Chem.* 12 (6): 1787–1791.

37 Yang, B., Myung, N., and Tran, T. (2021). 1D metal oxide semiconductor materials for chemiresistive gas sensors: a review. *Adv. Electron. Mater.* 7 (9): 2100271.

38 Morozov, V., Morozova, T., and Kallenbach, N. (1998). Atomic force microscopy of structures produced by electrospraying polymer solutions. *Int. J. Mass Spectrom.* 178 (3): 143–159.

39 Huang, Z., Zhang, Y., Kotaki, M. et al. (2003). A review on polymer nanofibers by electrospinning and their applications in nanocomposites. *Compos. Sci. Technol.* 63 (15): 2223–2253.

40 Gugulothu, D., Barhoum, A., Nerella, R. et al. (2019). Fabrication of nanofibers: electrospinning and non-electrospinning techniques. In: *Handbook of Nanofibers* (ed. A. Barhoum, M. Bechelany, and A.S.H. Makhlouf), 45–77. Cham: Springer International Publishing.

41 Park, J., Moon, J., Lee, S. et al. (2009). Fabrication and characterization of ZnO nanofibers by electrospinning. *Curr. Appl. Phys.* 9 (3, Supplement): S210–S212.

42 Wu, H. and Pan, W. (2006). Preparation of zinc oxide nanofibers by electrospinning. *J. Am. Ceram.* 89 (2): 699–701.

43 Zhang, Y., He, X., Li, J. et al. (2008). Fabrication and ethanol-sensing properties of micro gas sensor based on electrospun SnO_2 nanofibers. *Sens. Actuators, B* 132 (1): 67–73.

44 Yang, D., Kamienchick, I., Youn, D. et al. (2010). Ultrasensitive and highly selective gas sensors based on electrospun SnO_2 nanofibers modified by Pd loading. *Adv. Funct. Mater.* 20 (24): 4258–4264.

45 Lim, S., Hwang, S., Chang, D. et al. (2010). Preparation of mesoporous In_2O_3 nanofibers by electrospinning and their application as a CO gas sensor. *Sens. Actuators, B* 149 (1): 28–33.

46 Zheng, W., Lu, X., Wang, W. et al. (2009). A highly sensitive and fast-responding sensor based on electrospun In_2O_3 nanofibers. *Sens. Actuators, B* 142 (1): 61–65.

47 Guan, H., Shao, C., Wen, S. et al. (2003). Preparation and characterization of NiO nanofibres via an electrospinning technique. *Inorg. Chem. Commun.* 6 (10): 1302–1303.

48 Sahay, R., Suresh, P., Aravindan, V. et al. (2012). High aspect ratio electrospun CuO nanofibers as anode material for lithium-ion batteries with superior cycleability. *J. Phys. Chem. C* 116 (34): 18087–18092.

49 Zhou, T., Zang, Z., Wei, J. et al. (2018). Efficient charge carrier separation and excellent visible light photoresponse in Cu_2O nanowires. *Nano Energy* 50: 118–125.

50 Chevalier-César, C., Capochichi-Gnambodoe, M., and Leprince-Wang, Y. (2014). Growth mechanism studies of ZnO nanowire arrays via hydrothermal method. *Appl. Phys. A* 115 (3): 953–960.

51 Wang, H., Zhang, P., and Zang, Z. (2020). High performance $CsPbBr_3$ quantum dots photodetectors by using zinc oxide nanorods arrays as an electron-transport layer. *Appl. Phys. Lett.* 116 (16): 5.

52 Lin, H., Baltrus, J., Gao, H. et al. (2016). Perovskite nanoparticle-sensitized Ga_2O_3 nanorod arrays for CO detection at high temperature. *ACS Appl. Mater. Interfaces* 8 (14): 8880–8887.

53 Alfaro, C., Sanchez-Martinez, D., and Torres-Martínez, L. (2019). TiO_2 nanorods grown by hydrothermal method and their photocatalytic activity for hydrogen production. *Mater. Lett.* 237: 310–313.

54 Saravanakumar, B. and Kim, S. (2014). Growth of 2D ZnO nanowall for energy harvesting application. *J. Phys. Chem. C* 118 (17): 8831–8836.

55 Sheng, L., Liao, T., Kou, L. et al. (2017). Single-crystalline ultrathin 2D TiO_2 nanosheets: a bridge towards superior photovoltaic devices. *Mater. Today Energy* 3: 32–39.

56 Li, M., Chen, Y., Li, W. et al. (2017). Ultrathin anatase TiO_2 nanosheets for high-performance photocatalytic hydrogen production. *Small* 13 (16): 1604115.

57 Zhou, X., Li, X., Gao, Y. et al. (2019). Preparation and characterization of 2D ZnO nanosheets/regenerated cellulose photocatalytic composite thin films by a two-step synthesis method. *Mater. Lett.* 234: 26–29.

58 Wang, H., Cao, S., Yang, B. et al. (2020). NH_4Cl-modified ZnO for high-performance $CsPbIBr_2$ perovskite solar cells via low-temperature process. *Solar RRL* 4 (1): 1900363.

59 Cai, W., Li, H., and Zang, Z. (2021). One-volt, solution-processed InZnO thin-film transistors. *IEEE Electron Device Lett.* 42 (4): 525–528.

60 Cai, W., Li, M., Li, H. et al. (2022). Significant performance and stability improvement of low-voltage InZnO thin-film transistors by slight La doping. *Appl. Phys. Lett.* 121 (6): 062108.
61 Adamopoulos, G., Thomas, S., Wöbkenberg, P. et al. (2011). High-mobility low-voltage ZnO and Li-doped ZnO transistors based on ZrO_2 high-k dielectric grown by spray pyrolysis in ambient air. *Adv. Mater.* 23 (16): 1894–1898.
62 Isakov, I., Faber, H., Grell, M. et al. (2017). Exploring the leidenfrost effect for the deposition of high-quality In_2O_3 layers via spray pyrolysis at low temperatures and their application in high electron mobility transistors. *Adv. Funct. Mater.* 27 (22): 1606407.
63 Adamopoulos, G., Bashir, A., Thomas, S. et al. (2010). Spray-deposited Li-doped ZnO transistors with electron mobility exceeding 50 cm^2/Vs. *Adv. Mater.* 22 (42): 4764–4769.
64 Pattanasattayavong, P., Thomas, S., Adamopoulos, G. et al. (2013). P-channel thin-film transistors based on spray-coated Cu_2O films. *Appl. Phys. Lett.* 102 (16): 163505.
65 Lee, D., Chang, Y., Herman, G. et al. (2007). A general route to printable high-mobility transparent amorphous oxide semiconductors. *Adv. Mater.* 19 (6): 843–847.
66 Lee, S., Kim, J., Choi, J. et al. (2012). Patterned oxide semiconductor by electrohydrodynamic jet printing for transparent thin film transistors. *Appl. Phys. Lett.* 100 (10): 102108.
67 Kim, S., Kim, K., Hwang, Y. et al. (2016). High-resolution electrohydrodynamic inkjet printing of stretchable metal oxide semiconductor transistors with high performance. *Nanoscale* 8 (39): 17113–17121.
68 Jun, T., Song, K., Jeong, Y. et al. (2011). High-performance low-temperature solution-processable ZnO thin film transistors by microwave-assisted annealing. *J. Mater. Chem.* 21 (4): 1102–1108.
69 Cheng, H. and Tsay, C. (2010). Flexible a-IZO thin film transistors fabricated by solution processes. *J. Alloys Compd.* 507 (1): L1–L3.
70 Kim, Y., Heo, J., Kim, T. et al. (2012). Flexible metal-oxide devices made by room-temperature photochemical activation of sol–gel films. *Nature* 489 (7414): 128–132.
71 Xia, G., Zhang, Q., and Wang, S. (2018). High-mobility IGZO TFTs by infrared radiation activated low-temperature solution process. *IEEE Electron Device Lett.* 39 (12): 1868–1871.
72 Twyman, N., Tetzner, K., Anthopoulos, T. et al. (2019). Rapid photonic curing of solution-processed In_2O_3 layers on flexible substrates. *Appl. Surf. Sci.* 479: 974–979.
73 Rim, Y., Choi, H., Kim, K. et al. (2016). Effects of structural modification via high-pressure annealing on solution-processed InGaO films and thin-film transistors. *J. Phys. D: Appl. Phys.* 49 (7): 075112.

3

Metal Oxide Semiconductors for Diodes

3.1 Introduction

As a fundamental electronic component with a typical two-terminal structure, diodes enable current flow in only one direction. Over the past few decades, diodes have found numerous practical applications, including, but not limited to, UV optical sensing, gas sensing, light-emitting diodes (LEDs), and photovoltaics [1–4].

In the industry, Si and Ge remain the most commonly used semiconductor materials for diode fabrication. However, these materials possess limitations such as rigidity and opacity, which restrict their potential applications in flexible and transparent electronics. Alternative semiconductor materials, including III–V compound semiconductors, 2D materials, oxide semiconductors, and organic semiconductors, have gained attention for their potential use in diode fabrication. Among these materials, oxide semiconductors have emerged as promising alternatives to conventional semiconductors in next-generation printable electronics. This is due to their relatively low deposition temperature, suitable bandgap, transparency, flexibility, and cost-effectiveness.

The reported research on oxide-based diodes can be categorized into four main types, *p–n* heterojunction diodes, Schottky diodes, metal–insulator–semiconductor (MIS) diodes, and self-switching diodes (SSDs). In this chapter, the fundamental working principles of those four types of diodes will be introduced, followed by the discussions on their recent progress using oxide semiconductors, the potential applications, and the further directions toward commercialization.

3.2 *P–N* Heterojunction Diodes

P–N junction diodes, also known as rectifier diodes, are formed by interfacing a *p*-type and an *n*-type semiconductor material. This interface results in band bending, which modifies the electronic properties at the junction. Depending on the band alignment of *n*-type and *p*-type semiconductors, *p–n* junction diodes can be further classified into homojunction and heterojunction. However, achieving a *p–n* homojunction is challenging, especially for wide bandgap semiconductors due to doping

Metal Oxide Semiconductors: Synthesis, Properties, and Devices,
First Edition. Zhigang Zang, Wensi Cai, and Yong Zhou.

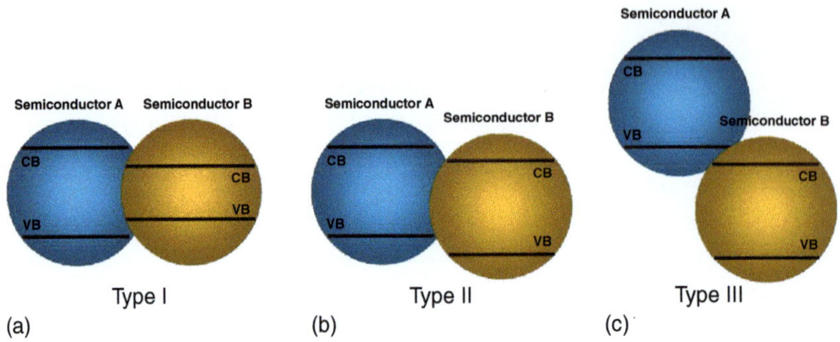

Figure 3.1 Different types of heterojunctions. (a) Type I, (b) type II and (c) type III.

asymmetry [5–7]. Heterojunctions, on the other hand, offer potential in the fabrication of bipolar devices with attempts that have already been reported.

As shown in Figure 3.1, heterojunctions can be further classified into three types, namely straddling gap (type I), staggered gap (type II), and broken gap (type III). In type I heterojunctions, the conduction-band minimum (CBM) and valence-band maximum (VBM) of one material are fully contained within the bandgap of the other material. In type II heterojunctions, there is a partial overlap between the bandgap of one material and the CBM of the other material. On the other hand, type III heterojunctions involve materials where the CBM and VBM of one material lie entirely outside the bandgap of the other material, with no overlap. Typically, type I heterojunctions are suitable for applications in LEDs and lasers, type II heterojunctions find applications in photovoltaic devices and photodetectors, while type III heterojunctions are primarily used in tunneling devices. It is worth noting that oxide semiconductors, which often have relatively large bandgaps, are challenging to form type III heterojunctions.

3.2.1 Representative Devices

The technology involving *p*–*n* heterojunction devices has several parameters such as reverse saturation current, diffusion potential, energy bandgap, forbidden gap, bias voltage, breakdown voltage (BV), and nature of the recombination centers, can be well characterized by the *I*–*V* characteristics. The specific nature and quality of semiconductor *p*–*n* junction can be understood by the simplest *I*–*V* characterization method.

Out of all potential oxide semiconductors, ZnO-based materials are the most frequently reported ones for forming oxide-based *p*–*n* heterojunctions. The study of *n*-type ZnO grown on other available and comparable *p*-type materials, such as Si, GaN, AlGaN, Cu_2O, NiO_x, and diamond, in realizing *p*–*n* diodes has been conducted for a long time and is widely reported [8]. In 2002, Hosono's groups reported a *p*–*n* heterojunction diode composed of *p*-$ZnRh_2O_4$ and *n*-ZnO [9]. These *p*–*n* heterojunction diodes had an abrupt interface and exhibited rectifying *I*–*V* characteristics with a threshold voltage of ~2 V, in good agreement with the bandgap of $ZnRh_2O_4$, suggesting a good carrier-blocking contact by the heterojunction formed by a narrow bandgap $ZnRh_2O_4$ and a wide bandgap ZnO. This behavior is similar to that of traditional *p–n* junctions and is not specific to the d-electron system.

Later in the same year, the same group also reported *p–n* heterojunction diodes on plastic substrates by combining $ZnRh_2O_4$ with IGZO [10]. Similar to *n*-ZnO, IGZO is an *n*-type oxide semiconductor. However, unlike the previous work where $ZnRh_2O_4$ was thermally annealed at 950 °C to convert the polycrystalline layer to an epitaxial single-crystalline layer, in this work, $ZnRh_2O_4$ was found to show a *p*-type conductivity even in an amorphous state.

Subsequently, *p*-type conductivity was also reported for $ZnCo_2O_4$ as an extension of this series to the $3d^6$ system. However, it is worth noticing that these materials, despite being used as *p*-type semiconductors for the *p–n* junction, are not used as *p*-channel materials in thin-film transistors (TFTs) due to the high concentrations of vacant *d* levels, which serve as hole traps in the bandgap.

More recently, NiO_x has garnered significant attention as a *p*-type oxide semiconductor in forming *p–n* heterojunction diodes, owing to their advantages including a wide bandgap of 3.6–4.3 eV, excellent chemical stability, low toxicity, and cost-effective processability. By sputtering NiO_x films on *n*-Si, *p–n* heterojunction diodes with Ni/*p*-NiO_x/*n*-Si/Ag structures were demonstrated by Yilmaz and coworkers [11], showing a homogeneous barrier height, an average effective barrier height, and an average ideality factor of 0.86 eV, 0.652 eV, and 1.948, respectively.

NiO_x-based heterojunction diodes were successfully achieved using oxide semiconductors only, forming heterostructures such as *n*-ZnO/*p*-NiO_x, *n*-IGZO/*p*-NiO_x, and *n*-Ga_2O_3/*p*-NiO_x. As early as in 2003, Ohta et al. prepared transparent tri-layered oxide films consisting of ZnO/NiO/ITO on a YSZ substrate and demonstrated *p–n* heterojunction diodes with a clear *I–V* characteristic, an ideality factor of ~2, and a forward threshold voltage of ~1 V [12]. The researchers also investigated the potential of these heterojunction diodes as UV detectors, given that both semiconductors used in the heterostructure had wide bandgaps. As shown in Figure 3.2, although the diodes showed a weak photoresponse at zero-bias voltage, a photoresponse up to ~0.3 A W^{-1} could be achieved at an applied reverse bias of −6 V under 360-nm light irradiation.

Later in 2010, Shao and coworkers [13] used a sol–gel process and an electrospinning technology to achieve *p*-NiO_x/*n*-ZnO heterojunction nanofibers with varying molar ratios of Ni to Zn. A comparative analysis revealed that these heterostructures exhibited higher photocatalytic activity for the degradation of rhodamine B dye under UV light irradiation, in contrast to those employing pure NiO_x or ZnO nanofibers. This was demonstrated to originate from the enhanced separation efficiency of photogenerated electron–hole pairs facilitated by the *p–n* heterojunction. Specifically, under UV irradiation, photogenerated electrons moved to ZnO while photogenerated holes moved to NiO_x. Subsequently, the holes were ultimately trapped by surface hydroxyl groups (or H_2O) on the catalyst surface to yield $OH^{\cdot}$ radicals. Concurrently, the dissolved oxygen molecules reacted with the electrons, yielding superoxide radical anions, $O_2^{\cdot -}$, which, upon protonation, generated hydroperoxy, $HO_2^{\cdot}$ radicals. These radicals further led to the generation of hydroxyl radical, $OH^{\cdot}$, a strong oxidizing agent for decomposing organic dye. Moreover, the

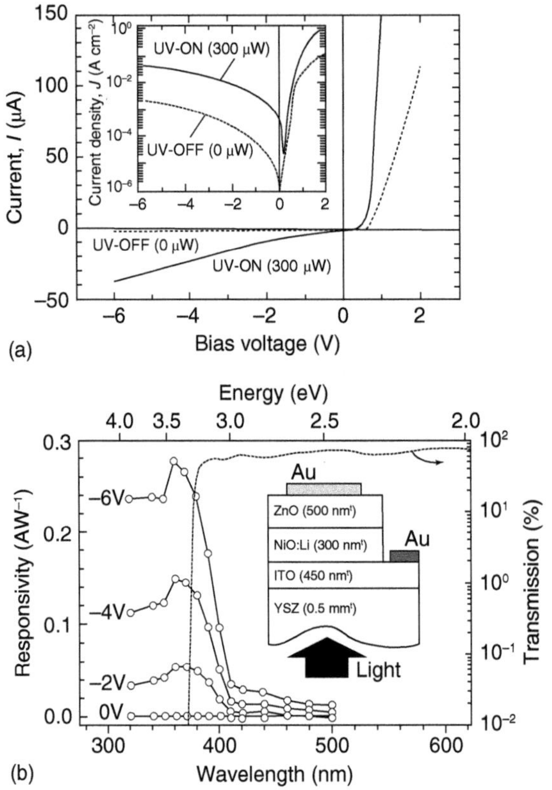

Figure 3.2 (a) Typical *I–V* characteristics of *n*-ZnO/*p*-NiO heterojunction diodes in the dark and under light illumination. (b) Special response of the diode at several reverse-bias voltages. Source: Reproduced from Ohta et al. [12]/with permission of AIP Publishing.

one-dimensional nanostructure of the heterojunction nanofibers facilitated their easy recyclability without compromising the photocatalytic activity, highlighting their significant potential for eliminating organic pollutants from wastewater.

Due to the challenging nature of achieving *p*-type Ga_2O_3, NiO_x has emerged as an effective material for forming *p–n* heterojunctions with Ga_2O_3, given their relatively similar bandgaps. In a recent study by Wang et al. [14], *p*-NiO_x/*n*-Ga_2O_3 heterojunction diodes were successfully demonstrated, which exhibited a large built-in potential and alleviated the peak electric field to enhance the BV. The diodes obtained a high on/off ratio of 10^9, a turn-on voltage of 2.6 V, a forward current of 1.3 kA cm^{-2}, a specific on resistance ($R_{on,\ sp}$) of 1.08 mΩ cm^2 at a bias voltage of 5 V, and a large BV of 1.22 kV. Such heterojunction diodes were also used as the gate in heterojunction field-effect transistors. By calculating the power figure-of-merit (P-FOM) through $\text{P-FOM} = BV^2/R_{on,\ sp}$, the highest P-FOM was found to be 1.38 and 0.39 GW cm^{-2} for heterojunction diodes and heterojunction field-effect transistors, respectively, suggesting the great potential in the high-power community.

3.2.2 Applications

P–N heterojunction diodes have been widely used in photodiodes in which a built-in electric field occurs, helping to separate photon-generated electron–hole

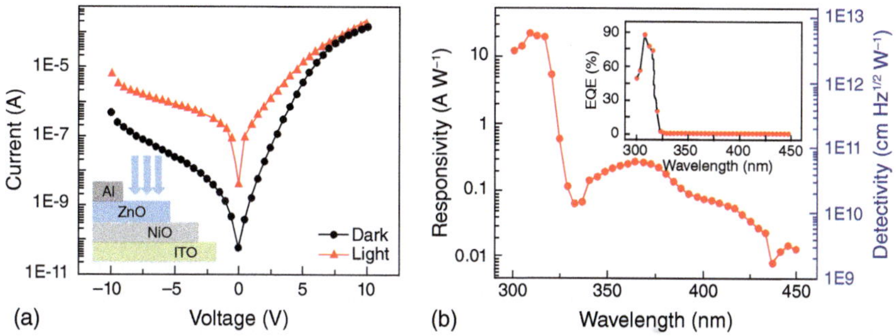

Figure 3.3 (a) *I*–*V* characteristics under both dark and UV illumination. (b) Responsivity, detectivity, and EQE as a function of wavelength. Source: Reproduced from Park et al. [18]/with permission of Royal Society of Chemistry.

pairs and excitons to move in opposite directions, enabling self-power under zero bias. Considering the wide bandgap of oxide semiconductors, oxide-based heterojunction photodiodes could also be used as photodetectors, especially for the UV region. Various combinations have been reported previously capable of working as *p–n* photodiodes, including, but not limited to, *n*-TiO_2/*p*-Cu_2O [15], *n*-ZnO/*p*-CuO_x [16, 17], *n*-ZnO/*p*-NiO_x [18, 19], and *n*-GZO/*p*-NiO_x [20]. Taking Wang's work as an example, thin-film heterojunction photodiodes made of sol-gel-deposited NiO and ZnO were reported [18]. As shown in Figure 3.3a, under dark conditions, the heterojunction photodiodes exhibited good rectifying behavior at room temperature, while showing an obvious photoresponse under illuminations and reverse biases. Besides, these photodiodes, operating in a wavelength range of 300–450 nm, also exhibited an exceptional performance in the UV region, including a maximum responsivity of 21.8 A W^{-1} and an equivalent external quantum efficiency (EQE) of 88% at 310 nm (Figure 3.3b).

Similarly, Tsai et al. reported heterojunction photodetectors using *p*-type NiO and *n*-type ZnO [21]. The energy band diagram of *p*-NiO/*n*-ZnO heterojunction confirmed the generation of a built-in electric field due to the Fermi level alignment, resulting in photodetectors with diode-like characteristics. Upon UV illumination, the current increased in both the forward and reverse bias directions, reaching a current sensitivity of 11.56 at a reverse bias voltage of −1 V.

Alternatively, oxide-based *p–n* heterojunction diodes have shown great potential as gas sensors, since they can enhance the gas-sensing properties of metal oxides toward targeted gases. This includes the enhanced gas-sensing properties for reducing gases (i.e. H_2, H_2S, CO, NH_3, and ethanol) and oxidization gases (i.e. NO_2). For example, enhancements in gas-sensing performance toward reducing gases such as H_2, toluene, and acetone were reported using In_2O_3-decorated CuO nanowires [22], SnO_2-decorated NiO nanostructure [23], SnO_2-decorated NiO foam [24], and Cu_xO-modified ZnO composite [25].

In terms of oxidization gases, NiO–SnO_2 nanocomposites (*p–n* junctions) were found to exhibit an improved gas-sensing performance toward NO_2 compared with pure NiO [26]. This improvement can be attributed to the higher Fermi level of

SnO_2 compared to NiO, allowing electrons to transfer from SnO_2 to NiO while holes diffuse from NiO to SnO_2. In an air atmosphere, the adsorption of oxygen molecules would capture electrons in SnO_2 and release holes to NiO, resulting in the formation of a thin depletion layer in SnO_2 and an accumulation layer in NiO. A potential barrier was then established between NiO and SnO_2, and the carriers in the sensing materials were primarily transported through the accumulation layer. Upon exposure to NO_2, the initial conductance significantly decreased, causing an increase in the equivalent hole concentration of the nanocomposites associated with effective electron transfer via the interface bonds at the heterojunction. The variation of contact potential in the nanocomposites before and after exposure to NO_2 also exerted a drastic effect on the transducer function for gas sensing. Thus, the sensing performance significantly improved compared with those using pure NiO.

However, despite showing a great potential in real applications, the formation of fully oxide-based *p–n* heterojunction diodes remains challenging, primarily due to the lack of high-performance *p*-type oxide semiconductors. Using alternative *p*-type materials has emerged as a strategy to realize heterojunctions, but limitations persist, such as defects and interface states induced by lattice mismatch and carrier blocking caused by band misalignment, and so on. Future advancements in this field are likely to focus on the development of new *p*-type oxide semiconductors and doping strategies to achieve high-performance *p–n* heterojunction diodes.

3.3 Schottky Diodes

Generally, achieving a low contact resistance is crucial for obtaining good Ohmic contacts and semiconductor devices with high operating speeds. In contrast, an ideal Schottky junction behaves like an ideal diode, allowing a high current flow in the forward-biased direction and exhibiting infinite resistance in the opposite direction [27]. The type of junction can be modified by varying electrode materials and doping levels of the semiconductor [28]. For example, for *n*-type oxide semiconductors, when metals with high work functions contact them, a Schottky barrier is formed at the interface. Schottky diodes are found ideal for rectifier circuits (i.e. power detectors, mixers, and frequency multipliers) operated at frequencies up to terahertz, which could easily find their practical applications in wireless communication systems and also energy-harvesting applications. By carefully engineering the Schottky barrier at the metal–semiconductor (MS) junction, a low voltage drop can be obtained. Their unipolar operation, involving only majority carriers, means that negligible storage of excess minority carriers occurs, leading to small capacitances that can further boost their operating frequency. In this subchapter, we provide a brief introduction to Schottky diodes, focusing on their working mechanisms, performance characterization, and presenting the most representative devices reported to date.

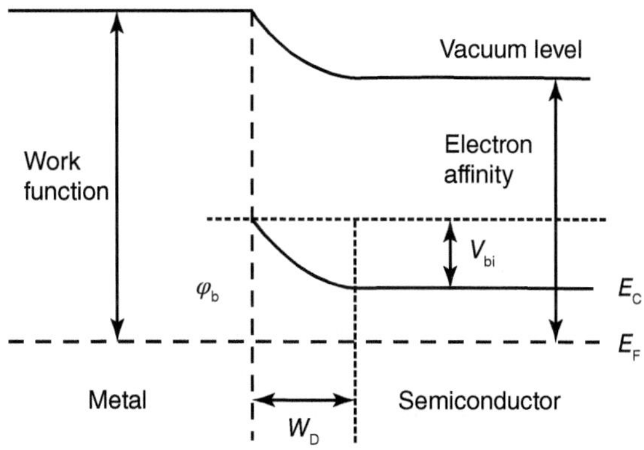

Figure 3.4 A schematic energy band diagram of metal/oxide semiconductor contact.

3.3.1 Working Mechanisms

Normally, Schottky diodes show a low turn-on voltage and a low junction capacitance, making them widely used in low voltage circuits for reverse current or discharge protection and high-frequency rectification. However, there have been relatively few studies conducted on oxide semiconductor-based Schottky diodes thus far.

As shown in Figure 3.4, the formation of a Schottky barrier requires the metal's work function to be larger than the semiconductor's electron affinity. The work function represents the minimum energy required to move an electron from the Fermi level to vacuum, while the electron affinity represents the energy gap between the conduction band and vacuum level. When the semiconductor gets in contact with the metal, the electrons at the interface transfer into the metal, resulting in a positively charged layer at the interface of the semiconductor. This leads to a reduction in the Fermi level within the semiconductor and a bending of the energy band at the interface. Figure 3.4 shows the barrier height, φ_b, and the built-in voltage, V_{bi}, which are caused by the band realignment. The Schottky barrier height (SBHM) signifies the energy required for an electron at the Fermi level in the metal to flow into the semiconductor. The built-in voltage describes the energy necessary for an electron to move from the semiconductor's CBM to the metal.

In theory, the current flow through metal to semiconductor follows the thermionic emission model, which can be described as

$$I = AA^{*}T^{2}exp\left(\frac{-q\varphi_{b}}{k_{B}T}\right)\left[exp\left(\frac{q(V-IR_{S})}{nk_{B}T}\right)-1\right] \tag{3.1}$$

where A is the area of the active area, A^* represents the Richardson constant, T is the temperature, φ_b is the barrier height, k_B is the Boltzmann constant, V is the applied bias, R_S is the series resistance, and n is the ideality factor. The depletion width can be obtained by

$$W_{D} = \sqrt{2\varepsilon_{s}qN_{bg}\left(V_{bi} - V - \frac{k_{B}T}{q}\right)} \tag{3.2}$$

where ε_s is the dielectric permittivity, N_{bg} is the background doping density, and V_{bi} is the built-in voltage. The capacitance per unit area of the depletion layer can be obtained by

$$C_D = \varepsilon_s / W_D \tag{3.3}$$

Thus, the relation between C_D and the applied bias can be expressed as

$$C_D^{-2} = \frac{2\left(V_{bi} - V - \frac{k_B T}{q}\right)}{q\varepsilon_s N_{bg}} \tag{3.4}$$

Here, if N_{bg} is a constant in the depletion layer, a straight line can be obtained by plotting C_D^{-2} as a function of voltage.

3.3.2 ZnO Schottky Diodes

As a simple binary oxide semiconductor, ZnO has been studied for several decades, including the uses in thin-film transistors, solar cells, photodetectors, and light-emitting devices. The investigation of SBHM on ZnO began with the pioneering work of Mead [29] in 1965, where they reported the barrier heights for several metals on a vacuum-cleaved ZnO surface, highlighting the potential of ZnO as an active layer in Schottky diodes.

In 2003, Park et al. [30] reported the fabrication of ZnO nanorod Schottky diode arrays. These ZnO nanorods were grown by a noncatalytic metalorganic vapor phase epitaxy with evaporated Au serving as the Schottky contacts on the vertically aligned ZnO. Later in 2006, Wang's group [31] reported ZnO nanobelt/nanowire Schottky diodes by forming ZnO across Au electrodes through a dielectrophoresis alignment process. The formation of Schottky diodes was attributed to the asymmetric contacts created during the dielectrophoresis aligning and was found to be dependent on the size of the nanobelts. As shown in Figure 3.5, a current of 0.5 μA was achieved at a forward bias of 1.5 V, and the diode was capable of withstanding applied voltages of up to 10 V. The diode also showed an ideality factor of ~3 with a current on/off ratio of 2000, suggesting the potential of single nanobelt/nanowire-based diodes for various applications.

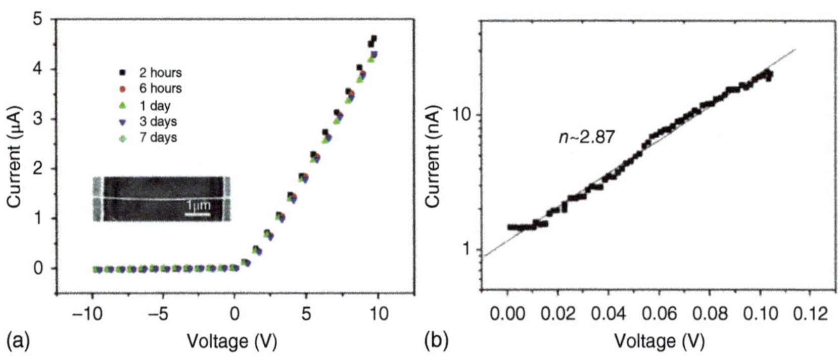

Figure 3.5 (a) I–V characteristics and (b) extracted ideality factor of ZnO Schottky diodes. Source: Reproduced from Lao et al. [31]/with permission of American Chemical Society.

ZnO-based Schottky diodes have potential applications as high-efficiency UV photodetectors, as reported by Nakano et al. [32], using PEDOT:PSS as the contact material for ZnO bulk single crystals. The fabricated devices showed a quantum efficiency as high as unity in ultraviolet regions, along with a visible rejection ratio of $\sim 10^3$ in the spectral response of the photodiode under zero-bias conditions. Furthermore, the photodiodes demonstrated a normalized detectivity of 3.6×10^{14} cm $Hz^{1/2}$ W^{-1} at a wavelength of 370 nm.

The emergence of next-generation flexible and transparent mobile electronics has led to increased research efforts in developing Schottky diodes with an ultra-high operating frequency. In 2016, Anthopoulos's group achieved a significant breakthrough by successfully fabricating ZnO-based coplanar Schottky nanodiodes on poly(ethylene terephthalate) (PET) substrates, achieving an operating frequency exceeding 400 MHz [33]. Utilizing adhesion lithography (a-Lith), they achieved electrode spacing of less than 20 nm in an unprecedented lateral architecture (Figure 3.6a). This architecture brought numerous benefits aside from its low surface area, including low device series resistance, negation of scattering at ZnO grain boundaries, and stable device performance. As shown in Figure 3.6b, despite having a large electrode width, the fabricated coplanar nanogap diodes operated at a low voltage of less than 3 V and while exhibiting a high current rectification ratio

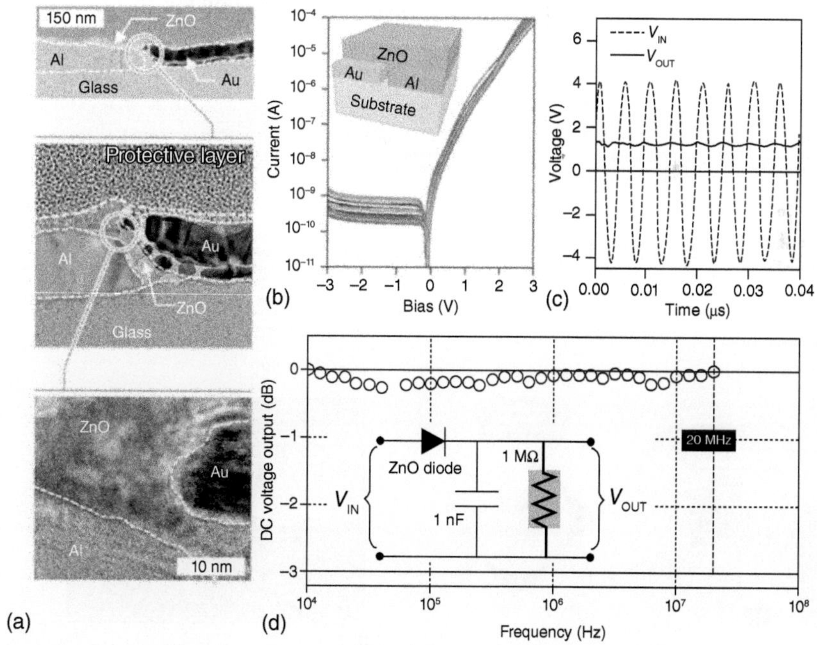

Figure 3.6 (a) Cross-sectional HRTEM images of a representative ZnO coplanar nano-Schottky diode. (b) I–V characteristics of 38 ZnO coplanar nano-Schottky diodes fabricated on the same substrate. (c) AC input and DC output signals of a ZnO diode-based rectifier circuit operated at 13.56 MHz. (d) Bode plot of the DC output signal vs AC input signal frequency up to 20 MHz with the inset showing the used DC rectifier circuit. Source: Reproduced with permission from Semple et al. [33] © 2016, JOHN WILEY & SONS, INC.

approaching 10^6. Besides, these diodes were also found to be suitable for integration into typical radio frequency identification (RFID) tags operating at 13.56 MHz (Figure 3.6c,d), underscoring the tremendous potential of the reported technologies.

Later, the same research group reported ZnO Schottky diodes with an exceptional intrinsic cut-off frequency exceeding 100 GHz, employing coplanar nanogap (~15 nm) electrodes to allow considerable reduction in capacitance [34]. The preparation of such Schottky diodes is shown in Figure 3.7a with the schematic diagram and SEM images shown in Figure 3.7b,c, respectively. In such devices, the active layer thickness was shrunk to nanogap channel length, while the device width remained at the macroscale, therefore effectively minimizing the device resistance and capacitance compared with conventional approaches, enabling the operation at ultrahigh frequencies.

Undeniably, ZnO has demonstrated great potential as active layers in Schottky diodes, offering advantages such as high operating frequencies and the ease of solution deposition, making them promising candidates for next-generation printable electronics. However, ZnO still has its limitations, especially regarding the requirement of crystallinity to achieve high mobility, which hinders large-area uniformity. Although Anthopoulos's work showcased the ability of ZnO Schottky diodes to achieve high operating frequencies while mitigating scattering at ZnO grain boundaries through structural modifications, further research is necessary, particularly at the material level, to advance commercialization efforts.

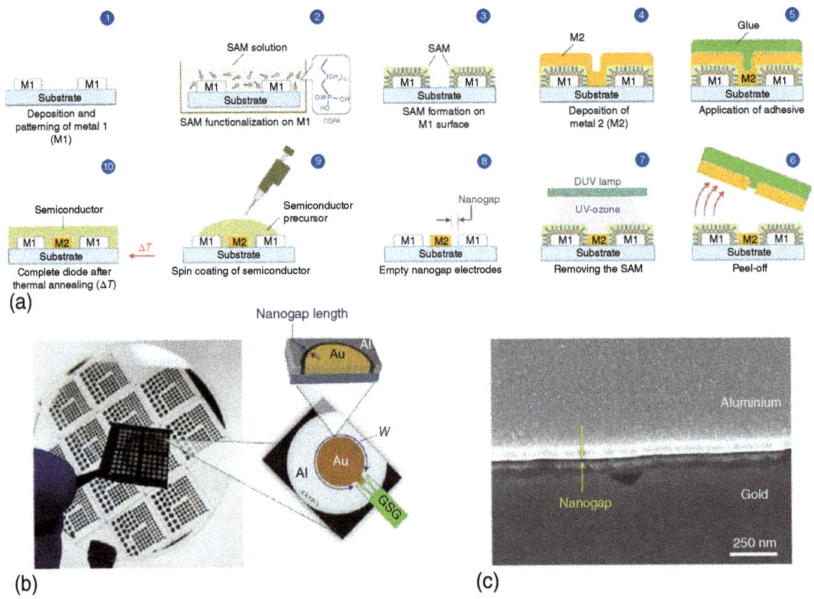

Figure 3.7 (a) Generic diagrams of adhesion lithography process steps for the fabrication of coplanar asymmetric Al/Au nanogap electrodes. (b) Photographs and corresponding schematic diagrams of adhesion-lithography-patterned Al/Au nanogap devices. (c) An SEM image showing the uniformity of the nanogap separating the two Al and Au electrodes. Source: Reproduced with permission from Georgiadou et al. [34] © 2020, Springer Nature.

3.3.3 IGZO Schottky Diodes

In comparison to ZnO, IGZO has the advantage of maintaining high mobility even in amorphous states, making it a promising candidate for mass production in the industry. There is now growing research focused on IGZO Schottky diodes, which have demonstrated excellent electrical properties including rectification ratios of $>10^7$, barrier heights of >0.9 eV, and ideality factors close to unity [35, 36]. Efforts have also been made to achieve flexibility and GHz operating frequencies, demonstrating the great potential in future flexible mobile technologies [36–38].

A representative flexible high-frequency IGZO Schottky diode was reported by Song's group in 2015 [37]. They studied a range of IGZO thicknesses to optimize the diode dimensions, as depicted in Figure 3.8a,b. Interestingly, they discovered an unexpected dependence of the diode speed on the IGZO thickness, with diodes using 100-nm-thick IGZO exhibiting the worst performance, possibly relating to an increased voltage drop across the diode, resulting in a reduced depletion width and an elevated effect series resistance. The dimensions of the diode's active area were also investigated, and by shrinking the dimensions, they were able to reduce capacitance while increasing contact resistance. By varying the diode dimensions, they achieved a high cut-off frequency of 6.3 GHz on PET substrates with an active area of 200 μm^2 (Figure 3.8c), a value beyond the critical benchmark speed of 2.45 GHz to satisfy the principal frequency bands of smartphones.

Replacing traditional metal contacts with transparent conductive materials could result in transparent Schottky diodes, beneficial for future fully transparent electronics. For example, So and coworkers reported transparent ITO/IGZO Schottky diodes achieved through gradient oxygen doping in the IGZO layer [39]. Wu and coworkers utilized transparent conducting polymer PEDOT:PSS to form Schottky contacts to IGZO, resulting in diodes with a high rectification ratio of $>10^5$ and an ideality factor of ~1.5 [40]. Semitransparent carbon paste was used as an alternative to noble metals by Huang et al. [41], showing increased barrier height and significantly improved rectified properties.

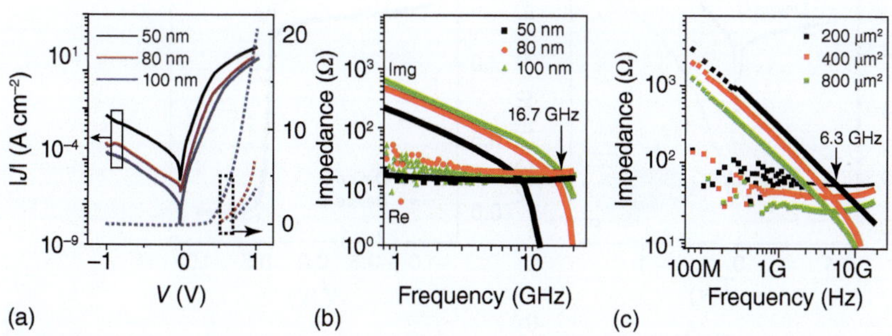

Figure 3.8 (a) J–V characteristics of IGZO Schottky diodes with different IGZO thicknesses. Resistance and reactance of IGZO Schottky diodes at different frequencies with (b) different IGZO thicknesses and (c) different active areas. Source: Reproduced from Zhang et al. [37]/with permission of Springer Nature.

The interfacial chemistry plays a crucial role in forming Schottky contacts on IGZO. The contact resistance between IGZO and metal decreases with improved band alignment between them, but the barrier height strongly depends on the oxygen treatment of the interface [42, 43]. In general, IGZO, with an electron affinity of 4.16 eV, forms an Ohmic contact with low work function metals such as Al (4.28 eV) and Ti (4.33 eV), while high work function metals such as Pd (5.12 eV), Pt (5.65 eV), and Au (5.1 eV) could form a Schottky contact with it. However, in reality, the situation is more complex. Metal contacts to oxide semiconductors have shown a quasi-ohmic behavior regardless of their work function, possibly due to the formation of conductive layers within the metal oxide [44]. This behavior is observed not only with reactive metal electrodes, but also with inert metals like Pt [42]. A possible explanation for this at Pt-IGZO interfaces is the reduction of In^{3+} to In^0, as suggested by Herman's work [45].

To investigate the influence of oxygen in the formation of Schottky barriers, Song and coworkers conducted a study where they sputtered IGZO and Pt under different conditions to create IGZO Schottky diodes [43]. As shown in Figure 3.9, while a poor Schottky junction was formed in devices sputtered in Ar, those with Pt or IGZO sputtered in Ar/O_2 mixed gases showed an improved performance, suggesting a determination of Schottky junction quality by the oxygen content at the Schottky interface. The presence of dangling bonds at the IGZO interface, caused by oxygen deficiencies, may contribute to a high density of interface traps, thereby pinning the Fermi level at a lower energy level. Depositing IGZO with 3% O_2 led to an improved barrier height and ideality factor by reducing the number of oxygen vacancies and dangling bonds. However, it slightly increased the resistivity of the IGZO film, resulting in higher series resistance. Devices with Pt sputtered in an Ar/O_2 mixture gas exhibited even better performance, with barrier height and ideality factor approaching ideal values, thanks to the reduced oxygen deficiencies.

The nonideal current–voltage (*J*–*V*) characteristics observed in IGZO Schottky diodes can be attributed to inhomogeneities in the barrier height. One of the

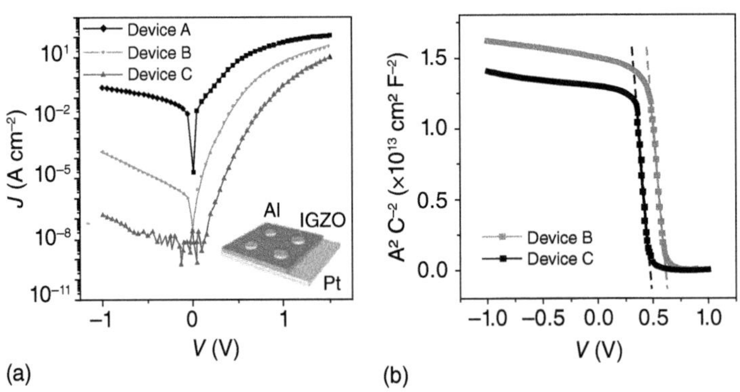

Figure 3.9 (a) *J*–*V* and (b) C^{-2}–*V* characteristics of IGZO Schottky diodes. Device A: IGZO and Pt sputtered in Ar, device B: IGZO in Ar/O_2 and Pt in Ar, device C: IGZO in Ar and Pt in Ar/O_2 mixture gas. Source: Reproduced from Zhang et al. [43]/with permission of AIP Publishing.

commonly reported causes of such inhomogeneity is local variations in interfacial oxygen content. Other factors include potential fluctuations in IGZO arising from the random distribution of In, Ga, and Zn elements [46] as well as variations in the metal work function resulting from the exposure of different metal crystal faces [47]. Numerous research groups have investigated this inhomogeneity through simulations and experiments, with a focus on improving the material properties and modifying the interfaces to address this challenge.

At present, most IGZO Schottky diodes operating at high to ultrahigh frequencies were deposited using sputtering. However, for the development of printable electronics, it is desirable to achieve high-performance IGZO using solution-based methods. Previous studies have demonstrated the deposition of high-performance IGZO thin-film transistors using methods such as spin-coating and inkjet printing [48]. However, unwanted organic residuals, high surface roughness, and remaining –OH groups/oxygen vacancies might act as defect states, restricting the attainment of high operating frequencies in solution-processed IGZO Schottky diodes.

In terms of structure modification, while in theory, the use of high work function metals could benefit the formation of Schottky contacts with IGZO, in reality, the performance of IGZO Schottky diodes is still limited, particularly by the barrier inhomogeneities. Thus, gaining a deeper understanding of barrier inhomogeneities in IGZO Schottky diodes is important to aid device design and help smoothing the path to wider applications.

3.3.4 Ga_2O_3 Schottky Diodes

Ga_2O_3, as a promising next-generation oxide semiconductor with an extremely large bandgap, has garnered numerous research attention particularly in recent years, in fields such as high-power devices and deep-UV detections. However, the *p*-type doping of Ga_2O_3 still faces difficulties, making the realization of fully Ga_2O_3-based bipolar devices quite challenging. Fortunately, substantial progress has been made in unipolar Ga_2O_3 Schottky diodes. For example, Ga_2O_3 Schottky barrier diodes were reported by Sasaki et al. [49], using $\beta-Ga_2O_3$ single-crystal substrates produced through a floating-zone method. The SBHM at the Pt/$\beta-Ga_2O_3$ interface was estimated to be 1.3–1.5 eV. Such devices also showed an ideality factor close to 1 and a reasonably high reverse BV of 150 V, which might be further improved by structure modifications.

Similar works were also reported by Long's and Tao's group in 2017, who studied Schottky barrier diodes with a Pt/$\beta-Ga_2O_3$/Ti structure and investigated their temperature-dependent current–voltage (*I*–*V*) characteristics [50]. Here, β-Ga_2O_3 single crystals with a (100)-orientation were prepared by an edge-defined film-fed growth (EFG) technique. As shown in Figure 3.10, the *I*–*V* measurements conducted at different temperatures, along with the thermionic emission modeling, suggested a good performance of the fabricated Schottky diodes at room temperature. Specially, the diodes exhibited a rectification ratio of 10^{10}, an ideality factor of 1.1, a SBHM of 1.39 eV, a threshold voltage of 1.07 V, an ON resistance of 12.5 mΩ cm^{-2}, a forward current density of 56 A cm^{-2} at 2 V, and an effective donor concentration of about

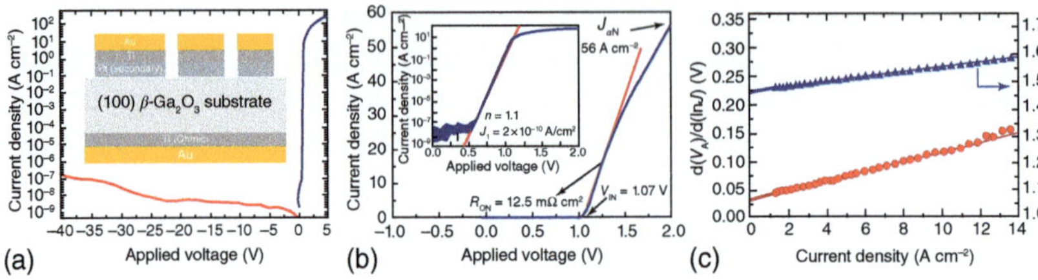

Figure 3.10 (a) $J-V$ characteristics of β−Ga_2O_3 Schottky diodes measured at room temperature. (b) Linear plot of the forward $J-V$ curve with the fitting line. (c) $d(V_A)/d(\ln J)$ and H as functions of current density with the fitting lines. Source: Reproduced from He et al. [50]/with permission of AIP Publishing/CC BY 4.0.

$2.3 \times 10^{14}\,cm^3$. Moreover, the estimated SBHM ranged from 1.3 to 1.39 eV across temperatures ranging from room temperature to 150 °C. Subsequent research by the same groups further demonstrated that the temperature-dependent ideality factor and the SBHM could be explained by the Gaussian distribution of barrier height inhomogeneity [51].

However, despite showing promising properties as Schottky barrier diodes for power devices, the development of Ga_2O_3 is still at an early stage. To date, most of the research on Ga_2O_3-based power Schottky barrier diodes has focused on the β-phase, and the achievement of other phases of Ga_2O_3 for Schottky diodes, especially in amorphous states, would be an interesting research direction in the future. Besides, the low thermal conductivity of Ga_2O_3 poses challenges in terms of self-heating effects. Thus, a better thermal management method of Ga_2O_3 is required to address the issue of poor thermal conductivity. Other potential improvements in the overall performance might also be done by optimizing the contact interface between Ga_2O_3 and metal, as well as device structure modification, similar to other Schottky diodes.

3.4 Metal–Insulator–Semiconductor Diodes

Instead of forming p–n junctions, metal–insulator–semiconductors (MIS) structures are also widely used, especially in the realizations of electroluminescence properties. The MIS structure involves the insertion of an insulating layer between the metal and semiconductor layers. This insulating layer, typically made of oxide, serves multiple purposes. It prevents the diffusion of metal into the semiconductor and also helps to alleviate the reduction of electric field in MIS diodes. The achievement of oxide-based MIS diodes can be dated back to 1978, when Shimizu et al. [52] reported ZnO-based MIS diodes using an insulating ZnO layer obtained by keeping the ZnO crystal in a 30% H_2O_2 solution. Until now, the working mechanisms of MIS diodes could rely on the formation of Schottky junctions between metal and semiconductor, behaving as ideal diodes with current primarily flowing only in the forward-bias direction. Alternatively, MIS diodes with an ultrathin insulator layer can function as MIS tunneling diodes. In this subchapter, we will mainly discuss these two types of MIS diodes.

3.4.1 MIS Schottky Diodes

Several parameters, such as the choice of metal materials and the doping level of the semiconductor, can affect the formation of Schottky junctions. MIS Schottky diodes show working mechanisms similar to normal Schottky diodes. However, compared to traditional MS structured Schottky diodes where the metal is directly deposited on the semiconductor, MIS Schottky diodes offer advantages of an enhanced barrier height, improved interface states, and reduced leakage current. To achieve an ideality factor close to unity, low series resistance and low reverse leakage current are required for MS Schottky diodes. However, their electrical properties are strongly influenced by various nonidealities, particularly the interface states. The presence of high interface trap density is known to be the main cause of nonideal current–voltage characteristics in MS diodes. In MIS diodes, the addition of an insulator layer on the semiconductor surface plays a crucial role in passivating dangling bonds and reducing anomalies in the diode's current–voltage characteristics by minimizing surface states. Moreover, for devices operating under high power, high frequency, and high temperature, MS junctions may easily suffer from issues of high leakage current and low BV, limiting device performance, reliability, and stability. By introducing an insulating layer between the metal and semiconductor, these issues can be somehow solved through the passivation of interface defects and the reduction of leakage current.

Similar to the equations given in Section 3.3.1, according to the thermionic emission theory, when applying a bias voltage $V \geq 3k_BT/q$ in the forward region, I–V characteristics for MIS Schottky diodes are given as follows:

$$I = AA^*T^2 exp\left(\frac{-q\varphi_b}{k_BT}\right)\left[exp\left(\frac{q(V-IR_S)}{nk_BT}\right)-1\right] \tag{3.5}$$

where A is the area of the active area, A^* represents the Richardson constant, T is the temperature, k_B is the Boltzmann constant, V is the applied bias, R_S is the series resistance, and n is the ideality factor. The ideality factor and saturation current I_0 can be determined using the slope and intercept of the semilog forward bias ln(I)–V characteristics with ideality factor can be calculated through

$$n = \frac{q}{k_BT}\left(\frac{d(V-IR_s)}{d(\ln(I))}\right) \tag{3.6}$$

For an ideal Schottky diode, the ideality factor should be unity. However, in MIS Schottky diodes, the reported ideality factor was found to be generally larger than 1, which might attribute to a change in the MS barrier height due to the interfacial layer, interface states, series resistance, and so on. The barrier height of a Schottky diode due to the presence of these interface states is strongly dependent on the electric field in the depletion region and hence the applied bias. The ideality factor for MIS Schottky diodes having interface states, N_{ss}, in equilibrium with semiconductor can then be expressed as

$$n(V) = 1 + \frac{\delta}{\varepsilon_i}\left[\frac{\varepsilon_s}{W_D} + qN_{ss}\right] \tag{3.7}$$

where ε_i, ε_S, δ, W_D, and N_{ss} are interfacial layer permittivity, semiconductor permittivity, thickness of the interfacial layer, depletion layer width, and density of interface states, respectively.

Actually, the use of oxide semiconductors as the semiconductor material in MIS Schottky diodes was relatively less, especially compared to other traditional semiconductors like GaN, GaAs, and *n*-Si. The larger bandgap of oxide semiconductors limits the choice of suitable insulators and metals for MIS structures. Additionally, the performance of metal insulator oxide semiconductor (MIOS) Schottky diodes is generally inferior to III–V semiconductor devices, primarily due to the polycrystalline nature of thin films, which introduces defects and grain boundaries. Although IGZO has shown reasonable mobility even in amorphous states, the reported mobility is still not comparable to that of single-crystalline Si or high-quality epitaxial III–V semiconductor films. Besides, oxide semiconductors-based Schottky diodes with a MS structure already show a reasonably high performance, making the addition of another insulator layer less beneficial and potentially complicating the deposition process. A summary of the previously reported MIOS Schottky diodes is shown in Table 3.1.

Alternatively, with a relatively larger bandgap than traditional semiconductors, oxide semiconductors can be used as the insulating layer in MIS diodes. For example, Lee et al. [57] reported GaN MIS diodes using insulating ZnO and metallic Pt, and studied their potential as hydrogen gas sensors. By inserting a 16-nm ZnO layer at the metal/semiconductor interface, Bilge et al. [58] reported Al/ZnO/*p*-Si diodes with a barrier height of 0.73 eV and an ideality factor of 1.44. Al/SnO_2/*p*-Si MIS Schottky diodes were reported by Tugluoglu and coworkers [59], focusing on the role of interface states and series resistance on the diode performance.

In 2017, Cicek and coworkers [60] replaced traditional SiO_2 with either organic polyvinyl alcohol (PVA) or inorganic ZnO interlayers in Au/*n*-GaAs Schottky diodes. While both diodes exhibited performance characteristics dependent on illuminations and applied bias voltages, ZnO-based MIS diodes showed a higher rectification ratio and hence a better rectification behavior compared to those using PVA (referred as MPS diodes). Additionally, MIS diodes also showed a higher reverse saturation current and sheet resistance, lower series resistance, and an ideality factor value much closer to 1 than MPS ones, despite having similar barrier heights. Notably, the ideality factor values for both diodes increased with increasing

Table 3.1 A summary of representative MIOS Schottky diodes.

Structure (M/I/S)	Insulator thickness (nm)	Operating voltage (V)	Ideality factor	Rectifying ratio	Reference
p^+-Si/Al_2O_3/IGZO	10	1.8	/	10^7	[53]
Au/MgO/ZnO	8	2	3.1	731	[54]
Au/MgO/ZnO	15	8	/	/	[55]
Si/SiO_2/NiO_x	100	20	/	2.2×10^5	[56]

illumination levels and were significantly greater than 1. These high ideality factor values can be ascribed to the intense distribution of interface states, thickness, and homogeneity of the interfacial layers between Au/n-GaAs, as well as laterally inhomogeneous barrier height and image strength reductions.

In 2020, Kaymaz et al. [61] reported Al/(ZnO-PVA)/*p*-Si diodes and investigated the effects of gamma-irradiation on their performance. Compared with diodes using only ZnO or PVA as the interfacial layer, the use of ZnO-PVA combined the advantage of both materials, including the high direct bandgap and large excitation binding energy of ZnO and nontoxic, semicrystalline, and water-soluble of PVA. Through capacitance/conductance–voltage measurements, they found that the resulting diodes exhibited high sensitivity to gamma-irradiation (Figure 3.11), making them promising candidates as gamma-ray detectors.

Similarly, Tuğluoğlu et al. [62] studied the effects of gamma-ray irradiation on the performance of Au/SnO_2/*n*-Si Schottky diodes. Both reverse and forward biases of *I*–*V* characteristics were studied before and after subjecting the diodes to 10 kGy ^{60}Co gamma-ray radiation to examine the current conduction mechanisms. While the forward bias $\ln(I_F)$–$\ln(V_F)$ characteristics confirmed the domination of space-charge-limited conduction (SCLC), the examination through $\ln(I_R)$–$V_R^{1/2}$ characteristics indicated the domination of Schottky emission at both 0- and 10-kGy irradiation levels.

Compared with MS Schottky diodes, which suffer from poor electrical performance due to strong Fermi-level pinning and metal-induced gap states presented at the interface, MIS Schottky diodes with an addition of interfacial layer could passivate the traps at the interface and reduce the leakage current. Such interfacial layers generally play a crucial role in determining the electrical properties of MIS Schottky diodes. While for high-frequency and high-power devices, they can effectively reduce leakage current, it might also cause drawbacks such as stability issues and drift effects. Currently, the report of using oxide semiconductors in MIS Schottky diodes was rare but did show potential. Considering that the ultimate device reliability and performance is intimately related to the surface conditions,

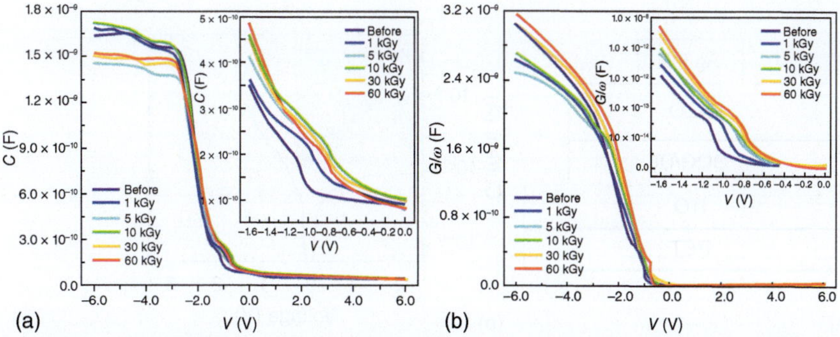

Figure 3.11 (a) *C*–*V* and (b) *G*/*ω*–*V* characteristics of Al/(ZnO-PVA)/p-Si-type Schottky diodes before and after irradiation. Source: Reproduced from Kaymaz et al. [61]/with permission of Springer Nature.

further research directions might be focused on the understanding and investigation of the device surface physics in oxide-based MIS Schottky diodes.

3.4.2 MIS Tunneling Diodes

In the case of MIS Schottky diodes, the rectifying behavior is primarily caused by the Schottky contact between the metal and the semiconductor, and the role of the inserted insulating layer is to improve the electrical properties by controlling the redistribution of interface states and/or reducing the chemical interaction at the interface. MIS tunneling diodes, on the other hand, rely on the tunneling of electrons in one bias condition, and hence the insulating layer in such a structure must be thin.

A typical example of MIS tunneling diodes was reported by Dominguez et al. [63], using ultrasonic spray-deposited Al-doped ZnO as the semiconductor and spin-on glass (SOG) as the insulator (Figure 3.12a). As shown in Figure 3.12b, when applying a negative voltage to ITO contacts, the current was independent of the applied voltage, exhibiting a good rectifying behavior. Applying a positive voltage to the ITO contacts enabled the tunneling of carriers through the insulator, resulting in an exponential increase in current with applied bias. The MIS diodes achieved an inverse saturation current lower than 100 pA and an on/off current ratio of 10^2 at |3| V.

Zhang et al. [64] conducted a simulation-based study to investigate the impact of oxide thickness on the performance of MIS tunneling diodes utilizing ZnO nanostructures with a MgO insulator layer. The study revealed the formation of a valley in the electrostatic potential at the ZnO/MgO interface, creating an energy barrier on the ZnO side. Electrons were required to overcome two barriers: the high and narrow MgO barrier and the barrier induced by the depletion region on the ZnO side of the ZnO/MgO interface. Increasing the MgO thickness deepened the potential valley, resulting in higher and wider barriers at the ZnO/MgO interfaces. Consequently, the current passing through the diodes rapidly decreased. By optimizing the MgO thickness, the authors achieved a rectifying ratio 169 times greater than that of typical MSM structures, with a MgO thickness of 1.5 nm.

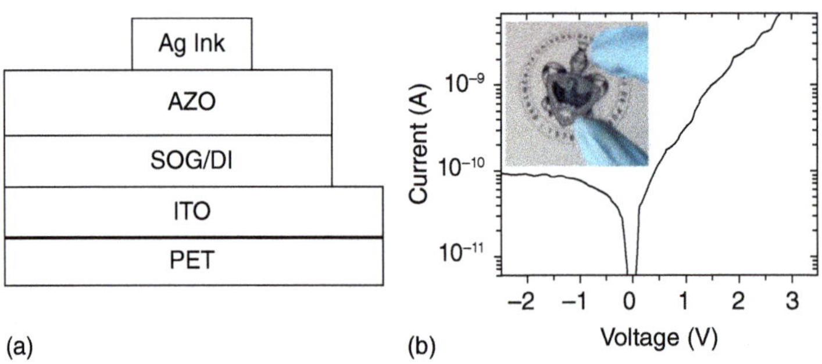

Figure 3.12 (a) A schematic diagram showing the diode structure. (b) Electrical performance of flexible MIS diodes. Source: Reproduced with permission from Dominguez et al. [63] © 2018, Elsevier.

The performance of MIS tunneling diodes has been found to be strongly influenced by the thickness of the insulator and the modulation of the SBHM. SBHM plays a significant role when the insulator is thin enough for the majority carriers to tunnel from the metal to the semiconductor, yet thick enough to retain a portion of the minority inversion layer. In general, a stronger SBHM allows more majority carriers to possess sufficient energy for injection from the metal to the semiconductor, resulting in a higher reverse saturation current in the MIS diodes. Therefore, the selection of the insulator thickness is crucial and may be influenced by factors such as the metal work function, semiconductor bandgap, substrate doping type, and so on.

3.4.3 Applications

While the direct report of MIS diodes is relatively fewer than MS ones, such a structure has shown significant potential, especially in photodetectors. Photodetectors with a *p*–*n* heterojunction offer a built-in electric field within the detector, which is advantageous for achieving self-powered devices [65]. However, they face challenges in terms of fabrication complexity, and the realization of *p*–*n* homojunctions remains difficult. In comparison, MIS-type photodetectors are more favorable as they offer additional advantages and improved detection performance over traditional photodetectors with MS structures. The addition of the insulator layer in MIS photodetectors blocks unwanted charge transfer. Shan's work [66] demonstrated that photodetectors with an Au/MgO/MgZnO structure exhibited higher responsivity than those without MgO (Figure 3.13a). Under dark conditions, the large barrier between the electron affinity of MgO and Au prevented most electrons from injecting into MgO under reverse bias, resulting in a lower dark current compared to MS structures. As shown in Figure 3.13b, under UV illumination, electron–hole pairs were generated in the MgZnO layer. While the generated electrons were collected by the Al contacts under reverse bias, photo-generated holes can be drifted toward the

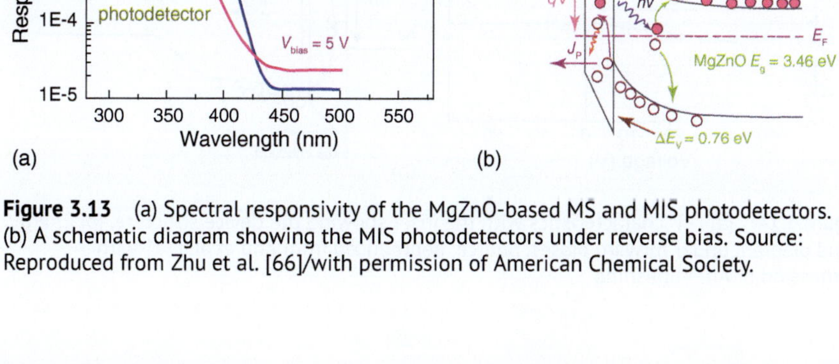

Figure 3.13 (a) Spectral responsivity of the MgZnO-based MS and MIS photodetectors. (b) A schematic diagram showing the MIS photodetectors under reverse bias. Source: Reproduced from Zhu et al. [66]/with permission of American Chemical Society.

Au contacts through MgO. These holes that tunneled through the MgO layer gained significant kinetic energy, leading to the generation of additional electron–hole pairs through the impact ionization process [67]. This multiplication process contributed to the exponential increase in responsivity as a function of the applied voltage.

Alternatively, MIS configurations are also widely utilized in LEDs. For example, ZnO LEDs with fair UV emission under continuous current injection were reported by Yang and coworkers [68], using an Au/SiO_x/ZnO structure. Shan and coworkers [69] demonstrated electrically pumped UV random lasing from ZnO nanocrystalline films using an Au/MgO/ZnO structure.

The relatively small bandgap of ZnO has presented challenges in achieving deep UV (DUV) light emission. Doping Mg into ZnO can increase the bandgap, but direct use of MgZnO may result in a low electron concentration. An electron layer, i.e. ZnO could be introduced to enhance the electron injection into the MgZnO layer, as reported by Shan and coworkers [70]. Nonetheless, the large conduction band offset between the two materials may also result in a substantial energy barrier and hence hinder the electron transfer. In an effort to address these challenges, Shan's group employed an Au/MgO/$Mg_{0.52}Zn_{0.48}O$/$Mg_xZn_{1-x}O$/n-ZnO structure, as shown in Figure 3.14a [71]. The corresponding band alignment of the device is shown in Figure 3.14b. With the assistance of a composition gradient $Mg_xZn_{1-x}O$ bridging layer, the injection process of electrons from n-ZnO to $Mg_{0.52}Zn_{0.48}O$ active layer was facilitated, resulting in pronounced DUV emission at approximately 276 nm.

The conduction mechanism of oxide semiconductors typically results in n-type oxide semiconductors exhibiting higher performance than p-type counterparts. Unlike oxide-based p–n heterojunction diodes, MIS diodes can be made without the need of p-type oxide semiconductors. However, despite that the introduction of an impact ionization process to produce holes could avoid the dependence on p-type doping of oxide semiconductors, the holes produced by this route are inefficient. More importantly, such devices often operate at high voltages, which might cause undesired heating effects. Further research is thus necessary to address these aforementioned challenges.

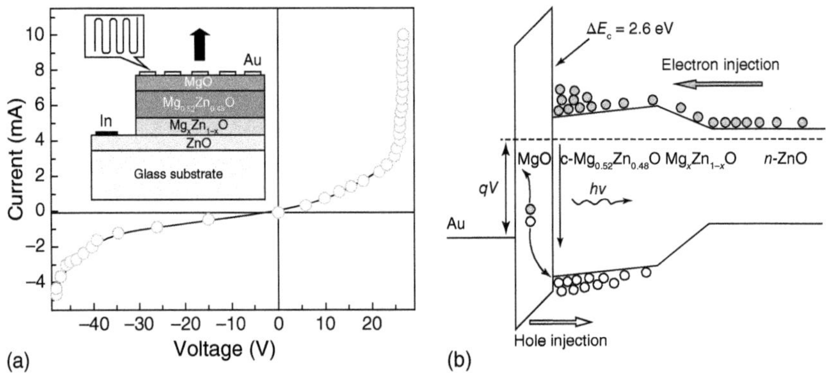

Figure 3.14 (a) I–V characteristics with the inset showing the device structure. (b) Energy band diagram under forward biases. Source: Reproduced from Zhu et al. [71]/with permission of AIP Publishing.

3.5 Self-Switching Diodes

The RF rectifier plays a crucial role in RFID tags. However, the conventional diodes discussed in the Sections 3.2–3.4 typically require a doping junction or energy barriers, necessitating complex multilayer sandwich structures, and hence a high degree of control process and complexity to achieve adequate RF performance by printing technologies [72–74]. SSDs are unipolar two-terminal nanodevices capable of functioning as room-temperature rectifiers at microwave and terahertz frequencies. Unlike traditional diodes, SSDs exhibit nonlinear behavior without the need for doping junctions or tunneling barriers. Moreover, the distinct operating principle enables SSDs and SSD-based simple circuits to be fabricated using a single lithography step, reducing manufacturing costs and time. The initial demonstration of this diode structure was carried out by Song et al. [75] using InGaAs, and later being applied to other semiconductors, including but not limited to III–V semiconductors [75, 76], graphene [77], organic semiconductors [78], and oxide semiconductors [79, 80].

A typical SSD structure is shown in Figure 3.15a. In such structures, the L-shape insulating trenches not only serve as the boundaries for a narrow semiconductor channel, but also create an asymmetrical structure, contributing to the nonlinear I–V characteristics of the device. Taking n-type semiconductors with negative surface charges as an example, depletion regions form at the wall of the trenches (i.e. insulator/semiconductor interface), caused by charges at the surface states, have made the effective channel width of SSDs smaller than the physical size. Due to these depletion regions, the SSD's effective channel is almost pinch-off under unbiased condition, as shown in Figure 3.15b. Applying a positive voltage ($V > 0$) to the right terminal reduces the depletion regions along the channel, hence opening the channel

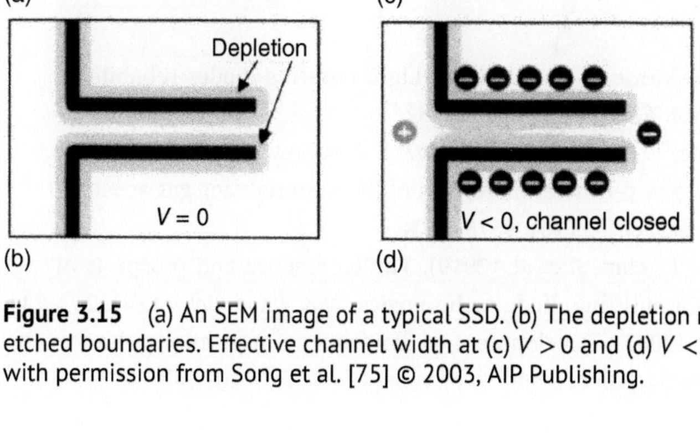

Figure 3.15 (a) An SEM image of a typical SSD. (b) The depletion region formed near the etched boundaries. Effective channel width at (c) $V > 0$ and (d) $V < 0$. Source: Reproduced with permission from Song et al. [75] © 2003, AIP Publishing.

for current flow across the device (Figure 3.15c). Alternatively, applying a negative voltage ($V < 0$) to the right terminal increases the depletion regions and eventually closes the channel (Figure 3.15d), preventing current flow across. This results in the diode-like nonlinear I–V characteristic of SSDs.

A representative work of oxide semiconductor-based SSDs was reported by Song's group [80] in 2011, employing polycrystalline ZnO thin films. The coating of PMMA and PVDF were found to further enhance the electric field coupling, maintaining the nonlinear behavior of the devices even as the current increased dramatically. Despite showing a low mobility, ZnO SSDs could operate at frequencies up to 51.5 MHz on glass substrates, mainly attributed to the greatly reduced parasitic capacitance enabled by the planar device structure and nanometer dimensions of the SSDs. This suggests the great potential of oxide-based SSDs for ultra-high frequency applications.

Similar works were also reported by Kimura et al. [81], who studied the rectification effects of ZnO-based transparent SSDs, showing that the rectifying characteristics could be controlled by the device's shape, while the cut-off voltage could be adjusted by reducing the electron concentration with a cost of a lower current. Later in 2013, the same group also reported ZnO SSDs with clear rectification on both glass and flexible substrates [79]. In this work, they found that the current strength was strongly affected by the channel length, and the addition of an HfO_2 film on SSDs shifted the turn-on voltage toward zero volt and increased the current strength.

In conclusion, since the initial report of SSDs in 2003, extensive research has been conducted on this field, aiming to improve the rectification performance and cut-off frequency of the device. With prudent considerations on the structural parameters and materials, SSDs were proven to have abilities working as rectifiers at high frequencies. The simplicity of the design and processes may offer cost-effective alternatives for high-frequency rectification in low-power devices. Although the utilization of such structures in oxide semiconductors is currently limited, they exhibit significant potential for next-generation printable electronics. Future research directions may focus on developing high-mobility oxide semiconductors to enable operation at ultrahigh frequencies.

References

1 Chang, M., Das, D., Varde, P. et al. (2012). Light emitting diodes reliability review. *Microelectron. Reliab.* 52 (5): 762–782.

2 Mathew, M., Shinde, P., Samal, R. et al. (2021). A review on mechanisms and recent developments in p-n heterojunctions of 2D materials for gas sensing applications. *J. Mater. Sci.* 56 (16): 9575–9604.

3 Kneissl, M., Seong, T., Han, J. et al. (2019). The emergence and prospects of deep-ultraviolet light-emitting diode technologies. *Nat. Photonics* 13 (4): 233–244.

4 Lo, S. and Burn, P. (2007). Development of dendrimers: macromolecules for use in organic light-emitting diodes and solar cells. *Chem. Rev.* 107 (4): 1097–1116.

5 Zhang, S., Wei, S., and Zunger, A. (2001). Intrinsic n-type versus p-type doping asymmetry and the defect physics of ZnO. *Phys. Rev. B* 63 (7): 075205.

6 Lany, S., Osorio-Guillén, J., and Zunger, A. (2007). Origins of the doping asymmetry in oxides: hole doping in nio versus electron doping in ZnO. *Phys. Rev. B* 75 (24): 241203.

7 Robertson, J. and Clark, S. (2011). Limits to doping in oxides. *Phys. Rev. B* 83 (7): 075205.

8 Özgür, Ü., Alivov, Y., Liu, C. et al. (2005). A comprehensive review of ZnO materials and devices. *J. Appl. Phys.* 98 (4): 041301.

9 Ohta, H., Mizoguchi, H., Hirano, M. et al. (2003). Fabrication and characterization of heteroepitaxial p-n junction diode composed of wide-gap oxide semiconductors p-$ZnRh_2O_4$/n-ZnO. *Appl. Phys. Lett.* 82 (5): 823–825.

10 Narushima, S., Mizoguchi, H., Shimizu, K. et al. (2003). A p-type amorphous oxide semiconductor and room temperature fabrication of amorphous oxide p–n heterojunction diodes. *Adv. Mater.* 15 (17): 1409–1413.

11 Grilli, M., Aydogan, S., and Yilmaz, M. (2016). A study on non-stoichiometric p-NiO_X/n-Si heterojunction diode fabricated by RF sputtering: determination of diode parameters. *Superlattices Microstruct.* 100: 924–933.

12 Ohta, H., Hirano, M., Nakahara, K. et al. (2003). Fabrication and photoresponse of a pn heterojunction diode composed of transparent oxide semiconductors, p-NiO and n-ZnO. *Appl. Phys. Lett.* 83 (5): 1029–1031.

13 Zhang, Z., Shao, C., Li, X. et al. (2010). Electrospun nanofibers of p-type NiO/n-type ZnO heterojunctions with enhanced photocatalytic activity. *ACS Appl. Mater. Interfaces* 2 (10): 2915–2923.

14 Wang, C., Gong, H., Lei, W. et al. (2021). Demonstration of the p-NiO_x/n-Ga_2O_3 heterojunction gate FETs and diodes with $BV^2/R_{on,sp}$ figures of merit of 0.39 GW/cm^2 and 1.38 GW/cm^2. *IEEE Electron Device Lett.* 42 (4): 485–488.

15 Tsai, T., Chang, S., Hsueh, T. et al. (2011). P-Cu_2O-shell/n-TiO_2-nanowire-core heterostucture photodiodes. *Nanoscale Res. Lett.* 6 (1): 575.

16 Hsueh, H., Chang, S., Weng, W. et al. (2012). Fabrication and characterization of coaxial p-copper oxide/n-ZnO nanowire photodiodes. *IEEE Trans. Nanotechnol.* 11 (1): 127–133.

17 Thirumoorthi, M., Dhavud, S., Ganesh, V. et al. (2022). High responsivity n-ZnO/p-CuO heterojunction thin film synthesised by low-cost silar method for photodiode applications. *Optic. Mater.* 128: 112410.

18 Park, N., Sun, K., Sun, Z. et al. (2013). High efficiency NiO/ZnO heterojunction UV photodiode by sol–gel processing. *J. Mater. Chem. C* 1 (44): 7333–7338.

19 Hakkoum, H., Tibermacine, T., Sengouga, N. et al. (2020). Effect of the source solution quantity on optical characteristics of ZnO and NiO thin films grown by spray pyrolysis for the design NiO/ZnO photodetectors. *Optic. Mater.* 108: 110434.

20 Liu, H. and Huang, Z. (2021). Investigation of p-NiO/n-$Ga_{0.3}Zn_{0.7}O$ heterojunction photodiodes for ultraviolet-b detection. *IEEE Sens. J.* 21 (19): 21486–21493.

21 Tsai, S., Hon, M., and Lu, Y. (2011). Fabrication of transparent p-NiO/n-ZnO heterojunction devices for ultraviolet photodetectors. *Solid-State Electron.* 63 (1): 37–41.

22 Li, X., Li, X., Chen, N. et al. (2014). Cuo-In_2O_3 core-shell nanowire based chemical gas sensors. *J. Nanomater.* 2014: 973156.

23 Gao, H., Zhao, L., Wang, L. et al. (2018). Ultrasensitive and low detection limit of toluene gas sensor based on SnO_2-decorated NiO nanostructure. *Sens. Actuators B Chem.* 255: 3505–3515.

24 Liu, T., Yu, Z., Liu, Y. et al. (2020). Gas sensor based on ni foam: SnO_2-decorated NiO for toluene detection. *Sens. Actuators B Chem.* 318: 128167.

25 Xue, X., Zhu, L., Yuan, K. et al. (2020). ZnO branched p-Cu_xO @n-ZnO heterojunction nanowires for improving acetone gas sensing performance. *Sens. Actuators B Chem.* 324: 128729.

26 Zhang, J., Zeng, D., Zhu, Q. et al. (2016). Enhanced room temperature NO_2 response of NiO–SnO_2 nanocomposites induced by interface bonds at the p–n heterojunction. *Phys. Chem. Chem. Phys.* 18 (7): 5386–5396.

27 Saha, B., Sarkar, K., Bera, A. et al. (2017). Schottky diode behaviour with excellent photoresponse in NiO/FTO heterostructure. *Appl. Surf. Sci.* 418: 328–334.

28 Digdaya, I., Adhyaksa, G., Trześniewski, B. et al. (2017). Interfacial engineering of metal-insulator-semiconductor junctions for efficient and stable photoelectrochemical water oxidation. *Nat. Commun.* 8 (1): 15968.

29 Mead, C. (1965). Surface barriers on ZnSe and ZnO. *Phys. Lett.* 18 (3): 218.

30 Park, W., Yi, G., Kim, J. et al. (2003). Schottky nanocontacts on ZnO nanorod arrays. *Appl. Phys. Lett.* 82 (24): 4358–4360.

31 Lao, C., Liu, J., Gao, P. et al. (2006). ZnO nanobelt/nanowire Schottky diodes formed by dielectrophoresis alignment across au electrodes. *Nano Lett.* 6 (2): 263–266.

32 Nakano, M., Makino, T., Tsukazaki, A. et al. (2008). Transparent polymer Schottky contact for a high performance visible-blind ultraviolet photodiode based on ZnO. *Appl. Phys. Lett.* 93 (12): 123309.

33 Semple, J., Rossbauer, S., Burgess, C. et al. (2016). Radio frequency coplanar ZnO Schottky nanodiodes processed from solution on plastic substrates. *Small* 12 (15): 1993–2000.

34 Georgiadou, D., Semple, J., Sagade, A. et al. (2020). 100 GHz zinc oxide Schottky diodes processed from solution on a wafer scale. *Nat. Electron.* 3 (11): 718–725.

35 Lee, D., Nomura, K., Kamiya, T. et al. (2011). Diffusion-limited a-IGZO/Pt Schottky junction fabricated at 200 °C on a flexible substrate. *IEEE Electron Device Lett.* 32 (12): 1695–1697.

36 Chasin, A., Nag, M., Bhoolokam, A. et al. (2013). Gigahertz operation of a-IGZO Schottky diodes. *IEEE Trans. Electron Devices* 60 (10): 3407–3412.

37 Zhang, J., Li, Y., Zhang, B. et al. (2015). Flexible indium–gallium–zinc–oxide Schottky diode operating beyond 2.45 GHz. *Nat. Commun.* 6 (1): 7561.

38 Zhang, J., Wang, H., Wilson, J. et al. (2016). Room temperature processed ultrahigh-frequency indium-gallium–zinc-oxide Schottky diode. *IEEE Electron Device Lett.* 37 (4): 389–392.

39 Ho, S., Yu, H., and So, F. (2017). Transparent indium-tin oxide/indium-gallium-zinc oxide Schottky diodes formed by gradient oxygen doping. *Appl. Phys. Lett.* 111 (21): 212103.

40 Chang, C., Hsu, C., and Wu, C. (2017). Rectified Schottky diodes based on PEDOT:PSS/InGaZnO junctions. *Org. Electron.* 48: 35–40.

41 Huang, C., Lin, P., Cheng, H. et al. (2019). Rectified Schottky diodes that use low-cost carbon paste/InGaZnO junctions. *Org. Electron.* 68: 212–217.

42 Chasin, A., Zhang, L., Bhoolokam, A. et al. (2014). High-performance a-IGZO thin film diode as selector for cross-point memory application. *IEEE Electron Device Lett.* 35 (6): 642–644.

43 Zhang, J., Xin, Q., and Song, A. (2016). High performance Schottky diodes based on indium-gallium-zinc-oxide. *J. Vac. Sci. Technol. A* 34 (4): 04C101.

44 Mosbacker, H., Strzhemechny, Y., White, B. et al. (2005). Role of near-surface states in ohmic-Schottky conversion of Au contacts to ZnO. *Appl. Phys. Lett.* 87 (1): 012102.

45 Flynn, B., Oleksak, R., Thevuthasan, S. et al. (2018). Interfacial chemistry-induced modulation of Schottky barrier heights: in situ measurements of the Pt–amorphous indium gallium zinc oxide interface using X-ray photoelectron spectroscopy. *ACS Appl. Mater. Interfaces* 10 (4): 4333–4340.

46 Kamiya, T., Nomura, K., and Hosono, H. (2010). Origin of definite hall voltage and positive slope in mobility-donor density relation in disordered oxide semiconductors. *Appl. Phys. Lett.* 96 (12): 122103.

47 Wilson, J., Zhang, J., Li, Y. et al. (2017). Influence of interface inhomogeneities in thin-film Schottky diodes. *Appl. Phys. Lett.* 111 (21): 213503.

48 Chen, R. and Lan, L. (2019). Solution-processed metal-oxide thin-film transistors: a review of recent developments. *Nanotechnology* 30 (31): 312001.

49 Sasaki, K., Higashiwaki, M., Kuramata, A. et al. (2013). Ga_2O_3 Schottky barrier diodes fabricated by using single-crystal β–Ga_2O_3 (010) substrates. *IEEE Electron Device Lett.* 34 (4): 493–495.

50 He, Q., Mu, W., Dong, H. et al. (2017). Schottky barrier diode based on β-Ga_2O_3 (100) single crystal substrate and its temperature-dependent electrical characteristics. *Appl. Phys. Lett.* 110 (9): 093503.

51 Jian, G., He, Q., Mu, W. et al. (2018). Characterization of the inhomogeneous barrier distribution in a Pt/(100)β-Ga_2O_3 Schottky diode via its temperature-dependent electrical properties. *AIP Adv.* 8 (1): 015316.

52 Shimizu, A., Kanbara, M., Hada, M. et al. (1978). ZnO green light emitting diode. *J. Appl. Phys.* 17 (8): 1435.

53 Lee, D., Park, J., Cho, N. et al. (2019). Verification of charge transfer in metal-insulator-oxide semiconductor diodes via defect engineering of insulator. *Sci. Rep.* 9 (1): 10323.

54 Hwang, J. and Hong, Z. (2021). Enhancing the Schottky-barrier height by inserting a thin MgO layer between Au and annealed-ZnO. *Mater. Res. Bull.* 144: 111478.

55 Shi, Z., Zhang, Y., Cui, X. et al. (2015). Photoluminescence performance enhancement of ZnO/MgO heterostructured nanowires and their applications in ultraviolet laser diodes. *Phys. Chem. Chem. Phys.* 17 (21): 13813–13820.

56 Park, J., Lee, D., Kim, K. et al. (2020). Rectification mechanism of a p-type oxide-based metal–insulator–oxide semiconductor–metal thin-film diode. *ACS Appl. Electron. Mater.* 2 (12): 3946–3952.

57 Lee, H., Huang, H., and Lee, C. (2011). Hydrogen sensing performances of Pt/i-ZnO/GaN metal–insulator–semiconductor diodes. *Sens. Actuators B Chem.* 157 (2): 460–465.

58 Kaymak, N., Efil, E., Seven, E. et al. (2019). Electrical characteristics analyses of zinc-oxide based mis structure grown by atomic layer deposition. *Mater. Res. Express* 6 (2): 026309.

59 Altındal, Ş., Karadeniz, S., Tuğluoğlu, N. et al. (2003). The role of interface states and series resistance on the i–v and c–v characteristics in Al/SnO_2/p-Si Schottky diodes. *Solid-State Electron.* 47 (10): 1847–1854.

60 Tan, S., Tecimer, H., and Çiçek, O. (2017). Comparative investigation on the effects of organic and inorganic interlayers in Au/n-GaAs Schottky diodes. *IEEE Trans. Electron Devices* 64 (3): 984–990.

61 Kaymaz, A., Uslu, T., Evcin, B. et al. (2020). Investigation of Gamma-irradiation effects on electrical characteristics of Al/(ZnO–PVA)/p-Si Schottky diodes using capacitance and conductance measurements. *J. Mater. Sci. Mater. Electron.* 31 (11): 8349–8358.

62 Akgül, F., Eymur, S., Akın, Ü. et al. (2021). Investigation of Schottky emission and space charge limited current (SCLC) in Au/SnO_2/n-Si Schottky diode with gamma-ray irradiation. *J. Mater. Sci. Mater. Electron.* 32 (12): 15857–15863.

63 Dominguez, M., Luna-Lopez, J., and Ceron, S. (2018). Low-temperature ultrasonic spray deposited aluminum doped zinc oxide film and its application in flexible metal-insulator-semiconductor diodes. *Thin Solid Films* 645: 278–281.

64 Yang, X., Gu, Y., Migliorato, M. et al. (2016). Impact of insulator layer thickness on the performance of metal–MgO–ZnO tunneling diodes. *Nano Res.* 9 (5): 1290–1299.

65 Wang, Y., Wu, C., Guo, D. et al. (2020). All-oxide NiO/Ga_2O_3 p–n junction for self-powered UV photodetector. *ACS Appl. Electron. Mater.* 2 (7): 2032–2038.

66 Zhu, H., Shan, C., Wang, L. et al. (2010). Metal–oxide–semiconductor-structured MgZnO ultraviolet photodetector with high internal gain. *J. Phys. Chem. C* 114 (15): 7169–7172.

67 Hudgins, J., Simin, G., Santi, E. et al. (2003). An assessment of wide bandgap semiconductors for power devices. *IEEE Trans. Power Electron.* 18 (3): 907–914.

68 Chen, P., Ma, X., and Yang, D. (2006). Fairly pure ultraviolet electroluminescence from ZnO-based light-emitting devices. *Appl. Phys. Lett.* 89 (11): 111112.

69 Zhu, H., Shan, C., Zhang, J. et al. (2010). Low-threshold electrically pumped random lasers. *Adv. Mater.* 22 (16): 1877–1881.

70 Zhu, H., Shan, C., Li, B. et al. (2011). Low-threshold electrically pumped ultraviolet laser diode. *J. Mater. Chem.* 21 (9): 2848–2851.

71 Zhu, H., Shan, C., Li, B. et al. (2011). Deep-ultraviolet light-emitting device realized via a hole-multiplication process. *Appl. Phys. Lett.* 99 (10): 101110.

72 Sun, J., Pal, B., Jung, B. et al. (2009). Solution-processed hybrid p–n junction vertical diode. *Org. Electron.* 10 (1): 1–7.

73 Steudel, S., Myny, K., Arkhipov, V. et al. (2005). 50 MHz rectifier based on an organic diode. *Nat. Mater.* 4 (8): 597–600.

74 Subramanian, V., Chang, P., Lee, J. et al. (2005). Printed organic transistors for ultra-low-cost RFID applications. *IEEE Trans. Compon. Packag. Technol.* 28 (4): 742–747.

75 Song, A., Missous, M., Omling, P. et al. (2003). Unidirectional electron flow in a nanometer-scale semiconductor channel: a self-switching device. *Appl. Phys. Lett.* 83 (9): 1881–1883.

76 Kasjoo, S. and Song, A. (2013). Terahertz detection using nanorectifiers. *IEEE Electron Device Lett.* 34 (12): 1554–1556.

77 Westlund, A., Winters, M., Ivanov, I. et al. (2015). Graphene self-switching diodes as zero-bias microwave detectors. *Appl. Phys. Lett.* 106 (9): 093116.

78 Majewski, L. and Song, A. (2016). 20 megahertz operation of organic nanodiodes. *Phys. Status Solidi B* 253 (8): 1507–1510.

79 Kimura, Y., Sun, Y., Maemoto, T. et al. (2013). Rectification effects of ZnO-based transparent nanodiodes on glass and flexible plastic substrates. *J. Appl. Phys.* 52 (6S): 06GE09.

80 Irshaid, M., Balocco, C., Luo, Y. et al. (2011). Zinc-oxide-based planar nanodiodes operating at 50 MHz. *Appl. Phys. Lett.* 99 (9): 092101.

81 Kimura, Y., Kiso, T., Higaki, T. et al. (2012). Rectification effects in ZnO-based transparent self-switching nano-diodes. In: *IEEE International Meeting for Future of Electron Devices*, IEEE, Kansai, 2012, pp. 1–2.

4

Metal Oxide Semiconductors for Transistors

4.1 Introduction

Transistors are considered the most groundbreaking invention in the twentieth century and are generally three-terminal devices in which current or voltage is applied between one pair to modulate the current through another pair. The first reports of transistors can be dated back to as early as 1947 at Bell Laboratories, where Ge was used as the active layer. By overcoming the surface states that blocked electric fields from penetrating the semiconductor material, Atalla and Kahng at Bell Labs successfully achieved the first insulated-gate field-effect transistor (FET) in 1959. Such transistors used silicon and its oxide in a sandwich structure with metal as the gate, oxide as the insulator, and Si as the semiconductor and hence are being named as metal–oxide–semiconductor field-effect transistors (MOSFETs). The invention of MOSFETs opened doors for the development of integrated circuits, microprocessors, and CPUs.

However, despite Si-based MOSFETs attracting considerable attention and showing immense potential and applications in our lives, their rigid nature of still limits their potential applications in future wearable electronics. Thin-film transistors (TFTs), which have operation and composition layers similar to well-known MOSFETs but with the advantage of being able to be deposited on a variety of substrates (i.e. flexible substrates) at a relatively low temperature, are considered promising alternatives. Materials such as low-temperature polycrystalline Si (LTPS), amorphous Si, and organic semiconductors have been studied as the active layer in TFTs. However, organic semiconductors, despite having a simple fabrication process and low cost, suffer from poor air stability and low mobility, which are crucial drawbacks. Similarly, although amorphous Si can be deposited by plasma-enhanced chemical vapor deposition (PECVD) at temperatures below 350 °C, the low obtained mobility of $\sim 1\ cm^2\ V^{-1}\ s^{-1}$ limits its potential applications. LTPS shows a higher mobility, but the performance might vary significantly even over a small area (i.e. on one substrate) due to the existence of grain boundaries, making it unsuitable for large-area electronics such as displays.

An alternative to them is oxide semiconductors, which show high mobility, low deposition temperature, and can be both flexible and transparent. Especially since the first report of flexible amorphous InGaZnO (IGZO) TFTs by Hosono

Metal Oxide Semiconductors: Synthesis, Properties, and Devices,
First Edition. Zhigang Zang, Wensi Cai, and Yong Zhou.

and coworkers in 2004 [1], oxide TFTs have garnered considerable attention in fields such as next-generation flat panel displays and biosensors, challenging traditional semiconductors not only in conventional circuit fields but also in paper electronics. In this chapter, we systematically review oxide semiconductor-based TFTs, including discussions on their device structures and mechanisms, typically reported *n*-type and *p*-type transistors with an emphasis on the critical barriers toward commercialization, and the major milestones in circuits that have already been achieved with this emerging and very promising technology.

4.2 Device Structures and Mechanisms

As a widely used type of transistor for oxide semiconductors, a TFT is a special kind of FET which is fabricated by depositing thin films of semiconductor, dielectric, and conductive materials onto a chosen substrate. The freedom to choose a suitable substrate means that, unlike Si MOSFETs, TFTs are not limited to a rigid form factor. There are four common TFT structures, namely top-gate top-contacts, top-gate bottom-contacts, bottom-gate top-contacts, and bottom-gate bottom-contacts, which are shown in Figure 4.1.

Unlike MOSFETs, which have *p*–*n* junctions as source-drain contacts, the contacts of a TFT are metal–semiconductor junctions. Typically, and often ideally, TFT source-drain contacts are ohmic, although in some cases this is not possible, and in other cases, Schottky contacts may be preferable.

With the source connected to the ground and a positive voltage being applied to the drain side, electrons are attracted from the source to the drain in transistors. Transistors have three stages of operation depending on applied voltages, which

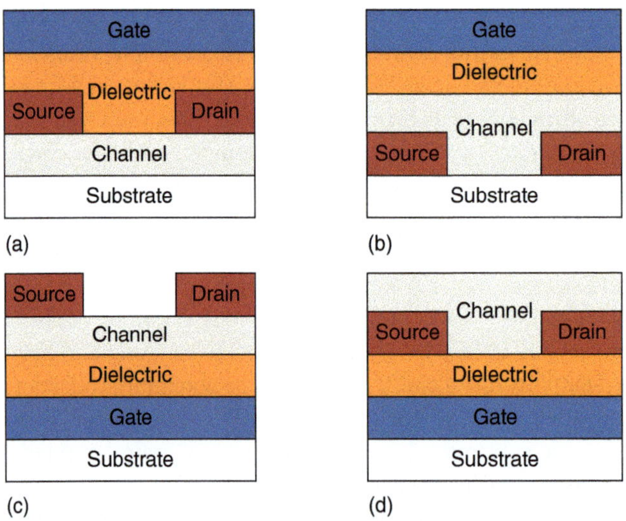

Figure 4.1 Schematic diagrams of basic TFT structures. (a) Top-gate top-contacts. (b) Top-gate bottom-contacts. (c) Bottom-gate top-contacts. (d) Bottom-gate bottom-contacts.

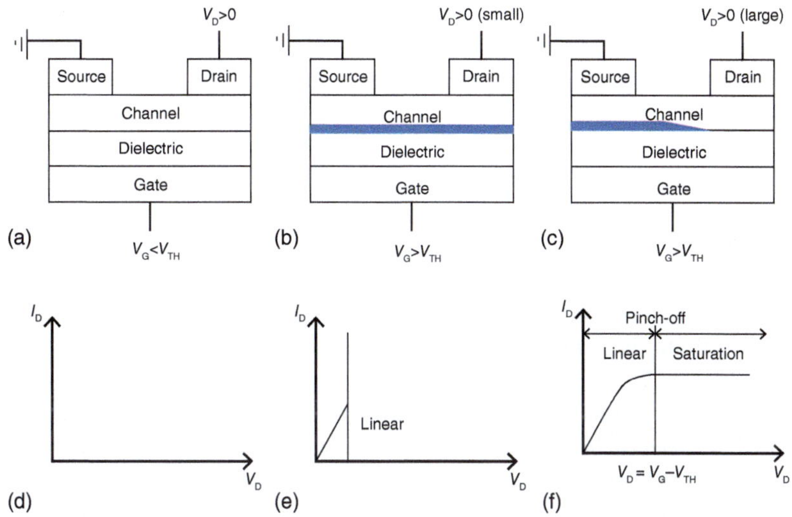

Figure 4.2 Cross-sectional views of an ideal *n*-type TFT in (a) cut-off, (b) linear, and (c) saturation regions. The corresponding output curves of the TFT in (d) cut-off, (e) linear, and (f) saturation regions.

are cut-off, linear, and saturation regions. An ideal case of *n*-type TFT (bottom-gate top-contacts) operation is shown in Figure 4.2, including both the cross-sectional view and the corresponding output curves.

When the gate voltage, V_G, is smaller than the threshold voltage, V_{TH}, no drain current, I_D, flows between the drain and source electrodes since there are no electrons in the channel. This case is shown in Figure 4.2a,d and is known as the cut-off region.

The channel of a TFT is formed due to charge accumulation, unlike the inversion process commonly used in MOSFET, where a field-effect induces an *n*-type region in a *p*-type semiconductor. In the case of TFTs, the gradual channel approximation, which is typically used to describe Si MOSFETs, is employed as a simple and reasonably accurate model for TFT behavior. However, it is important to note that charge transport in TFTs is more complex and depends on the specific materials and geometry employed. For an *n*-type semiconductor, the application of a positive V_G leads to the accumulation of electrons at the semiconductor–dielectric interface and the drain current can be modeled as

$$I_D = \frac{W}{L}\mu C\left[(V_G - V_{TH})V_D - \frac{V_D^2}{2}\right] \tag{4.1}$$

where W and L are the channel width and length, C is the gate capacitance per unit area, μ is the mobility, V_{TH} is the threshold voltage, and V_D is the drain voltage.

When V_G is greater than V_{TH}, carriers accumulate at the interface between the semiconductor and the gate dielectric layer, forming a conductive channel layer and allowing I_D to flow between the source and drain electrodes. At this time, the state of the transistor mainly depends on the difference between V_D and gate overvoltage

$(V_G - V_{TH})$. In the case where V_D is sufficiently low, i.e. $V_G - V_{TH} \gg V_D$, the accumulation of electrons can be assumed to be uniformly across the whole area of the channel. This case is shown in Figure 4.2b,e and is known as the linear region. In this time,

$$I_D = \frac{W}{L}\mu C[(V_G - V_{TH})V_D] \tag{4.2}$$

As V_D increases, the accumulation of charge near the drain decreases, causing a nonlinear relationship between current and voltage. When $V_D = V_G - V_{TH}$, I_D in the ideal case becomes independent of V_D and the TFT is said to be in saturation. Under these conditions, the potential of the drain is large enough to attract electrons directly to the drain edge. This leads to a depletion of the accumulated channel beneath the drain edge and is known as channel pinch-off. Further increasing the V_D cannot push this pinch point backward toward the source electrode, since the electric field needs to be balanced. Therefore, the resistance of the channel from the source to the pinch point remains constant, and I_D remains constant as well. This case is known as the saturation region and is shown in Figure 4.2c,f. The equation for the saturation region is expressed as

$$I_D = \frac{W}{2L}C\mu(V_G - V_{TH})^2 \tag{4.3}$$

It is worth mentioning that in Eq. (4.2), an ideal case is considered, which assumes no contact resistance in TFTs. However, in reality, the influence of contact resistance can be serious, especially for devices with a short channel length. By taking contact resistance, R_{SD}, into consideration, Eq. (4.2) should be extended as

$$I_D \approx \frac{W}{L}C\mu(V_G - V_{TH})(V_D - 2I_D R_{SD}) = \frac{I_{D0}}{1 + 2\mu C\frac{W}{L}(V_G - V_{TH})R_{SD}} \tag{4.4}$$

where I_{D0} is the ideal drain current with no contact resistance. The contact resistance can be extracted through the transfer characteristics at the linear region, using

$$R_{total} = \frac{V_D}{I_D} = R_{ch} \times L + 2R_{SD} = \frac{L}{C\mu W(V_G - V_{TH})} + 2R_{SD} \tag{4.5}$$

where R_{ch} is the channel resistance. Contact resistance can also be extracted using the transmission line method, through

$$R_{total} = R_{sh}\frac{L}{W} + 2R_{SD} \tag{4.6}$$

where R_{sh} is the channel sheet resistance. By plotting the relationship between R_{total} and L at a fixed gate voltage, contact resistance can be extracted from the intercept at $L = 0$ and R_{sh}/W can be extracted from the slope.

The most important characteristics for TFTs are mobility, current on-off ratio, turn-on voltage, threshold voltage, and subthreshold swing, as shown in Figure 4.3. Generally, TFTs with good electrical properties will maximize mobility and current on-off ratio, minimize subthreshold swing, and have a threshold voltage near zero-volt.

Mobility (μ): an intrinsic material property that affects the efficiency of carrier transport in a material. Indeed, according to Ohm's law, current increases proportionally with mobility. Therefore, higher mobilities are desirable in TFTs as they

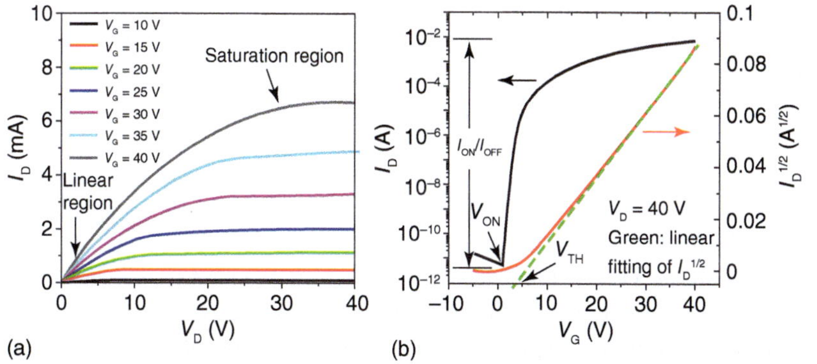

Figure 4.3 Typical (a) output and (b) transfer characteristics of *n*-type TFTs. The dashed green line in (b) is the linear fitting of $I_D^{1/2}$ (red line), from which the threshold voltage can be obtained.

allow for reduced TFT dimensions without sacrificing performance or enable faster charging of capacitances, resulting in faster operation.

The mobility of a TFT is determined by various scattering mechanisms in the accumulation channel, which forms near the interface between the semiconductor and the dielectric. The scattering mechanisms most relevant to TFT operation include, but not limited to, grain boundary scattering, interface roughness, and Coulomb scattering from dielectric charges.

Generally, mobility can be calculated in two different regions: saturation and linear regions. In saturation regions ($V_D > V_G - V_{TH}$),

$$\mu_{sat} = \frac{\left(\frac{d\sqrt{I_D}}{dV_G}\right)^2}{\frac{1}{2}C\frac{W}{L}} \tag{4.7}$$

While in linear regions, the output conductance (g_d) at low V_D can be used to determine the effective mobility,

$$\mu_{eff} = \frac{g_d}{C\frac{W}{L}(V_G - V_{TH})} \tag{4.8}$$

and the transconductance (g_m) at low V_D can be used to obtain the field-effect mobility,

$$\mu_{FE} = \frac{g_m}{C\frac{W}{L}V_D} \tag{4.9}$$

For each method, there are advantages and drawbacks. The calculation of μ_{eff} is less commonly used because it requires the determination of threshold voltage and is more sensitive to contact resistance. Although the calculation of field-effect mobility also suffers from sensitivities to contact resistance, it does not require the knowledge of threshold voltage, and can be easily calculated from the derivative of transfer characteristics. Saturation mobility, although it does not require threshold voltage and is less sensitive to contact resistance, only describes saturation states, where the channel is pinched-off.

On–off ratio (I_{ON}/I_{OFF}): the ratio between maximum and minimum drain current. The off-current is normally related to the gate leakage current and the resistivity of the channel layer. In display technologies, for example, TFTs with higher on-currents (and with sufficiently high mobilities) enable the realization of simpler, lower-cost displays because the driver/control circuitry can be included on the same panel as the pixels. A low off-current, ideally zero, means less power dissipations where one switch of TFT can keep the pixel on for longer, thus reducing the required refresh rate of a display.

Turn-on voltage (V_{ON}): the value of V_G at which I_D starts to increase in the I_D–V_G plot. V_{ON} roughly tracks the flat-band condition; thus when V_G is below (above) V_{ON}, an *n*-type (*p*-type) TFT can be considered as fully off.

Threshold voltage (V_{TH}): in different operating regions, V_{TH} can be extracted by linear extrapolation of V_G when $I_D = 0$ from either an I_D–V_G plot at low V_D (linear region) or an $I_D^{1/2}$–V_G plot at high V_D (saturation region), as shown in Figure 4.3b. For an ordinary MOSFET, V_{TH} is considered as the V_G at which a conducting channel is accumulated at the semiconductor–dielectric interface. However, as mentioned above, the mobility of a TFT can vary with V_G as a result of trap states, making the extraction methods inaccurate and decoupling V_{TH} from its physical meaning. Despite this, V_{TH} is still a commonly used figure of merit for TFTs.

Subthreshold swing (*SS*): represents the necessary V_G needed to increase I_D by one decade. Such a value represents the ease of switching between the on and off states and can be calculated through

$$SS = \left(\left. \frac{\mathrm{d}\log(I_D)}{\mathrm{d}V_G} \right|_{\max} \right)^{-1} \tag{4.10}$$

where the derivative is the gradient extracted from the $\log(I_D)$–V_G curve. Normally, for a good transistor, *SS* needs to be $\ll 1\ \mathrm{V\,dec^{-1}}$ as a smaller value of *SS* offers higher speed and implies a lower V_{TH}, and thus a reduced power consumption. For an ideal device without traps, the theoretical minimum is limited by thermal effects to a value of $\sim 60\ \mathrm{mV\,dec^{-1}}$ at room temperature.

Often the current–voltage relation in the subthreshold region cannot be perfectly fitted with an exponential, as shown in Figure 4.3b. To gain a deeper insight into trap states, the V_G dependence of subthreshold swing should be taken into account. In practice, the lowest (i.e. best) value of *SS* is often given so due care should be taken when comparing devices, especially as *SS* is also dependent upon operating conditions and equipment resolution.

4.3 *N*-Type TFTs

4.3.1 History Overview

The first oxide semiconductor-based TFT was reported by Klasens and Koelmans in 1964, using SnO_2 as a semiconductor, Al as source/drain and gate electrodes, and an anodized Al_2O_3 gate dielectric [2]. In 1968, ZnO single crystals were used as semiconductors for fabricating *n*-type TFTs, but the obtained electrical performance was poor [3].

Oxide TFTs with good electrical properties started to appear in 2003, with reports on ZnO TFTs showing a respectable performance comparable to or even surpassing that of amorphous Si and organic-semiconductor-based TFTs, mostly in terms of mobility [4–6]. In 2013, radio-frequency sputtered polycrystalline ZnO TFTs showed a high mobility of >50 $cm^2 V^{-1} s^{-1}$ [7]. However, owing to the polycrystalline structure of ZnO with large-area uniformity limited by grain boundaries, applications were limited.

In 2003, Hosono and coworkers [8] reported the fabrication of a single-crystalline oxide semiconductor, $InGaO_3(ZnO)_5$. The film was first deposited by pulsed laser deposition (PLD) to form a 2-nm-thick ZnO epitaxial layer at 700 °C on a (111) single-crystal yttria-stabilized zirconia (YSZ) substrate as the template and then a deposition of $InGaO_3(ZnO)_5$ layer at room temperature. After that, the bilayer film was covered with a YSZ plate and subjected to thermal annealing at 1400 °C for 30 minutes in an atmospheric electric furnace. In such semiconductors, InO_2^- and $Ga(ZnO)_5^+$ blocks were stacked alternately along the (0001) axis, with the carrier doping being controlled by the amount of Ga. The doping of Ga^{3+} ion into the $GaO(ZnO)_5^+$ block was found not to generate carriers in $InGaO_3(ZnO)_5$ since Ga^{3+} ion not only substituted the Zn^{2+} tetrahedral sites but also took trigonal-bipyramidal coordination sites, which kept the local electroneutrality. The In_2O_3 layer worked as the blocking layer for oxygen outdiffusion, which suppressed the generation of oxygen vacancy. By gating with an 80-nm-thick HfO_2 layer, the TFTs exhibited a high current on/off ratio of ~10^6 and a mobility of ~80 $cm^2 V^{-1} s^{-1}$, surpassing those oxide TFTs reported before this work.

While the above-mentioned $InGaO_3(ZnO)_5$ TFTs showed a high mobility, a high deposition temperature of 1400 °C was required to obtain the high-quality single-crystalline film, limiting their potential applications especially in flexible electronics. In the next year, the same group reported a remarkable breakthrough on *n*-type oxide TFTs with the deposition of amorphous IGZO through PLD at room temperature and demonstrated high-performance flexible transistors [1]. In comparison to Si, which shows a significant mobility drop from 1000 $cm^2 V^{-1} s^{-1}$ in single crystal to 1 $cm^2 V^{-1} s^{-1}$ in amorphous states, IGZO could maintain a high mobility of 10 $cm^2 V^{-1} s^{-1}$ even in its amorphous state, making it a promising material for future displays. The origin of the high mobility was attributed to the electronic orbital structure of the material. As shown in Figure 4.4, direct overlap among neighboring metal *ns* orbitals occurred, creating a conduction path for free carriers that remained largely unaffected even in the amorphous phase. The conduction band in IGZO primarily formed through the overlap of In 5*s* orbitals, which exhibited isotropic properties. The spherical symmetry of the 5*s* orbitals rendered IGZO insensitive to structural deformation, allowing the semiconductor to maintain a high mobility even in the amorphous state.

Two years after that, Hosono's group also reported the first IGZO TFT deposited using magnetron sputtering, which had now become the most commonly used method in industry for depositing oxide films in large-area devices [9]. By adjusting

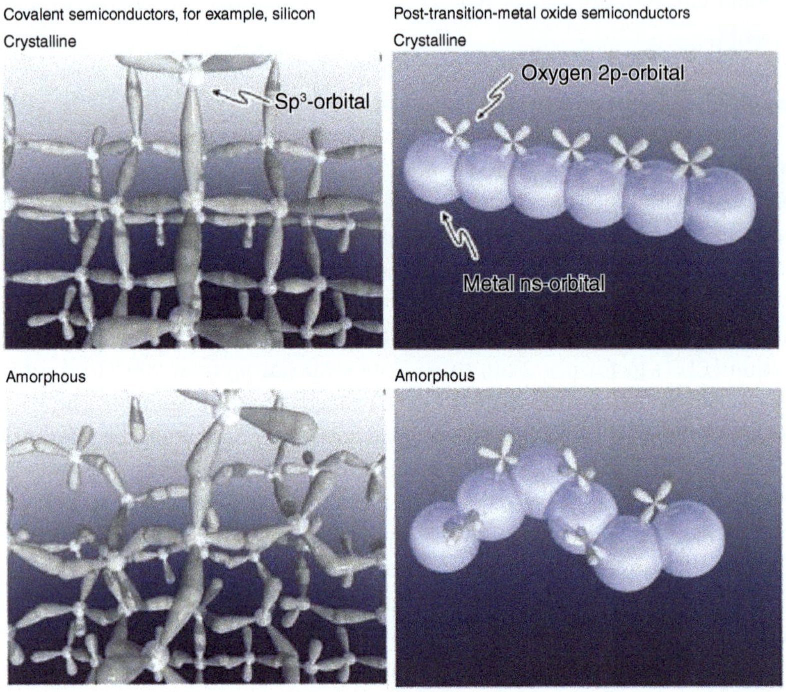

Figure 4.4 Schematic orbital structures of the CBM states in Si and an ionic oxide semiconductor. Source: Reprinted with permission from Ref. [1]. © 2004 Springer Nature.

the mixing ratio of sputtering gases, $O_2/(Ar + O_2)$, from ~3.1% to 3.7%, the conductivity of the IGZO films could be controlled from $\sim 10^{-3}$ to 10^{-6} S cm^{-1}. Under optimized conditions, the top-gated TFTs showed a mobility of 12 cm^2 V^{-1} s^{-1}, a current on/off ratio of $\sim 10^8$, and a subthreshold swing of 0.2 V dec^{-1}. Such sputtered IGZO TFTs also showed excellent reproductivity, making them promising candidates for the commercial production of flexible electronics.

Currently, numerous research groups are actively working on oxide semiconductor-based thin-film electronics, with the number of published papers gradually increasing, reaching more than 200 papers per year especially since 2011. Through analysis of those papers, IGZO is the most extensively studied *n*-type oxide semiconductor in TFTs, with sputtering being the most commonly employed deposition method. To realize practical applications, oxide TFTs with low power consumption and high stability are required. Also, the growing interest in printable electronics has led to an increased focus on solution-processed TFTs. A detailed analysis of these aspects will be given in the upcoming subsections.

4.3.2 Composition

Binary oxide semiconductors, such as ZnO, SnO_2, and In_2O_3, are fundamental *n*-type oxide semiconductors commonly used in TFTs. As early as in the 1960s, oxide TFTs using those binary semiconductors were reported despite showing an unsatisfactory performance. With the development over past few decades, ZnO TFTs with a reasonable performance were reported, but they generally require a high deposition temperature for crystallization, and the presence of grain boundaries can significantly degrade the uniformity and reproductivity over large areas. In_2O_3 TFTs were reported to show a high electron mobility even at a low deposition temperature, but the high conductivity of such films might result in TFTs being hard to turn off.

In general, the combination of those binary compounds with other metal oxides to form multicomponent amorphous oxides could lead to an enhanced TFT performance. Unlike in crystalline materials where carrier transport is often limited by grain boundaries, in amorphous oxides, carrier transport is mostly governed by the potential barriers located around the CBM. The structural randomness also makes the carrier transport be easily surpassed in properly processed films by increasing V_G. The relative proportion of cations can directly control the background carrier concentration, N, within a broad range, thereby affecting the transistor performance.

Out of all compound oxides, InZnO and ZnSnO are the most commonly reported ternary oxides. It was previously reported that IZO TFTs with In/(In + Zn) atomic ratios between 0.5 and 0.8 had properties closer to the predominant binary compound [10]. At In/(In + Zn) = 0.8, both I_D and μ_{FE} are similar to In_2O_3 TFTs, while at In/(In + Zn) = 0.5 the properties move toward the one of ZnO TFTs. An intermediate performance was obtained at In/(In + Zn) = 0.67, suggesting that N and μ_{FE} get higher with the increment of In content. However, IZO might have their properties changed around the best performance composition, making the composition control difficult. Moreover, these TFTs are often challenging to achieve clear on and off states within a small V_G range due to the presence of a large number of free carriers. To reduce the number of free carriers, metal elements can be added as carrier suppressors in the film to mitigate the presence of oxygen vacancies. For example, a considerably better device performance was obtained using IGO instead of IZO, owing to the strong bonds between Ga and O, which suppressed the generation of free carriers in the film [11]. Researchers therefore put more attentions on quaternary oxides, i.e. IGZO, as the channel layer. As shown

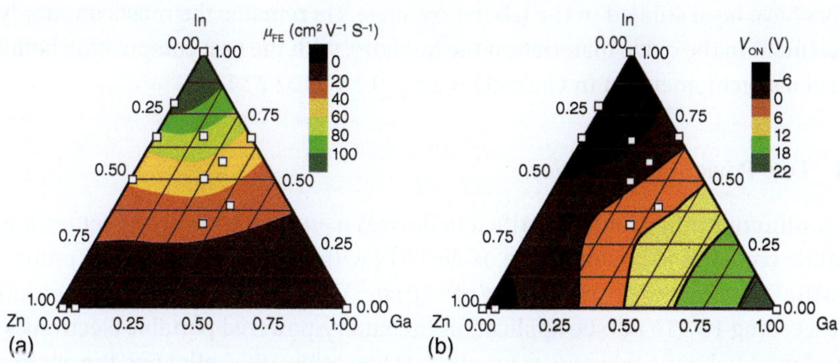

Figure 4.5 (a) μ_{FE} and (b) V_{ON} obtained for TFTs with different oxide compositions in the IGZO system. Source: Reprinted with permission from Ref. [11]. © 2012 John Wiley and Sons.

in Figure 4.5, V_{ON} shifted toward more positive with the increase of Ga content, owing to decrease of background N as a result of the stronger bond between Ga and O. However, the incorporation of Ga into the IZO TFTs could also lower μ_{FE} due to the less fraction of empty traps that are needed to be filled before E_F reaching CBM or above for materials with a higher background N.

The bonding distortion energies for In–O and Zn–O are 346 and 284.1 kJ mol^{-1}, respectively. Therefore, to be effective carrier suppressors, doping/alloying elements need to exhibit higher binding energies with O compared to In–O and Zn–O. In addition to well-reported Ga, elements such as Al, La, Hf, Zr, and B have been reported as capable of tuning the turn-on voltage and enhancing the stability of IZO TFTs. In a recent study by our group, we reported that the introduction of a third cation doping element, specifically Al, in solution-processed IZO TFTs, resulted in a shift of the V_{ON} toward 0 V, which is ideal for circuit applications. Also, both bias stress and long-term stability were significantly improved, mainly due to the higher bonding energy of Al–O than that of In–O and Zn–O.

However, it is worth mentioning that while doping/alloying the aforementioned metal elements can enhance lattice formation and tune carrier concentration and transport properties, they can also reduce oxygen vacancies and free carriers within the oxide semiconductors, thereby resulting in reduced mobility. Marks and coworkers [14] systematically studied the doping effects of Ga^{3+}, Sc^{3+}, Y^{3+}, and La^{3+} – materials with similar chemical properties but increasing ionic radii – on the performance of IZO TFTs. They found that only Ga^{3+} worked at a high doping concentration in increasing the M–O–M content, while other three with larger ionic radii than In^{3+} and Zn^{2+} worked as effective amorphicizing agents but not strong oxygen binders. As such, Ga-doped IZO TFTs showed an impressively stability in both $\Delta\mu$ and ΔV_{TH}, with a maximum deviation of only 5% and 1.6 V, respectively, while they were in the range of 10–22% and 17–20 V for other three metals-doped TFTs.

To conclude, in n-type oxide TFTs, cations such as In and Sn are used to reduce the film resistivity and increase the carrier concentration. Zn, on the other hand, is generally added to maintain the amorphous states of the oxides since ZnO has a high crystalline temperature. To achieve high stability, although various carrier suppressors have been studied in the laboratory stage, Ga remains the most commonly used element in the oxide materials in the industry, with the best composition being found at a stoichiometry of In:Ga:Zn:O = 1 : 1 : 1 : 4 or 2 : 2 : 1 : 7.

4.3.3 Low Power Consumption

With continuous improvement verified in devices using oxide semiconductors, it is now quite common to obtain n-type oxide TFTs with remarkable electrical properties, including a high electron mobility of ~10 cm^2 V^{-1} s^{-1} and a large current on/off ratio exceeding 10^7. To pursue applications in battery-powered portable electronics, it is preferred to have a power consumption as low as possible, allowing the devices to work longer without recharging. One possible way to reduce power consumption is by lowering the operating voltage. For a fixed charge density, this can be achieved by increasing the specific capacitance of the gate dielectrics. The capacitance, C, of a conventional dielectric can be expressed as

$$C = \frac{\varepsilon_0 \varepsilon_r A}{d}, \tag{4.11}$$

where ε_0 and ε_r are the permittivity of vacuum and dielectric layer, respectively, A is the effective area, and d is the thickness of the dielectric layer. To increase C, the easiest way to do is to decrease the dielectric thickness. By reducing the thickness of SiO_2 to 8.64 nm, Zhang et al. [15] reported IGZO TFTs with a low operating voltage of 2 V and a current on/off ratio up to 10^8. However, at ultra-thin thicknesses, SiO_2 might easily suffer from leakage and uniformity problems, resulting in a reduced device performance.

Alternatively, high gate capacitance could be achieved using high dielectric constant (high-κ) materials, which allow for the same capacitance to be obtained as with the traditional low-κ dielectrics but with a greater thickness, thereby addressing the leakage issue. Until now, a variety of oxide high-κ dielectrics, such as Al_2O_3, HfO_2, ZrO_2, and rare-earth oxides have been studied [16–19]. For example, Song and coworkers [19] deposited ultra-thin Al_2O_3 through atomic layer deposition and reported IGZO TFTs with a low operating voltage of 1 V and a high current on/off ratio of 10^7. Fortunato and coworkers [18] combined solution combustion synthesis and ultraviolet treatment to prepare HfO_x, achieving IGZO TFTs with a low operating voltage of 2 V. Chang et al. [20] utilized a 200-nm-thick Ta_2O_5 layer as the gate dielectric, resulting in TFTs with a low operating voltage of 3 V and a current on/off ratio of 10^5. By using ScO_x thin films as gate dielectrics, Shan and coworkers [21] achieved a high gate capacitance of 460 nF cm^{-2} and a permittivity of 12.1, leading to IZO TFTs with a low operating voltage of 1.5 V and a high current on/off ratio of 2.7×10^7.

With the dielectric constant being proportional to the gate capacitance and the dielectric thickness inversely proportional to it, to maximize the gate capacitance, a high dielectric constant and a low dielectric thickness should be achieved simultaneously. However, despite thinner films having a higher capacitance and being beneficial for low-voltage operation, they may also experience a higher leakage current due to the reduction in film thickness. Conversely, although a thicker film could reduce the leakage current, the capacitance might be too low to operate the device within a small voltage range. Therefore, several groups have attempted to find the balance point between film thickness and gate capacitance.

Song's group, for example, used anodization to deposit ultra-thin Al_xO_y, and achieved one-volt-operated IGZO TFTs on both rigid and flexible substrates by modifying the dielectric thickness through anodization voltages [12, 22]. Anodization is a simple, solution-based method, which can deposit pinhole-free, homogeneous films with an ultra-thin thickness at room temperature. Three anodization voltages were examined by them: 1.54, 2.30, and 3.08 V, corresponding to a thickness of 2, 3, and 4 nm for Al_xO_y, respectively. As shown in Figure 4.6a, while all the devices showed a low operating voltage of 1 V, decreasing the anodization voltage from 2.30 to 1.54 V caused an increase in subthreshold swing and interface trap density owing to the presence of a greater proportion of native aluminum oxide formed before anodization. Although devices gated with Al_xO_y anodized at 2.30 and 3.08 V showed a similar interface trap density, increasing the anodization voltage led to a decrease of gate capacitance and hence a lower on-current and mobility. With the use of the 2.3 V anodized Al_xO_y, they also demonstrated flexible IGZO TFTs, which showed a low operating voltage of 1 V and a high current on/off ratio of $>10^5$ even under a curvature radius of 11 mm, demonstrating the promising features of the reported methods in low-power flexible electronics. By substituting Al_xO_y with anodized HfO_x and studying the effects of anodization voltages on the device performance, the same group also reported ultra-thin HfO_x-gated flexible IGZO TFTs, which showed a low operating voltage of 0.5 V with a high current on/off ratio of $>10^5$ (Figure 4.6b) [13].

Besides, some electronically insulating but ion-conducting electrolytes can be used to form an electric-double layer at dielectric/channel interface, resulting in a high gate capacitance and hence a low operating voltage. Those electrolytes include polymer electrolytes, polyelectrolytes, ionic liquids, ion gels, and solid-state electrolytes, and have been widely employed for both organic and oxide TFTs to achieve a low operating voltage [23]. For example, Xin and coworkers [24] reported flexible IGZO TFTs with a low operating voltage of 1 V thanks to the utilization of a high capacitance ($4.5\,\mu F\,cm^{-2}$ at 20 Hz) poly(styrenesulfonic acid sodium salt) (PSSNa) as the gate dielectric. Yuan et al. [25] reported electrolyte-gated ZnO transistors using an ionic liquid *N, N*-diethyl-*N*-(2-methoxyethyl)-*N*-methylammonium bis(trifluoromethylsulphonyl-imide) (DEME-TFSI) as the dielectric layer, showing a low operating voltage of 3 V owing to the high gate capacitance of $34\,\mu F\,cm^{-2}$. Using photo-crosslinkable ion gels (SEAS and EMIM-TFSI) and stencil-based printing techniques, Frisbie's group [26] reported ZnO electrolyte-gated transistors with a low operating voltage of 1 V.

The increased requirements for the integration with other electronic components have led to a focus on the development of solid-state electrolytes. In this regard, PECVD can be employed using SiH_4 and O_2 as reactive gases. The incorporation

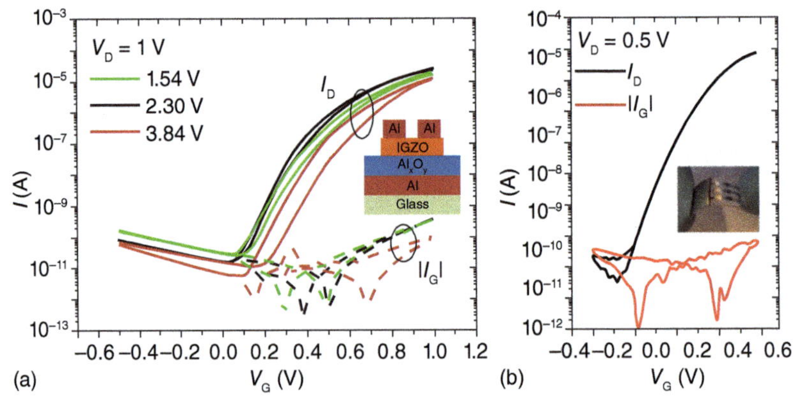

Figure 4.6 Transfer characteristics of IGZO TFTs gated with (a) Al_xO_y and (b) HfO_x. Source: Reprinted with permission from Ref. [20]. © 2018 IEEE, 2019 American Chemical Society.

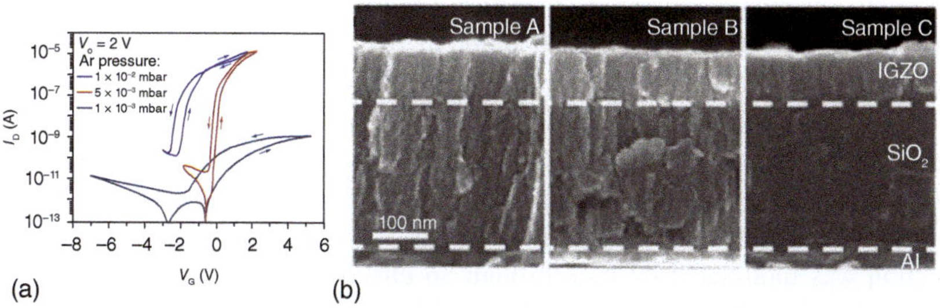

Figure 4.7 (a) Transfer characteristics and (b) corresponding cross-sectional SEM images of SiO_2 electrolyte-gated TFTs. Source: Reprinted with permission from Ref. [34]. © 2017 Springer Nature.

of reaction by-products related to –OH groups into the SiO_2 network during this process leads to the formation of porous SiO_2. When applying an electric field, the bond between oxygen and hydrogen might break, allowing the oxygen atoms to combine with mobile protons and form three-coordinate oxygen centers (Si–OH^+–Si) since free protons cannot exist in SiO_2. According to the polarity of the external electric field, the protons can hop between neighboring oxygen centers and move to the dielectric/channel or dielectric/gate interface to form an electric-double layer. With the use of such PECVD-deposited porous SiO_2, oxide TFTs with a low operating voltage of less than 1.5 V were deposited on both rigid and flexible substrates, as reported several times by Wan's group [27–30] and Dou's group [31–33].

Apart from those deposited through PECVD, porous SiO_2 have also been reported by Song's group, who achieved porous SiO_2 using radio-frequency magnetron sputtering in a pure Ar ambient [34, 35]. Unlike PECVD-based SiO_2, which has a known source of proton, the protons inside the sputtered SiO_2 are generated by absorbed water molecules. To support this theory, the authors explored the relationship between Ar pressure during SiO_2 deposition and the performance of the resulting TFTs, as shown in Figure 4.7. The cross-sectional SEM images suggested that decreasing the sputtering power decreases the granularity of the SiO_2 layer. A highly granular structure was capable of absorbing water molecules, which benefited the proton conduction. The absorbed water molecules could be ionized into H^+ and OH^- and the protons were driven to the dielectric/channel interface under a positive gate bias. Due to the lack of granular-like microstructure in SiO_2 sputtered at the lowest Ar pressure (sample C), the SiO_2 layer could not absorb enough water molecules to form an effective electric-double layer and hence the devices were hard to turn on and off fully within a low voltage range. Since sputtering is one of the most common large-area thin-film deposition methods in industry, this fabrication method might have potential applications in low-cost, low-power portable electronics applications.

Despite the above-mentioned methods for achieving a low operating voltage, low power consumption can also be achieved by Schottky barrier (SB) TFTs. By adjusting the height of the Schottky barrier formed between a semiconductor and contacts, Nathan and Lee [36] reported IGZO TFTs with an ultralow power of less than 1 nW, a switching voltage of less than 1 V, and a very high intrinsic gain of >400. To accomplish this, they used different oxygen-gas partial pressures, P_{OX}, relative to argon to deposit more compensated (MC, $P_{OX} = 15\%$) and less compensated (LC, $P_{OX} = 4\%$) IGZO films. Unlike conventional Ohmic contact-based TFTs, where the drain current followed a linear response at low drain voltages, SB TFTs showed a nonlinear response of the drain current at $V_{DS} < V_{tran}$ (transition voltage), suggesting a forward-biased Schottky diode at the drain junction. The electron collection was modulated by V_{GS} through an effective Schottky barrier lowering at the drain side. With the increase of V_{DS} to larger than V_{tran}, I_{DS} became firmly saturated due to the formation of a reverse biased Schottky diode at the source. With channel length, L, far larger than the depletion width at the drain, W_D, a negligible image-charge effect from the drain to the source was resulted. Consequently, in the saturation region of SB TFTs, I_D showed minimal change with the channel width.

Low-power devices were also reported by Wu and coworkers [37] in 2019, using an ultra-thin ITO as the channel layer. They conducted a systematic study on the effects of dielectric and channel layer thickness, as well as the channel length. As shown in Figure 4.8, the reduction of channel layer thickness to 4 nm increased the bandgap, making ITO capable of functioning as a semiconductor layer instead of the commonly reported contact layers in TFTs. Besides, the reduction of equivalent of thickness (EOT) of dielectric layers to 0.8 nm using La-doped HfO_2 increased the gate capacitance, thereby reducing the operating voltage and subthreshold swing. The ultra-thin channel layer and the small EOT combined with the low permittivity of ITO (~4) ensured a sufficiently small nature length of 1.8 nm, and hence negligible short channel effects in a 40-nm device. Thus, they achieved TFTs with a short-channel immunity at a channel length of 40 nm, showing a low subthreshold swing of 66 mV dec^{-1}, an off-current of <100 fA μm^{-1}, a high current on/off ratio reaching 10^9, all of which were highly promising for ultralow power applications.

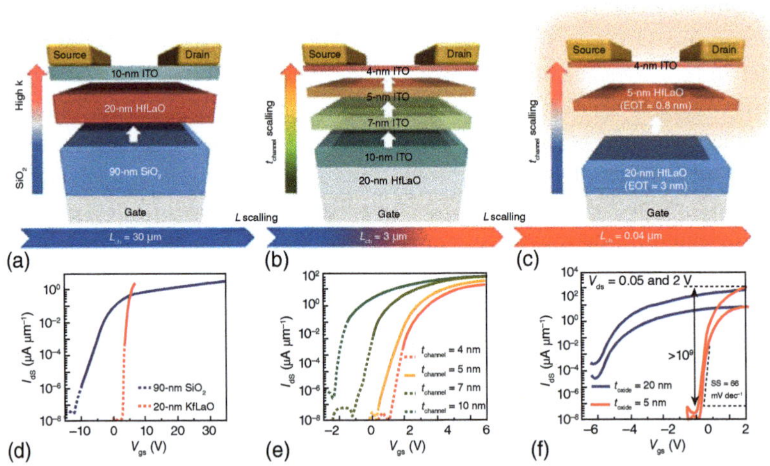

Figure 4.8 Electrical performance of ITO TFTs. (a-c) Device structures and (d-f) corresponding transfer characteristics. Source: Reprinted with permission from Ref. [37]. © 2019 Springer Nature.

4.3.4 Stability

The long-term stability and reliability of TFTs are crucial factors for mass production and commercialization. Until now, numerous groups have studied the stability of oxide TFTs and found a general right shift in the threshold voltage of *n*-type oxide TFTs under positive bias stress. This can be explained by several factors, including the trapping of positive charges (i) in the gate dielectric layer, (ii) at dielectric/channel interface, (iii) in the bulk channel, and (iv) at back-channel surface as well as (v) the creation of acceptor-type deep traps and (vi) a decrease of donor concentration. Among these contributions, (vi) is unlikely the reason since the defect density is difficult to decrease during a degradation test.

The charge trapping at dielectric/channel interface can be reduced by inserting an interfacial layer to passivate traps at the dielectric surface. While in organic TFTs, it is quite common to use a self-assembled monolayer (SAM) at the dielectric/channel interface to reduce the interface traps and increase the overall performance, such a treatment was rarely reported in oxide TFTs due to the potential damage to ultra-thin organic SAMs during the deposition of oxide semiconductors. In 2020, Song's group [38] studied the effectiveness of an SAM treatment on gate dielectrics (anodized Al_xO_y or HfO_x) in TFTs with sputtered IGZO channel layers, and found that by carefully controlling the sputtering power, the damage to the SAM could be minimized and effectively controlled. While the as-deposited dielectric layers might contain defects (i.e. –OH groups and adsorbed water molecules) at the surface, these traps could be passivated using SAMs, resulting in a reduction of trap density by more than 50%. As shown in Figure 4.9a, the untreated devices showed an increase of *SS* after the bias stress, possibly due to stress-induced traps at the dielectric/channel interface. However, treating the devices with *n*-octadecyltrichlorosilane (OTS) resulted in a negligible change in *SS*, suggesting that the creation of trap states at the dielectric/channel interface induced by the gate bias stress was negligible (Figure 4.9b). The treated devices also showed a smaller threshold voltage shift and were more resilient to carrier mobility degradation compared to the untreated ones, regardless of whether Al_xO_y or HfO_x was used as the dielectric, demonstrating the broad applicability of such methods in enhancing the bias stress stability of oxide TFTs.

Unlike Si or organic semiconductors based TFTs, the channels of oxide TFTs are quite sensitive to water and oxygen molecules in air. Under a positive gate bias, adsorbed oxygen can capture electrons from the conduction band and form various oxygen species, such as O^{2-}, O^-, or O_2-, resulting in a lower effective gate bias being applied to the TFTs and a larger positive voltage being required to turn the device on. Also, the adsorption/desorption process of water on the active layer might create/remove traps in the active layer, causing instabilities. As reported in [39], some positively charged water molecules on the surface of oxides would be desorbed upon the application of a positive gate bias, causing a simultaneous positive shift of threshold voltage and a decrease of subthreshold swing.

To address the instability issue caused by the traps at the back-channel surface, efforts have been made by passivating the top surface with a dense and gas-tight

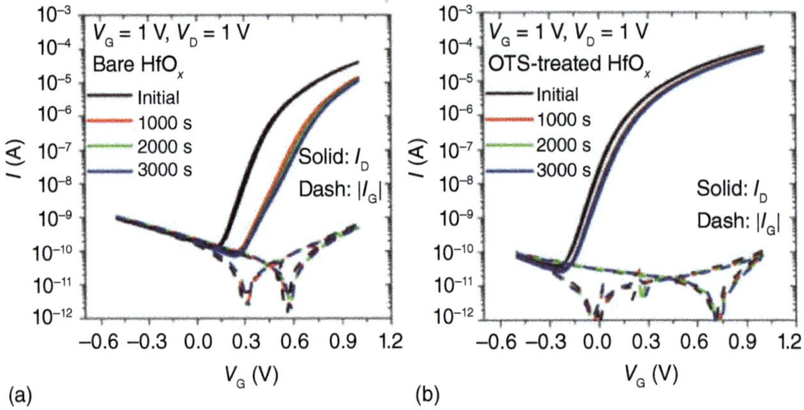

Figure 4.9 Transfer characteristics of IGZO TFTs gated with (a) bare HfO_x and (b) OTS-treated HfO_x before and after being positively biased at $V_G = 1$ V for 3000 seconds. Source: Reprinted with permission from Ref. [38]. © 2020 John Wiley and Sons.

layer. For example, in 2020, Song and coworkers [40] reported that by treating the top surface of IGZO TFTs with OTS, the bias stress stability of the devices could be significantly enhanced, as shown in Figure 4.10. The schematic in Figure 4.10c indicates that the OTS treatment facilitated the desorption of weakly adsorbed water molecules, transforming the surface of IGZO from hydrophilic to hydrophobic, and resulting in reduced interaction of impurity species and effects of adsorbed polar molecules on the surface of the film. Also, as a dense, chemically stable barrier layer, the formed OTS layer could prevent further adsorption of additional oxygen and water molecules from the ambient atmosphere. Similarly, Peng and coworkers [41], Chen and coworkers [42], and Lee and coworkers [43] also reported significant enhancement of bias stress stability in oxide TFTs by capping the top channel surface with an SAM. Other passivation layers, including organic PMMA, SU-8, and CYTOP, and inorganic insulator layers were also previously reported to improve the stability of oxide TFTs [44–46].

The instability under negative bias stresses with light illumination (NBIS) is another important consideration that hampers the realization of fully transparent oxide-based electronics. Oxide TFTs show high photosensitivity to light with photon energies smaller than their bandgaps. The defects near the valence band maximum (VBM), called subgap states (SBGs), are generally resulted by the oxygen vacancies, weakly bonded oxygen and hydrogen impurities, and are considered the main source of photo-instability [47, 48]. Thus, to improve the NBIS stability of oxide TFTs, one can try to eliminate the SGSs or widen the energy difference between the SGSs and the conduction band minimum (CBM) [49]. Compared with the first method, it is more practical to widen the bandgap if the location of the SGSs is pushed down with the VBM. As shown in Figure 4.11, lowering the energy level of VBM (E_{VBM}) is critical to suppress the NBIS instability while maintaining a relatively high mobility. Hosono and coworkers [49] examined the electronic structures of various

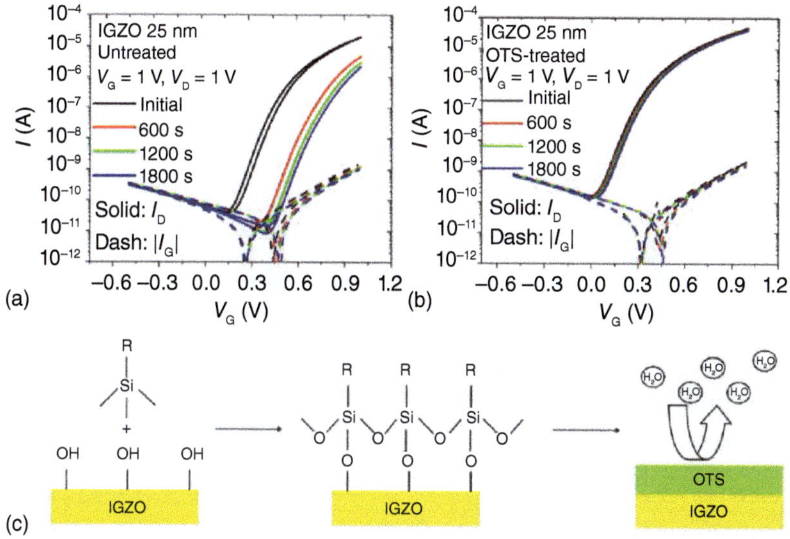

Figure 4.10 Transfer characteristics of IGZO TFTs (a) without and (b) with OTS treatment under a positive bias stress. (c) A schematic diagram showing the OTS formation on the top surface of IGZO. Source: Reprinted with permission from Ref. [39]. © 2020 American Chemical Society.

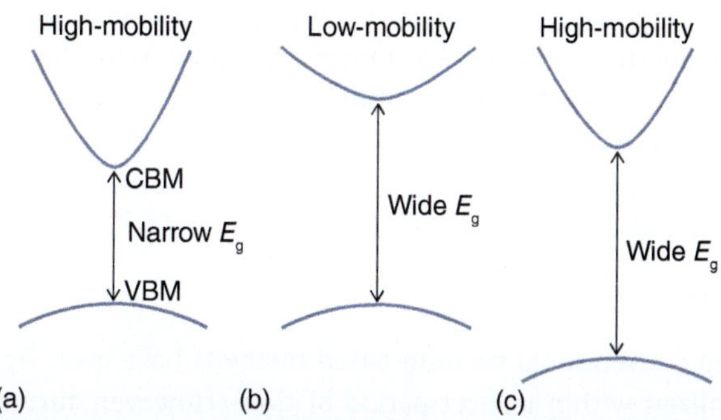

Figure 4.11 The excepted electronic structures and electrical properties of oxide semiconductors with (a) a high electron mobility and a narrow bandgap, (b) a low electron mobility and a wide bandgap, and (c) a proposed electronic structure with a suitable photostability and a high mobility. Source: Reprinted with permission from Ref. [49]. © 2019 AIP Publishing.

oxide semiconductors and found significant variations in their ionization potentials with changes in metal cations. Figure 4.12 indicates that oxide semiconductors tend to have a deeper energy level of CBM (E_{CBM}) with the increase of In content, resulting in an increased mobility. However, this would also cause a degradation of photostability. A comparison between the energy level of a-IZO and a-ZGO clearly indicates the contribution of GaO_x in deepening the E_{VBM} owing to its intrinsic

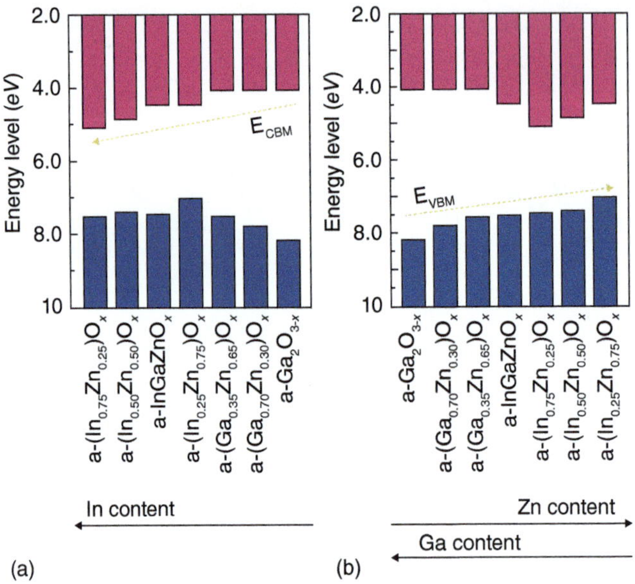

Figure 4.12 Energy levels of different oxide semiconductors arranged in order of (a) CBM level and (b) VBM level, respectively. Source: Reprinted with permission from Ref. [49]. © 2019 AIP Publishing.

properties. Thus, the author concluded that to achieve a high photostability, In/Zn should be avoided and GaO_x should be included as much as possible. With the use of a-$Zn_{0.3}Ga_{0.7}O_x$ as the channel layer, they successfully demonstrated TFTs with an excellent NBIS stability and a mobility of ~9 $cm^2\ V^{-1}\ s^{-1}$, mainly due to the deep E_{VBM} of 7.8 eV and the wide bandgap of ~3.8 eV.

4.3.5 Solution-Based TFTs

Oxide TFTs fabricated using conventional vacuum-based methods have been well established and commercialized within a short period of time. However, current methods generally involve sequential deposition, including patterning and etching of selected layers, which requires multiple photolithography and vacuum deposition processes, resulting in high manufacturing costs. In contrast, direct printing offers possibilities of depositing thin films using a direct additive patterning process, which can significantly reduce fabrication costs. However, despite the exploration of organic semiconductors for low-cost electronics over the years, their relatively low mobility and reliability issues still limit their applications in fields such as TFTs for backplane drivers of displays. The growing demand for printable electronics has thus shifted attention toward fabricating oxide TFTs using solution-processed technologies such as spin-coating, inkjet printing, and gravure printing.

In general, there are two main routes for depositing metal oxides from solution: nanoparticle-based processes and metal salt-based sol–gel chemistry. However, nanoparticle-based processes are seldom used in high-performance oxide TFTs due to the presence of voids within the nanoparticle film and poor dielectric/channel interface states. Among the metal salt-based approaches, sol–gel chemistry is widely reported. It involves dissolving metal salts in an alcohol or aqueous solvent, followed by coating the precursor solution and a post-annealing process to facilitate the formation of M–O–M structures, remove byproducts and solvents, and densify the resulting oxide films.

As early as 2007, Chang et al. [50] already reported solution-processed ZTO and IZO TFTs using a metal salt sol–gel route, achieving a high mobility of $16\,cm^2\,V^{-1}\,s^{-1}$, comparable to vacuum-deposited counterparts. Street's work in 2014 showed that the device performance highly relied on the precursor materials, composition, and processing variables [51]. By carefully controlling these factors, a high mobility of 30 and $20\,cm^2\,V^{-1}\,s^{-1}$ for IZO and IGZO TFTs, respectively, could be achieved by them.

In 2021, Zang's group studied the effects of annealing temperature on the performance of spin-coated IZO TFTs [52]. They studied several annealing temperatures ranging from 275 to 400 °C and found that the optimal device performance was achieved at an annealing temperature of 350 °C. They concluded that lower temperatures resulted in insufficient formation of M–O bonds, leading to film defects, while higher temperatures made the devices excessively conductive, affecting their turn-off characteristics. By combining the spin-coated IZO TFTs with a solution-processed ultra-thin Al_xO_y layer, they demonstrated 1-volt-operated TFTs with a high mobility approaching $10\,cm^2\,V^{-1}\,s^{-1}$ and a high current on/off ratio exceeding 10^5. These device performance metrics were comparable to those achieved using vacuum-based methods, highlighting the significant potential of printable, low-power electronics.

However, traditional sol–gel-based methods typically require a high annealing temperature above 300 °C to cure the oxide film, which limits their practical application in flexible electronics. Consequently, various approaches, including deep-ultraviolet (DUV) irradiation, combustion chemistry, and aqueous routes, have been investigated to lower the annealing temperature to less than 300 °C, as summarized in Table 4.1.

Among these methods, DUV irradiation is the most commonly employed technique for reducing the post-annealing temperature. It enables the preparation of stable oxide semiconductors with improved performance through a photochemical activation process involving rapid chemical condensation and gradual structural rearrangement and densification (Figure 4.13a) [53]. By subjecting the as-spun films to a high-density DUV treatment system (a mercury lamp with a peak emission wavelength of 184.9 nm (10%) and 253.7 nm (90%)) under N_2 purging, Park's group reported film and TFT properties comparable to those thermally annealed at 350 °C (Figure 4.13b) [53]. Such DUV-derived TFTs showed high mobilities of 14 and $7\,cm^2\,V^{-1}\,s^{-1}$ on glass and polymer substrates, respectively. Also, the authors demonstrated seven-stage ring oscillators on a flexible substrate with an oscillation frequency exceeding 340 kHz, suggesting the great potential of this method in future flexible electronics.

Table 4.1 Representative low-temperature (<300 °C), solution-processed oxide TFTs.

Methods	Channel layer	T (°C)	μ (cm^2 V^{-1} s^{-1})	I_{ON}/I_{OFF}	Reference
DUV irradiation	In_2O_3	150	11.29 ± 1.62	10^8	[53]
	IZO	150	4.43 ± 0.59	10^8	[53]
	IGZO	150	8.76 ± 0.98	10^8	[53]
Combustion synthesis	In_2O_3	250	3.37	10^7	[54]
	ZTO	250	1.76	10^7	[54]
	IZO	250	0.91	10^6	[54]
Microware irradiation	IZO	150	0.1	10^6	[55]
	ZnO	140	0.72	10^7	[56]
Aqueous route	ZnO	180	11	$>10^4$	[57]
	In_2O_3	250	19	10^7	[58]
	In_2O_3	200	3.14	$>10^9$	[59]
Plasma assisted	In_2O_3	140	1.3	$>10^7$	[60]
	ZnO	140	1.4	10^7	[61]
	IGZO	250	3.8	10^6	[62]

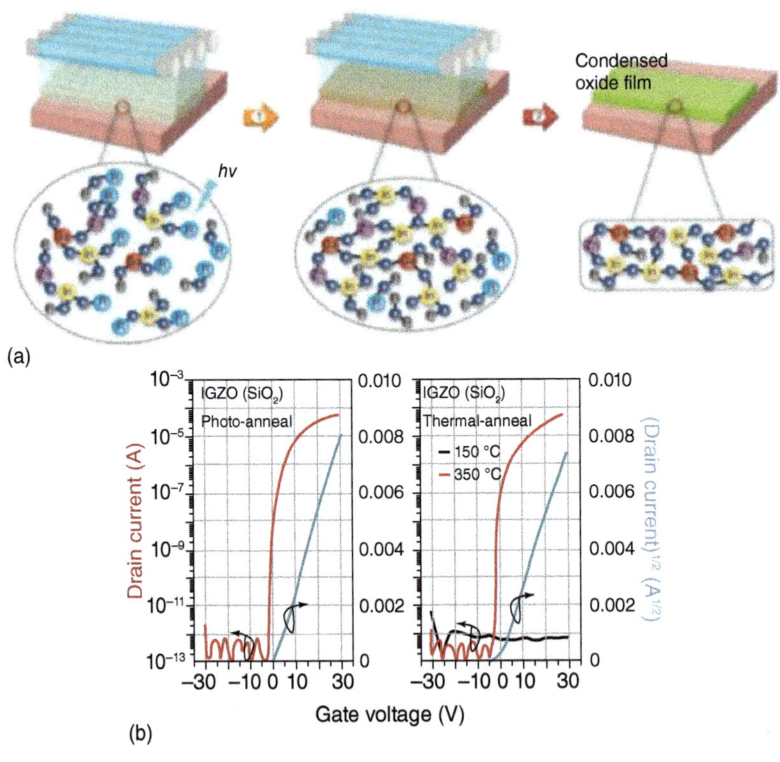

Figure 4.13 (a) Schematic diagrams of the condensation mechanism of DUV-irradiated oxide precursors. (b) Transfer characteristics of IGZO TFTs prepared by DUV irradiation or thermal annealing. Source: Reprinted with permission from Ref. [53]. © 2012 Springer Nature.

In 2015, the same group conducted a study to investigate the underlying physicochemical mechanism of DUV irradiation in various sol–gel oxide films [63]. They discovered that DUV irradiation could facilitate the rapid formation of oxide thin films, including densification, polycondensation, and impurity decomposition, within a short time frame of five minutes through *in-situ* radical-mediated reactions. Utilizing this method, they successfully demonstrated the rapid fabrication of all-solution-based oxide TFT circuitry, exhibiting stable and reliable electrical performance. Subsequently, the group reported an upgraded process known as the DUV light irradiation-water treatment-DUV irradiation (DWD) method [64]. In this process, water treatment was employed to induce the preferred adsorption of water molecules at oxygen vacancies and subsequent hydroxide formation in the bulk a-IGZO films. The second DUV irradiation step performed on the water-treated devices enabled the more complete formation of M–O–M lattice with reduced oxygen vacancies in the oxide films. As a result, oxide films with passivated imperfections (e.g. –OH groups and oxygen vacancies) were obtained, leading to a 30% improvement in TFT performance compared to the ordinary DUV method.

Furthermore, the deposition temperature of oxide semiconductors can be reduced by employing a combustion synthesis approach that incorporates fuels such as acetylacetone and urea into the precursor solutions. Marks and coworkers [54] utilized acetylacetone as a fuel and metal nitrates as oxidizers to develop a self-generated combustion chemistry process, which effectively lowered the annealing temperature required for oxide film formation. As shown in Figure 4.14, this process could generate heat autonomously, converting the precursors into metal oxide lattices and reducing the need for external annealing at high temperatures. By employing a solution-processed Al_2O_3 dielectric, In_2O_3 TFTs prepared using the combustion method exhibited a mobility of $6\,cm^{-2}\,V^{-1}\,s^{-1}$ even at an annealing temperature as low as 200 °C. Using similar approaches, the same group and several other groups have also reported the synthesis of various oxide semiconductors and even high-κ oxide dielectrics [14, 65–68].

In addition to the material and process modifications mentioned above, structural engineering techniques have also been reported to demonstrate the ability to reduce annealing temperatures. One of the most effective and extensively studied methods is the utilization of multistacked active layers, which includes both homojunction

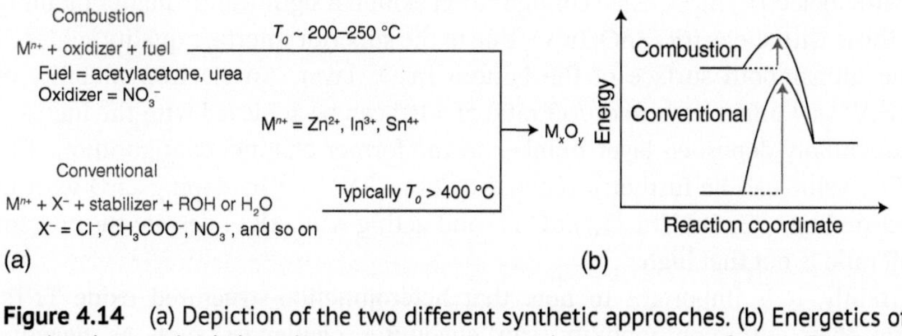

Figure 4.14 (a) Depiction of the two different synthetic approaches. (b) Energetics of combustion synthesis-based process versus conventional processes. Source: Reprinted with permission from Ref. [53]. © 2012 Springer Nature.

and heterojunction active layers. For instance, Kim *et al.* [69] successfully optimized the electrical performance of IGZO TFTs by adjusting the thickness of homojunction dual active layers, resulting in an enhanced mobility compared to single-channel configurations. However, a high annealing temperature of 450 °C was required in their work.

Compared with homojunction active layers, despite heterojunction active layers facing challenges related to interface properties, they also offer advantages in terms of easy controlling of electrical performance, including the reliability and chemical stability of TFTs. Kim and coworkers [70] conducted notable research on heterojunction dual active layered TFTs, where they utilized AIZO/IZO dual channels deposited at low temperatures and successfully achieved an optimized mobility of 5.62 $cm^2\ V^{-1}\ s^{-1}$ by varying the In/Zn ratio and IZO film thickness.

Recent advancements have showcased high-performance oxide TFTs utilizing heterojunction active layers, where the presence of an electron confinement layer at the interface between two active layers has led to a significant enhancement in mobility. The improved carrier mobility in this context is closely linked to the disparity in Fermi energy levels between the two active layers and the confinement of electrons within the potential well. One representative work related to solution-processed oxide TFTs using the behavior of electron confinement was reported by Yang and coworkers [71], who incorporated an ultra-thin ITZO layer between IGZO and the dielectric layer, resulting in a substantial boost in mobility. The band alignment confirmed that the observed increase in mobility was due to the control of effective electron flow induced by the barrier between IGZO and ITZO.

The aforementioned research presents exciting prospects for achieving both low-temperature processing and high mobility, prompting researchers to focus on process modifications. Anthopoulos's group, for instance, has made significant contributions in this field [72–75]. As early as in 2017, they reported the fabrication of solution-processed In_2O_3/ZnO TFTs using a sub-second photonic process with a high-power xenon flash lamp, achieving an impressive mobility of 36 $cm^2\ V^{-1}\ s^{-1}$ at a low temperature of 200 °C [72]. Later in 2019, they also reported the development of TFTs based on multilayer channels composed of alternating layers of In_2O_3 and ZnO at temperatures less than 200 °C and found that the TFT performance depended highly on the exact layer configuration and the number of In_2O_3/ZnO heterojunctions in the channel, as shown in Figure 4.15 [74]. Heterojunction channels with dielectric/In_2O_3/ZnO configuration exhibit a significantly higher mobility than those with dielectric/ZnO/In_2O_3 due to the superior interface quality achieved by the ultrasmooth surface of the bottom In_2O_3 layer. An electron mobility of 13 $cm^2\ V^{-1}\ s^{-1}$ and a current on/off ratio of $>10^8$ can be achieved with the increase of sequentially deposited layer number in the former channel configuration. The mobility value can be further promoted to 25 $cm^2\ V^{-1}\ s^{-1}$ by doping ZnO with Li (structure In_2O_3/ZnO:Li/In_2O_3/ZnO:Li) and gating with AlO_x, despite the current on/off ratio is not that high.

Certainly, it is important to note that heterojunction-structured oxide TFTs, despite their ultrahigh mobility, can encounter challenges such as negative threshold voltage or difficulties in turning off, which are not desirable for circuit applications. This behavior is attributed to the excessive carrier concentration in

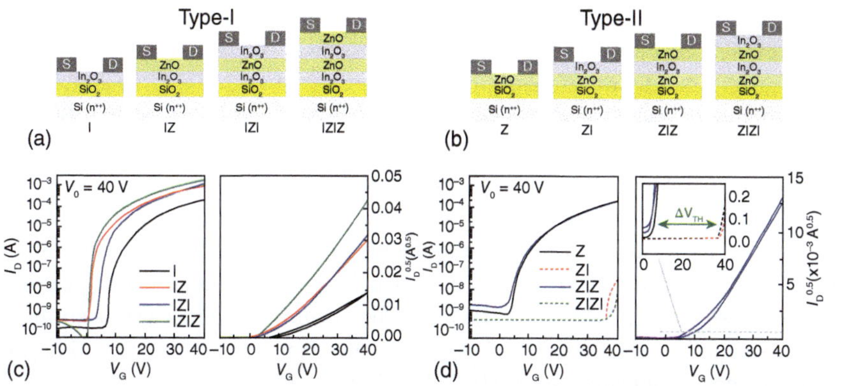

Figure 4.15 Schematics of (a) Type-I and (b) Type-II transistor architectures. Transfer characteristics and corresponding $I_D^{0.5}$ versus V_G plots for (c) Type-I and (d) Type-II TFTs. Inset: Zoom in region of the plot highlighting the difference in threshold voltage between Type-II TFTs.

heterojunction-structured oxide semiconductors. To address this issue, it is crucial to carefully select materials and explore strategies such as bandgap modification through doping or other potential methods. Further investigation and research are needed to find effective solutions to overcome these challenges and optimize the performance of heterojunction-structured oxide TFTs for practical circuit applications.

4.4 *P*-Type TFTs

The advancement of *n*-type oxide TFTs has indeed sparked interest in developing high-performance *p*-type oxide TFTs. The realization of high-performance *p*-type oxide TFTs is essential for enabling highly compact complementary circuits with low power consumption, which is crucial for the development of next-generation wearable electronics. However, *p*-type oxide semiconductors face challenges due to the carrier conduction path being formed from the oxygen *p* asymmetric orbitals. This limitation significantly restricts the carrier mobility in *p*-type oxide TFTs. Consequently, *p*-type oxide TFTs generally exhibit poor electrical performance, which poses a major obstacle in achieving high-performance complementary circuits. Nevertheless, recent research efforts have been focused on *p*-type oxide semiconductors, including Cu-based oxides, tin monoxide, and nickel oxide. These materials have shown promise in overcoming the limitations and have demonstrated *p*-type oxide TFTs with device performance comparable to their *n*-type counterparts. In the following discussion, we will primarily focus on these three promising *p*-type oxide semiconductors.

4.4.1 Copper Oxides

Copper oxides have two well-known forms, namely Cu_2O and CuO, both of which are *p*-type oxides owing to the existence of negatively charged Cu vacancies. Cu_2O has a cubic crystal structure, consisting of two interpenetrating diamond-like oxygen and copper networks with Cu atoms inserted between two consecutive body centered cubic (bcc)-arrayed oxygen layers. Each oxygen atom in the structure is surrounded by a tetrahedron of Cu atoms, and each Cu atom is twofold coordinated. As for CuO, each Cu atom forms four coplanar bonds with oxygen in the tenorite structure. Cu_2O has been reported to show a Hall mobility exceeding $100\,cm^2\,V^{-1}\,s^{-1}$, which is related to their band structure. Although most metal oxides with valence band formed mainly by the localized and anisotropic O 2*p* orbitals show a low hole mobility due to hopping conduction, here the very top of the VBM in Cu_2O is composed mainly of the Cu *d* states, resulting in a less-localized hole-transport pathway. Similarly, the *p*-type behavior of CuO has also been theoretically studied in relation to the presence of Cu vacancies. In this subsection, we review the most promising studies on TFTs using copper oxides.

The first *p*-type Cu_2O TFTs were reported by Hosono and coworkers [76] in 2008, prepared by PLD on MgO substrates. However, such devices were fabricated at a relatively high temperature of 700 °C and showed a low field-effect mobility of $0.26\,cm^2\,V^{-1}\,s^{-1}$ with a current on/off ratio of just 6, which was not suitable for practical applications. The main reason for the low mobility here was the subgap traps formed by the extra defects, i.e. oxygen vacancies and/or secondary CuO phase.

A big breakthrough was achieved two years later by Fortunato et al. [77], who reported Cu_2O TFTs deposited by an industry-compatible RF sputtering method at room temperature followed by a low-temperature post-annealing treatment. The Cu_2O phase composition was adjusted by tuning the oxygen partial pressure, and the annealing of the as-sputtered film at 200 °C for 10 hours successfully increased the Hall mobility from 0.65 to $18.5\,cm^2\,V^{-1}\,s^{-1}$ and the grain size from 8.3 to 15.72 nm. As a result, they achieved TFTs with a field-effect mobility of $1.2\times10^{-3}\,cm^2\,V^{-1}\,s^{-1}$ and a current on/off ratio of 200.

Another significant achievement in *p*-type Cu_2O TFTs was reported by Zou et al. [78], who deposited Cu_2O by PLD. The top-gated Cu_2O TFTs with high-κ HfON gate dielectrics showed a dramatic current on/off ratio of 3×10^6 and a saturation mobility of $4.3\,cm^2\,V^{-1}\,s^{-1}$, which was already comparable to those of *n*-type oxide TFTs and was of great importance for pursuing complementary circuits. Such an improved TFT mobility was considered mainly due to the decreased scattering of both ionized defects and grain boundaries for Cu_2O channels, which was also revealed by the low subthreshold swing of $0.18\,V\,dec^{-1}$.

Oxygen vacancies are considered the general reason affecting the performance of TFTs as they can interfere with the carrier flow by acting as hole-trapping sites. This happens particularly in the fabrication of Cu_2O from the reduction process of CuO. Recently, Kim et al. [79] reported a performance enhancement from a material science perspective by doping Cu_2O with Ga. Since Ga has a high oxygen affinity, the doping of it could decrease the oxygen vacancies and improve the hole conduction, as shown in Figure 4.16a. Compared with the undoped TFTs, the optimized doped TFTs showed an increase of mobility from 0.46 to $0.74\,cm^2\,V^{-1}\,s^{-1}$, a decrease of *SS* from 12.5 to $7.72\,V\,dec^{-1}$, and an increase of current on/off ratio by approximately two orders of magnitude (Figure 4.16b).

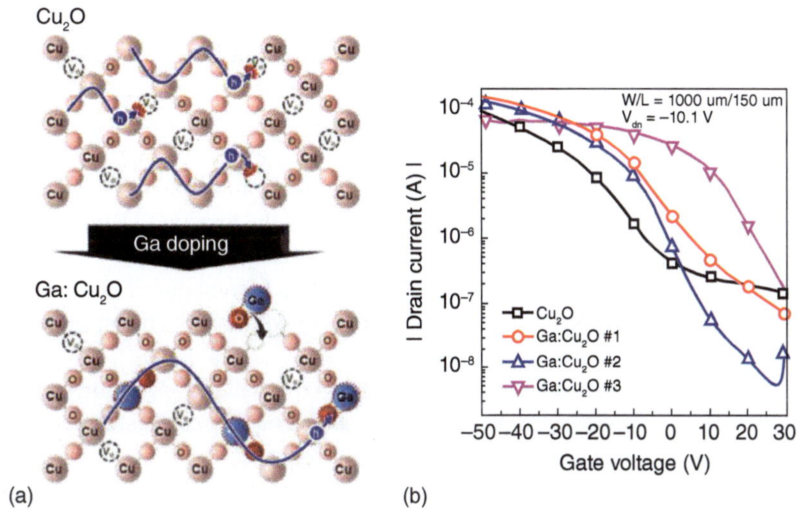

Figure 4.16 (a) Schematic diagrams showing the effects of Ga doping on hole conduction in Cu_2O. (b) Transfer characteristics of Ga:Cu_2O TFTs with different doping amounts. Source: Reprinted with permission from Ref. [79]. © 2020 American Chemical Society.

Apart from depositing through most reported vacuum-based methods, Cu_2O films with high quality were also reported by solution-based processes. In 2013, Cho and coworkers [80] first reported Cu_2O thin films deposited by sol–gel spin coating and found that an N_2 post-annealing treatment before the O_2 annealing could significantly improve the film morphology. The resulted *p*-type Cu_2O TFTs showed a field-effect mobility of $0.16\,cm^2\,V^{-1}\,s^{-1}$ and a current on/off ratio of ~100. However, this work still needed a high annealing temperature, making it unsuitable for flexible substrates. In the same year, Anthopoulos et al. [81] changed to deposit Cu_2O by spray pyrolysis, using a substrate temperature of 275 °C and a post annealing at 200 °C in air. The use of such films as the active layers has successfully resulted in TFTs with a field-effect mobility of $1\times10^{-2}\,cm^2\,V^{-1}\,s^{-1}$ and a current on/off ratio of $\sim4\times10^3$, demonstrating the potential of depositing high-performance Cu_2O using simple solution-processing techniques and encouraging development for *p*-type oxides.

Notably, although Cu_2O films show high Hall mobility, the field-effect mobility of the TFTs is still not satisfactory. Although numerous strategies have been done in material strategies to increase the grain sizes and decrease the effects from grain boundaries, a significant gap still exists between Hall and field-effect mobility, limiting the potential applications of Cu_2O TFTs. Recently, Jo's work suggested that effects of the contact resistance and interface traps might be the main reasons for this large difference [82]. Thus, better structure design strategies and deposition processes are required to further improve the TFT performance. For example, one can optimize the dielectric/channel interface to reduce the number of defective trap states. Treatments at the channel/contacts interface to reduce the energy barriers and resulting contact resistance might be also helpful.

4.4.2 Tin Monoxide

As early as in the 1960s, tin oxide was used as the channel layers in oxide TFTs, showing a reasonable device performance. However, the early research primarily focused on using SnO_2 as *n*-type semiconductors. Actually, the unique electronic structure of SnO_x with the VBM formed by hybridized orbitals of localized oxygen (2*p*) and spatially spread Sn metal (5*s*) makes it a great candidate for *p*-type oxide semiconductors [83]. In 2008, Hosono's group reported the first SnO TFTs, showing a Hall mobility of 2.4 $cm^2\ V^{-1}\ s^{-1}$ and a field-effect mobility of 1.3 $cm^2\ V^{-1}\ s^{-1}$. However, owing to the large hole density in the channel, the current on/off ratio of their TFTs was unsatisfactory. Nevertheless, the work opened the doors of using SnO in high-performance *p*-type oxide TFTs [84].

In 2010, Fortunato et al. [85] reported SnO TFTs deposited by magnetron sputtering. This marked a milestone in this field since sputtering is a commercially viable technique and is compatible with existing TFT technologies. The as-deposited films were amorphous while they changed to a polycrystalline structure after the post-annealing treatment, exhibiting a maximum Hall mobility of 4.8 $cm^2\ V^{-1}\ s^{-1}$. The XRD patterns suggested that although the films were dominated by α-SnO, some metallic tin phases also existed. As shown in Figure 4.17, the bottom-gate TFTs with a 30-nm-thick SnO channel and Au/Ti contacts showed a high field-effect mobility of 1.2 $cm^2\ V^{-1}\ s^{-1}$ and a current on/off ratio of approximately 10^3, demonstrating the great potential of sputtering for depositing high-performance *p*-type SnO. Since this work, sputtering has become the most common method for depositing high-performance SnO. Here, the recent works of SnO TFTs deposited by sputtering are summarized in Table 4.2.

The sensitive native of SnO has led a lot of research being conducted to investigate the fabrication process of SnO TFTs. For example, Caraveo-Frescas et al. [120] investigated the effects of main sputtering parameters, i.e. the oxygen partial pressure (O_{PP}) and total pressure, on the performance of *p*-type SnO, as shown in Figure 4.18. The XRD patterns suggested that the phase formation strongly depended on the O_{pp} and total sputtering pressure. *P*-type SnO with a polycrystalline phase only

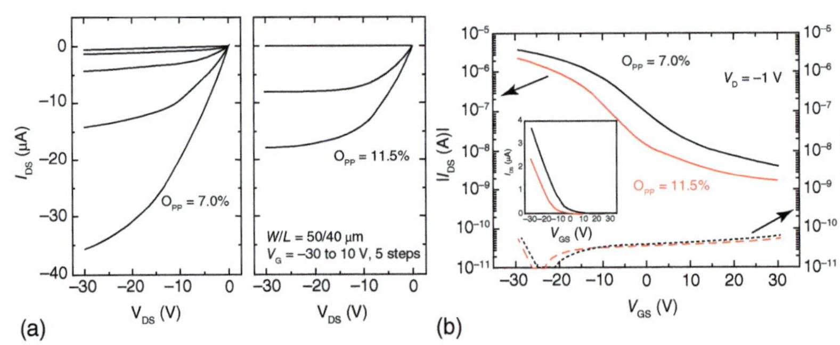

Figure 4.17 Electrical performance of SnO TFTs with (a) the output characteristics and (b) the transfer characteristics. Source: Reprinted with permission from Ref. [85]. © 2010 AIP Publishing.

Table 4.2 A summary of representative SnO TFTs deposited by sputtering.

S/D	*d* (nm)	Annealing *T* (°C)	Time (min)	Con	V_{TH} (V)	μ ($cm^2 V^{-1} s^{-1}$)	I_{ON}/I_{OFF} ($\times 10^3$)	*SS* ($V dec^{-1}$)	Year	Reference
Ti/Au	30	200	60	Air	−5	1.2	1	/	2010	[85]
Pt	50	300	120	N_2	30	0.24	0.1	/	2010	[86]
Ni/Au	8	150	60	Air	1.4	1.3 ± 0.2	0.1	6.9	2011	[87]
Au/Ni	/	200	/	/	/	4.6	70	/	2011	[88]
Al/Au	8	150	30	Air	1.4	1.3	0.1	6.9	2013	[89]
Ni	25	250	60	Vac	/	1.8	1.3	/	2013	[90]
ITO	30	250	60	Air	2.5	0.24	1	2	2014	[91]
Ni	25	230	60	Air	3.6	0.59	3.1	9.1	2014	[92]
Mo	30	260	60	Vac	/	0.43	0.67	—	2014	[93]
Ni/Au	24	200	120	Air	32.5	1.36	1.55	/	2014	[94]
ITO	15	225	30	Air	3.5	0.33	1	2.5	2014	[95]
Ni	10	200	100s	/	4.1	1.2	25	1.1	2014	[96]
Ni	10	250	100s	/	0.24	1.8	100	0.55	2015	[97]
Ni	10	200	100s	/	/	1.4	200	1.1	2015	[98]
Mo	25	250	60	Vac	/	1.16	0.23	/	2015	[99]
Ni	10	200	100s	/	3.7	0.8	26	1.3	2016	[100]
Ni	10	250	100s	Vac	2.5	0.65	22	1	2016	[101]
Ni	15	300	30	O_2	1.95	/	10	0.683	2016	[102]
ITO	14	225	30	Air	5	0.06	10	1.6	2016	[103]
Ni	8	200	30	O_2	1.05	2.14	>10	/	2016	[104]
Pd	27	225	60	Air	/	0.73	/	8.6	2016	[105]
Pt	27	225	60	Air	20.11	0.51	0.102	28.7	2017	[106]
Ti	20	225	60	Air	−5.05	1.19	22	1.5	2017	[107]
Ti	20	225	120	Air	−12	0.81	38	6.65	2018	[108]
ITO	25	250	30	Air	2.7	1	~0.1	3.24	2018	[109]
Ti/Au	20	225	120	Air	−4.8	0.7	26	1.4	2018	[110]
Ni	12	250	100s	/	−1.3	0.8	8.1	1.7	2019	[111]
ITO	30	300	60	Air	8.3	1.2	7.3	6.9	2019	[112]
Ti/Au	20	225	120	Air	−1.54	0.87	35.5	1.45	2019	[113]
Ni	8	200	/	/	−0.92	2.13	~10	~0.1	2019	[114]
Ni	8	200	30	N_2	−2.1	2.1	30	0.174	2020	[115]
ITO	16	180	30	Air	6.3	5.4	1.9	3.5	2020	[116]
Ni	20	300	/	Air	/	1.4	0.31	/	2021	[117]
Ni	7	200	45	N_2	/	4.4	200	0.526	2021	[118]
Al	10	300	30	O_2	−2.65	0.26	31	0.48	2021	[119]

d: channel thickness; *T*: temperature; μ: mobility; V_{TH}: threshold voltage; *SS*: subthreshold swing.

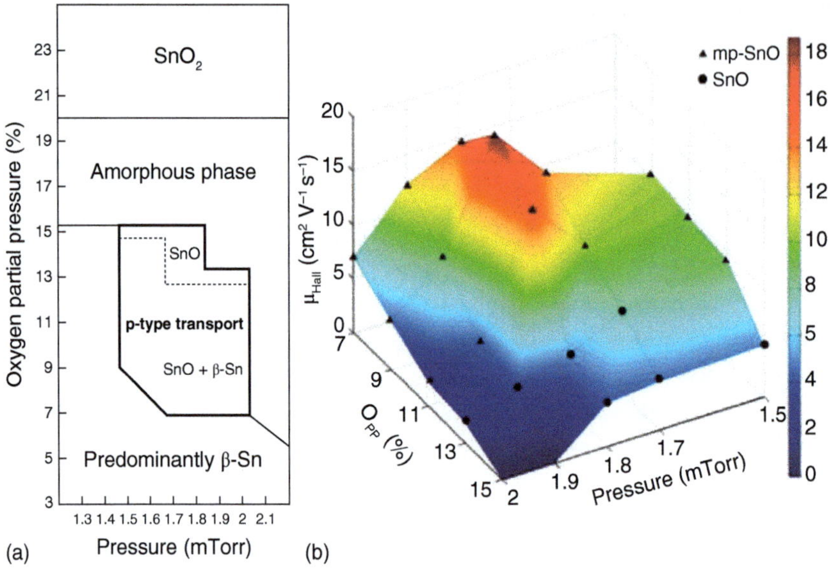

Figure 4.18 (a) Phase-stability and (b) Hall mobility as a function of total pressure and oxygen partial pressure. Source: Reprinted with permission from Ref. [120]. © 2013 American Chemical Society.

appeared in a narrow O_{pp} window between 7% and 15% and pressures between 1.5 and 2 mTorr. They also found that phase-pure SnO did not result in the highest mobility phase, but instead, well-controlled amounts of residual metallic tin could substantially increase the hole mobility, reaching a peak hole mobility of 18.7 cm^2 V^{-1} s^{-1} for films containing 3% metallic Sn. The DFT calculations also suggested that defects generated under Sn-rich growth conditions gave a more metallic character to the valence band of SnO, resulting in higher mobility. Using SnO with highest mobility, they achieved champion TFTs with a high mobility of 6.75 cm^2 V^{-1} s^{-1}, a threshold voltage of −1 V, and a current on/off ratio of 6×10^3.

Later, Zhong et al. [102] studied the impacts of gate dielectrics and oxygen annealing conditions on SnO TFTs and obtained TFTs with a high hole mobility of 3.33 cm^2 V^{-1} s^{-1}, a low threshold voltage of 1.95 V, and a current on/off ratio of $\sim 10^4$ using SiO_2 as a dielectric layer and a post-annealing treatment of 30 minutes. Cho et al. [92] studied the effects of air annealing and achieved an increased current on/off ratio, a decreased subthreshold swing, and an improved mobility after annealing in air at 230 °C for one hour. However, further increasing the annealing temperature would cause a decrease in on-current and the existence of ambipolar behavior due to the transition from SnO to SnO_2. They also studied the effects of channel thickness on the performance of SnO TFTs and found that the required annealing temperature for SnO TFTs decreased with the decreasing of channel thickness.

Apart from sputtering, other methods including e-beam evaporation [121], atomic-layer deposition [122–124], and solution-processing were also reported capable of fabricating SnO TFTs with a reasonable performance. For example, Kim et al. [123] demonstrated ALD-deposited SnO TFTs with a high current on/off ratio of 2×10^6, a subthreshold swing of $1.8\,V\,dec^{-1}$, and a field-effect mobility of $1\,cm^2\,V^{-1}\,s^{-1}$. In 2012, Hahn and coworkers [125] reported the first *p*-type SnO TFTs deposited by a solution process. In this work, $SnCl_2 \cdot 2H_2O$ was first dissolved in methanol and spin-coated on Si/SiO_2 substrates in a N_2-filled glove box. By exposing the films to NH_4OH exposure for one hour, the states of the film have been changed to $Sn_6O_4(OH)_4$ and a further annealing at 450 °C for two hours under N_2 and H_2 mixed gases successfully formed SnO. The films changed to polycrystalline after the treatment, and the obtained TFTs showed a mobility of $0.13\,cm^2\,V^{-1}\,s^{-1}$ and a current on/off ratio of 85.

Same as other metal-oxide semiconductors, the electrical properties of SnO TFTs might be degraded due to the interaction of the SnO surface with air. In 2014, Han et al. [96] studied the effect of ambient conditions on the electrical properties of SnO TFTs and found the electrical performance decreased a lot after being exposed to water. They also investigated the effects of passivation layers on SnO TFTs and presented that the devices passivated by SU-8 showed much improved long-term reliability and bias-stress stability under air environment. Similarly, Xin's group [108] studied the effects of inorganic (Al_2O_3) and organic (PMMA) passivation layers on the performance of SnO TFTs. By varying the channel thickness, they found that while PMMA passivation was more effective in reducing the shallow traps states, the passivation with Al_2O_3 could reduce deep traps in the SnO channel. The champion devices with a 20-nm-thick SnO and Al_2O_3 passivation showed a high current on/off ratio of 3.8×10^4, a mobility of $0.81\,cm^2\,V^{-1}\,s^{-1}$, and a small subthreshold swing of $6.65\,V\,dec^{-1}$, all of which were significantly enhanced compared with the uncapped devices.

It is worth mentioning that unlike other oxide semiconductors, SnO has been demonstrated to show an ambipolar behavior due to its valence/conduction-band structure and the energy-band configuration. For example, Nomura et al. [127] fabricated ambipolar SnO TFTs with saturation mobilities of $\sim 0.81\,cm^2\,V^{-1}\,s^{-1}$ for *p*-channel and $\sim 5 \times 10^{-4}\,cm^2\,V^{-1}\,s^{-1}$ for *n*-channel. As shown in Figure 4.19, Liang et al. [126] demonstrated the ambipolarity of SnO TFTs using a small bandgap, e-beam evaporated polycrystalline SnO as the channel layer and obtained a saturation field-effect mobility of $\sim 0.16\,cm^2\,V^{-1}\,s^{-1}$ for *p*-channel and $\sim 0.63\,cm^2\,V^{-1}\,s^{-1}$ for *n*-channel, as well as a linear field-effect mobility of $\sim 0.32\,cm^2\,V^{-1}\,s^{-1}$ for *p*-channel and $\sim 1.02\,cm^2\,V^{-1}\,s^{-1}$ for *n*-channel, respectively. Such an ambipolarity and balanced hole and electron mobilities of SnO TFTs have also enabled the realization of CMOS-liked inverters by SnO only, as reported by the same group, showing a maximum gain of over 30 under a supply voltage of 40 V. In 2018, Xin's group found that the Sn in the film might be detrimental to the SnO film quality and promote the ambipolar conduction when it was segregated from the film to enhance the stoichiometric balance [128]. Based on this, they sputtered the Sn target at a high power of 200 W and post-annealed the resulting films at a relatively low temperature of 150–200 °C, successfully achieving SnO TFTs with an ambipolar behavior.

Although SnO thin films show a lower Hall mobility than Cu_2O, the average device field-effect mobility is higher, with tens of works already existing showing a field-effect mobility of more than $1\,cm^2\,V^{-1}\,s^{-1}$. This means, compared with Cu_2O,

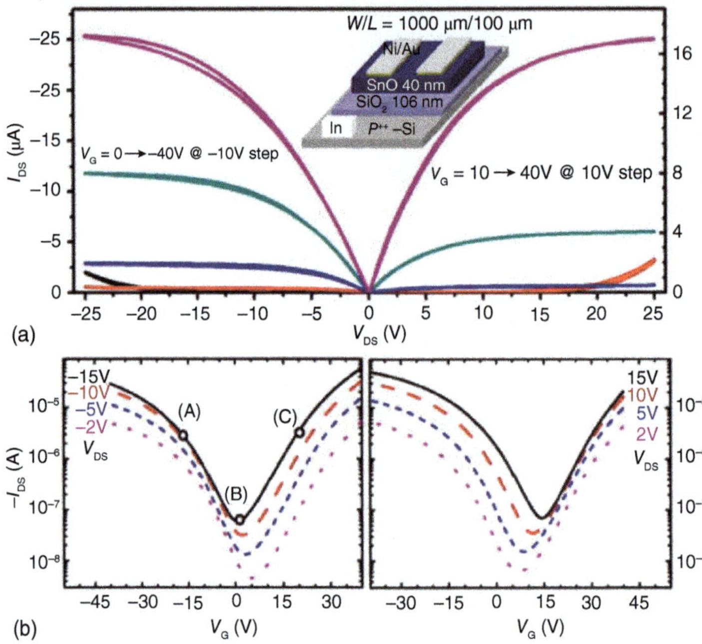

Figure 4.19 (a) Output and (b) transfer characteristics of ambipolar SnO TFTs. Source: Reprinted with permission from Ref. [126]. © 2012 AIP Publishing.

SnO TFTs can have a better matching with *n*-type oxide TFTs in complementary circuits. However, the current on/off ratio and the subthreshold swing reported so far for SnO TFTs was less than expected. Further work including dielectric/channel interface treatment to reduce the trap density, top encapsulation, and the use of higher-quality dielectrics might help.

4.4.3 Nickel Oxide

Apart from copper oxide and tin monoxide, nickel oxide is also a *p*-type oxide semiconductor that has gained significant attention in TFT-related applications. Similar to SnO and Cu_2O, the hole conductivity of NiO_x can be controlled by the formation of nickel vacancies. The first realization of NiO_x TFTs was reported by Iwasa and coworkers [129], who used single-crystal NiO as the channel and achieved a field-effect mobility of 1.6×10^{-4} cm^2 V^{-1} s^{-1} and a current on/off ratio of 130. Later in 2013, Zhang and coworkers [130] prepared NiO_x films by thermally oxidizing the e-beam evaporated Ni thin films and achieved NiO_x TFTs with a field-effect mobility of 5.2 cm^2 V^{-1} s^{-1} and a current on/off ratio of 2.2×10^3.

Interestingly, unlike Cu_2O and SnO films where vacuum-deposited ones generally show a higher performance, NiO_x films with a high quality are mainly deposited by solution-based methods, i.e. spin-coating, ink-jet printing, and solution combustion [131–137]. Nowadays, solution-processed NiO_x films with a wide bandgap, a high work function, and a low electron affinity have been widely used in emerging

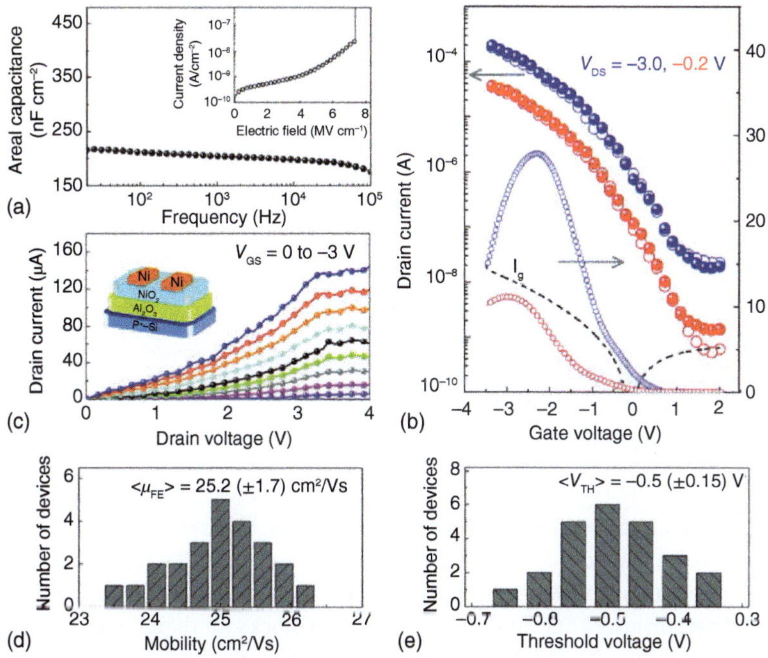

Figure 4.20 Electrical performance of NiO_x TFTs. (a) Al_2O_3 quality. (b) Transfer characteristics. (c) Output characteristics. Statistic diagrams of (d) mobility and (e) threshold voltage. Source: Reprinted with permission from Ref. [133]. © 2016 Royal Society of Chemistry.

optoelectronics and *p–n* heterojunctions [132]. As for TFTs, Shan's group have done a lot of work on solution-processed NiO_x TFTs, including, but not limited to, the modification of precursor solutions, post annealing conditions, and gate dielectric materials. In 2016, they studied the effects of post-annealing temperatures on the performance of NiO_x films and TFTs, and found that the optimized condition was achieved at an annealing temperature of 250 °C [131]. By further changing the gate dielectrics from SiO_2 to high-κ Al_2O_3, they found that the mobility could be enhanced to 4.4 $cm^2\ V^{-1}\ s^{-1}$ while decreasing the operating voltage from 30 to 2 V. In the same year, NiO_x TFTs with a high field-effect mobility of 25 $cm^2\ V^{-1}\ s^{-1}$ were demonstrated by the same group (Figure 4.20), which was the highest mobility reported so far for solution-processed NiO_x TFTs [133]. This value was surprising, considering that NiO_x TFTs based on traditional SiO_2 gate dielectrics showed a mobility of less than 1 $cm^2\ V^{-1}\ s^{-1}$. Such an enhancement was considered due to the reduction of Fevel level pinning and dielectric/channel interface traps. It is worth mentioning that the hole mobility of NiO_x is quite low, which might be even further suppressed by the defects in the film, limiting the resulted field-effect mobility.

Similar to other oxide semiconductors, doping is considered an effective method in improving the device performance. By doping the NiO_x with Li, Cu, and Sn, the *p*-type conductivity can be improved by valence band dispersion, resulting in an enhanced overall TFT performance [134–136]. However, the reported field-effect mobilities of those works were still not satisfactory, in the order of 1 $cm^2\ V^{-1}\ s^{-1}$. To further enhance the TFT performance, one should carefully control the processing and post-processing methods and conditions to reduce defects such as metal interstitials, valance state impurities, and grain boundaries. Besides, Shan's work already demonstrated that the device performance was highly dependent on the quality of dielectric layers and dielectric/channel interfaces; therefore, developing high-quality dielectric layers to reduce the interface traps might be helpful for mobility increasing.

4.5 Circuit Applications

4.5.1 Oxide NMOS/PMOS

Ever since the first report of oxide TFTs, numerous efforts have been made on developing oxide-based circuits including inverters, flip-flops, ring oscillators, and logic circuits. Due to the absence of high-performance *p*-type oxide TFTs, until now, most reported oxide circuits are NMOS implemented only by *n*-type oxide TFTs. Out of all the circuit types, inverters are fundamental building blocks for digital logic circuits and several analog circuits in current modern electronics. The schematics of equivalent circuit diagrams of inverters are shown in Figure 4.21, which include both an enhancement mode and a depletion mode depending on which mode the load TFT works. As can be seen in Figure 4.21a, the enhancement-mode inverters consist of two enhancement-mode TFTs with a positive threshold voltage for both load and drive TFTs. The gate of load TFT is connected to the drain terminal, which is also known as diode-connect of diode-load, making them work as a nonlinear resistor and operate in the saturation region. At a low input voltage, the drive TFT is in the cut-off region with no steady state flowing in the circuit and the load TFT pulls the output up toward the supply voltage. As for the depletion-mode inverters, the load devices work in depletion mode with a negative threshold voltage and have the gate and source terminals being connected together, as shown in Figure 4.21b. The load

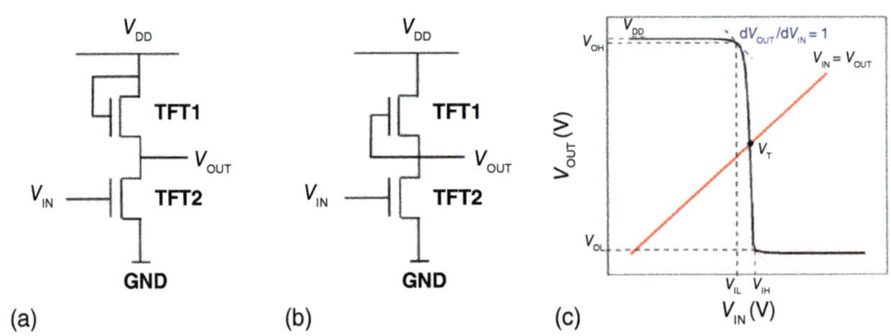

Figure 4.21 Schematic diagrams of NMOS inverters using (a) enhancement-mode and (b) depletion-mode TFTs as the load TFTs. TFT1 is a load transistor and TFT2 is a drive transistor. (c) Representative voltage transfer characteristics.

TFTs are always on since the gate bias is 0 V regardless of the input voltage, and hence could also be called zero-V_{GS}. Generally, depletion-mode inverters show a better performance including a sharper transition region, larger noise margins, and smaller overall layout area than enhancement-load inverters.

Typical voltage transfer characteristics of inverters are shown in Figure 4.21c, from which important parameters including threshold voltage, V_T, noise margin high, NM_H, and noise margin low, NM_L, transition width, ΔV_{IN}, and voltage gains can be calculated. V_T is defined as $V_{DD}/2$ and is ideally to be $V_{IN}/2$. The noise margins and transition width can be calculated using $NM_H = V_{OH} - V_{IH}$, $NM_L = V_{IL} - V_{OL}$, and $\Delta V_{IN} = V_{IH} - V_{IL}$, where V_{OH}, V_{IH}, V_{IL}, and V_{OL} are output and input voltages at high and low levels, respectively. The voltage gain is defined as dV_{OUT}/dV_{IN} and is generally considered a benchmark to indicate the quality of inverters.

Table 4.3 summarizes previously reported *n*-type oxide inverters, detailing their device structures and electrical performance. Three main conclusions can be drawn here: (i) While the first inverter reported by Wager et al. showed only a voltage gain of 1.5 using IGO TFTs in 2006, with significant development over past decades, voltage gains approaching 500 were reported. (ii) Out of all the reported *n*-type inverters, IGZO and ZnO channels are the most reported ones, mainly due to their high mobility. (iii) Compared with enhancement-mode inverters, depletion-mode ones show a better performance especially in terms of the voltage gain and full swing and are hence more preferable.

The realization of depletion-load inverters requires the fabrication of stable enhancement-mode and depletion-mode TFTs on a single chip. To achieve this, one can fabricate two enhancement-mode TFTs first and convert one to depletion-mode by post treatments, such as selective capping as reported by Cao and coworkers [154] and local laser annealing as reported by Wang and coworkers [149]. Besides, one can also directly fabricate depletion-mode TFTs using methods such as controlling the channel thickness and/or dielectric thickness, using a top-gate/dual-gate structure, and controlling the channel compositions [142, 148, 163, 166, 167]. For example, Song's group used a multi-step anodization to prepare Al_xO_y gate dielectrics and achieved load and drive IGZO TFTs with different dielectric thicknesses, resulting in IGZO inverters with a voltage gain of 34 at a low supply voltage of 1 V [163]. Moreover, applying bias stresses and/or light irradiations to the load TFT are also potential ways to achieve depletion-load inverters [145, 162, 168].

Until now, there are several remarkable progresses being achieved in *n*-type oxide inverters. For example, in 2016, Nathan and Lee [36] reported SB IGZO TFTs operated in a deep subthreshold region and achieved a depletion-load inverter with a gain larger than 220 and a low power consumption of less than 150 pW at a low supply voltage of 2 V, suggesting the potential in ultra-low power electronics. In 2007, Alshareef and coworkers [159] reported the replacement of ITO gate electrodes with ALD-deposited SnO_2 in ZnO TFTs, and achieved inverters with a high voltage gain of 382. With a low propagation stage delay of 150 ns, 11-stage ring oscillators were also demonstrated by them using the ZnO inverters, showing a high oscillation frequency of 303 kHz. In 2019, Wu's group reported ITO channel inverters and achieved

Table 4.3 A summary of NMOS-type inverters.

Year	Type of load TFT	Channel materials	Gain	V_{DD} (V)	NM_H/NM_L (V)	Reference
2006	Enhancement	IGO	1.5	30	/	[138]
2007	Enhancement	IGZO	1.7	18	/	[139]
2008	Depletion	ZnO	4	7	/	[140]
2008	Depletion	ZnO	22	2	/	[141]
2009	Depletion Enhancement	IGZO	37.4 1.6	20 20	8.2/8.4 7.5/0.3	[142]
2009	Depletion	ZTO	10.6	10	2.09/5.9	[143]
2009	Depletion	IGZO/IZO bilayer	15	5	3.5/1.05	[144]
2011	Depletion Enhancement	IGZO	4.18 2.81	10 10	/	[145]
2012	Depletion	IGZO	16.9	10	6/1.7	
2012	Depletion	SiInZnO/IGZO	22.5	10	/	[146]
2013	Depletion	ZnO	130	5	/	[147]
2013	Depletion	ZnO	23	100	/	[148]
2014	Depletion Enhancement	IGZO	20.5 1.5	20 20	/	[149]
2015	Depletion	SiZnSnO/ZTO	25.22	10	1.2/1.2	[150]
2016	Depletion	In_2O_3	45	20	/	[151]
2016	Depletion	ZnO	248	25	/	[152]
2016	Depletion	IGZO	14	5	3.53/0.6	[153]
2016	Depletion	IGZO	220	2	/	[36]
2016	Depletion	IGZO	45.9	15	/	[154]
2016	Depletion Enhancement	IGZO	24 4.1	1 1	/	[155]
2016	Depletion	IZO/IGZO	43.9	2	/	[156]
2016	Depletion	SiZnSnO	14.2	15	6.1/2.3	[157]
2017	Depletion	ZnO	48	10	3.13/1.15	[158]
2017	Depletion	ZnO	382	25	/	[159]
2018	Enhancement	ZTO	3.5	10	/	[160]
2018	Depletion	IGZO	40	10	/	[161]
2019	Depletion	SIZO	12.33	5	/	[162]
2019	Depletion	IGZO	34	1	0.46/0.48	[163]
2019	Depletion	ITO	478	2.5	/	[37]
2020	Enhancement	ITO	7.13	10	/	[164]
2021	Depletion	SnO_2	6.3	3	/	[165]
2021	Depletion	In_2O_3/IGZO	112	10	/	[166]

a high voltage gain of 478 at a supply voltage of 2.5 V, which was higher than most 2D materials and oxide semiconductors based inverters reported so far [37].

Apart from inverters, the demonstration of NAND and NOR gates is crucial for real commercialization. In 2012, Muth and coworkers [169] demonstrated transparent IGZO NAND and NOR gates with an operating frequency up to 5 kHz. In 2019, Lee et al. [170] reported NMOS-based NAND and NOR logic gates by simply changing the metal capping layer on the channel layer. Due to the different work function of metal capping layer on the channel layer, they found capping of different metal layers on SZTO TFTs could result in different contact properties and hence a different TFT performance. By capping the depletion-mode TFTs with Al and enhancement-mode TFTs with Ti/Al, they achieved NAND and NOR gates with output and input characteristics being driven in the same manner as the truth table, suggesting the great potential in next-generation thin-film logic circuits.

Although NMOS shows advantages such as higher density and higher speed operations over PMOS owing to the higher carrier mobility of *n*-type devices, some specific applications could still be found for PMOS. For example, LTPS-based PMOS inverters were proposed for reducing the manufacturing cost and improving the yields of larger-area displays. Hence, the development of *p*-type oxide circuits is still quite useful. Fully oxide PMOS circuits were reported by Myny and coworkers [171], who reported both enhancement and depletion-load PMOS inverters using SnO as the channel layer. The SnO TFTs showed a mobility of ~1.6 $cm^2 V^{-1} s^{-1}$, a threshold voltage of −0.3 V, a current on/off ratio of 10^3–10^4, and a contact resistance of 148 Ω cm, securing the successful operation of PMOS devices. A 19-stage ring oscillator was also demonstrated by them using the PMOS inverters, showing a stage delay as low as 300 ns. The use of such 19-stage ring oscillators as the clock signal in an 8-bit RFID code generator comprising 294 *p*-type oxide TFTs enabled the resulted devices to show a data rate as high as 12.2 $kb s^{-1}$, demonstrating the great potential of *p*-type oxide circuits.

The above-mentioned works clearly show the great potential of oxide semiconductors in next-generation circuit-related applications. However, several issues still exist and need to be solved for proceeding this technique to next level. For NMOS circuits, although an excellent performance has already been achieved in *n*-type oxide TFTs, further research is still required for pursuing higher voltage gains, more balanced noise margins, and most importantly, a lower power consumption. As for PMOS circuits, their development is largely limited by the poor device performance of *p*-type oxide TFTs and further improvements either from material strategies or from device structure strategies are required to enhance the *p*-type oxide TFT performance.

4.5.2 Oxide CMOS

All the circuit applications reported in Section 4.5.1 are made with only *n*-type or *p*-type oxide TFTs. Compared with them, complementary circuits offer several advantages especially a lower power dissipation and a higher density of logic functions on a chip.

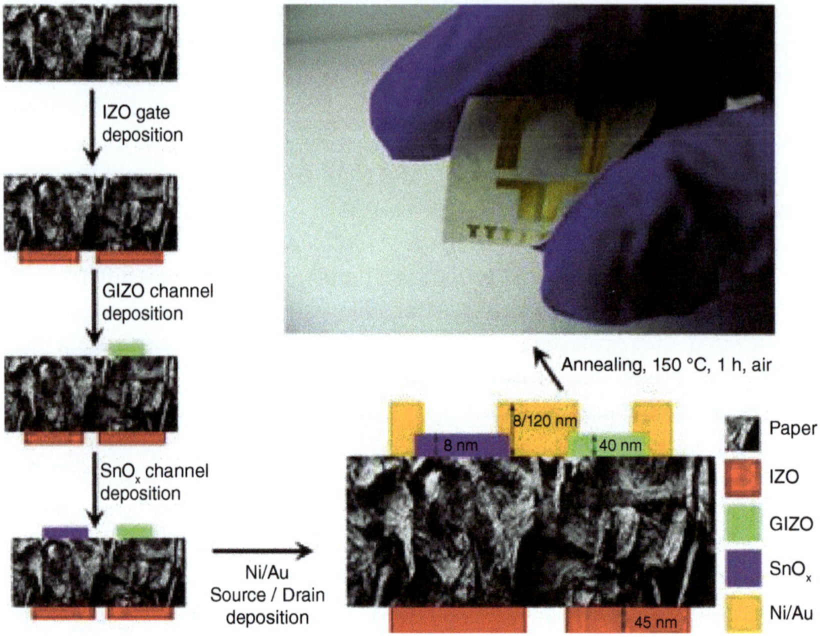

Figure 4.22 Schematic diagrams and real images of the complementary circuits. Source: Reprinted with permission from Ref. [87]. © 2011 John Wiley and Sons.

In 2011, Fortunato and coworkers [87] first reported CMOS inverter circuits layered on a flexible and recyclable fiber-based paper substrate, which could simultaneously work as the gate dielectrics. As shown in Figure 4.22, such inverters used IGZO as the n-type channel and SnO_x as the p-type channel, prepared at a relatively low temperature of 150 °C. Although the inverters showed a relatively low gain owing to the hard saturation of p-type devices, the work still opened doors for light weight, low cost, and fully recyclable complementary circuits.

From an industry point of view, a high-voltage gain is desirable for achieving large noise margins. This highly relies on the achievement of comparable performance for n-type and p-type TFTs. With the performance improvement being achieved in p-type oxide TFTs especially in recent years, this becomes possible. In 2018, Song's group reported complementary inverters based on n-type IGZO (mobility of $10.05\,cm^2\,V^{-1}\,s^{-1}$, V_{TH} of 5 V, SS of $0.86\,V\,dec^{-1}$, and current on/off ratio of 5.28×10^5) and p-type SnO (mobility of $1.19\,cm^2\,V^{-1}\,s^{-1}$, V_{TH} of −5.05 V, SS of $1.5\,V\,dec^{-1}$, and current on/off ratio of 2.2×10^4), achieving an extremely high gain of 112 at $V_{DD} = 10\,V$ [107]. Such devices also showed a noise margin high of 5.22 V and a noise margin low of 3.54 V, indicating that the inverters could withstand a high noise level. Based on these inverters, the same group also demonstrated 3-stage ring oscillators, NAND, and transmission gates, showing the great potential in large-area transparent/flexible electronics.

One clear advantage of oxide CMOS over traditional Si-based CMOS is that the oxide ones can be made flexible. For example, Cheng's group reported CMOS inverters based on SnO and ZnO TFTs on PI substrates [103], achieving a voltage gain up to 12 at $V_{DD} = 12\,V$. They also studied the bending effects on the inverter performance and found that the voltage gains slightly degraded under tensile strain but showing negligible effects under compressive strain. Five of such inverters were then cascaded to form flexible ring oscillators, achieving an oscillation frequency of ~18.4 kHz at $V_{DD} = 12\,V$. Recently, the same group also reported flexible inverters by monolithically integrating a top-gated IGZO TFT and a bottom-gated SnO TFT with HfO_2 capping [172], resulting in devices with a high static voltage gain of 370 and balanced noise margins at a supply voltage of 10 V. The high performance of their inverters was also demonstrated to be maintained under mechanical bending strains and after a long-term storage.

Although all-oxide-based complementary circuits have made remarkable progresses, especially in recent years, the unbalanced device performance of *n*- and *p*-type TFTs still makes the obtained performance unsatisfactory yet, which is mainly due to the absence of high-performance *p*-type TFTs. The development of high-performance *p*-type oxide TFTs that exhibit comparable performance with *n*-type ones is indispensable for approaching high-performance complementary circuits and would undoubtedly guide the oxide TFT technologies in a broad range of next-generation electronic device applications.

References

1 Nomura, K., Ohta, H., Takagi, A. et al. (2004). Room-temperature fabrication of transparent flexible thin-film transistors using amorphous oxide semiconductors. *Nature* 432 (7016): 488–492.

2 Klasens, H. and Koelmans, H. (1964). A tin oxide field-effect transistor. *Solid-State Electron.* 7 (9): 701–702.

3 Boesen, G. and Jacobs, J. (1968). ZnO field-effect transistor. *Proc. IEEE* 56 (11): 2094–2095.

4 Hoffman, R., Norris, B., and Wager, J. (2003). ZnO-based transparent thin-film transistors. *Appl. Phys. Lett.* 82 (5): 733–735.

5 Carcia, P., McLean, R., Reilly, M. et al. (2003). Transparent ZnO thin-film transistor fabricated by RF magnetron sputtering. *Appl. Phys. Lett.* 82 (7): 1117–1119.

6 Masuda, S., Kitamura, K., Okumura, Y. et al. (2003). Transparent thin film transistors using ZnO as an active channel layer and their electrical properties. *J. Appl. Phys.* 93 (3): 1624–1630.

7 Nilsen, C., Jin, J., Luo, Y. et al. (2013). Sputtered ZnO thin-film transistors with carrier mobility over 50 cm^2/vs. *IEEE Trans. Electron Devices* 60 (10): 3424–3429.

8 Nomura, K., Ohta, H., Ueda, K. et al. (2003). Thin-film transistor fabricated in single-crystalline transparent oxide semiconductor. *Science* 300 (5623): 1269–1272.

9 Yabuta, H., Sano, M., Abe, K. et al. (2006). High-mobility thin-film transistor with amorphous $InGaZnO_4$ channel fabricated by room temperature RF-magnetron sputtering. *Appl. Phys. Lett.* 89 (11): 112123.

10 Fortunato, E., Barquinha, P., and Martins, R. (2012). Oxide semiconductor thin-film transistors: a review of recent advances. *Adv. Mater.* 24 (22): 2945–2986.

11 Kim, Y., Kim, T., Avis, C. et al. (2016). Stable and high-performance indium oxide thin-film transistor by Ga doping. *IEEE Trans. Electron Devices* 63 (3): 1078–1084.

12 Cai, W., Park, S., Zhang, J. et al. (2018). One-volt IGZO thin-film transistors with ultra-thin, solution-processed Al_xO_y gate dielectric. *IEEE Electron Device Lett.* 39 (3): 375–378.

13 Cai, W., Brownless, J., Zhang, J. et al. (2019). Solution-processed HfO_x for half-volt operation of InGaZnO thin-film transistors. *ACS Appl. Electron. Mater.* 1 (8): 1581–1589.

14 Hennek, J., Smith, J., Yan, A. et al. (2013). Oxygen “getter” effects on microstructure and carrier transport in low temperature combustion-processed a-InXZnO (X = Ga, Sc, Y, La) transistors. *J. Am. Chem. Soc.* 135 (29): 10729–10741.

15 Zhang, Y., Yang, H., Peng, H. et al. (2019). Self-aligned top-gate amorphous InGaZnO TFTs with plasma enhanced chemical vapor deposited sub-10 nm SiO_2 gate dielectric for low-voltage applications. *IEEE Electron Device Lett.* 40 (9): 1459–1462.

16 Wang, G., Liu, D., Fan, S. et al. (2021). High-k erbium oxide film prepared by sol-gel method for low-voltage thin-film transistor. *Nanotechnology* 32 (21): 215202.

17 Liu, A., Liu, G., Zhu, H. et al. (2014). Fully solution-processed low-voltage aqueous In_2O_3 thin-film transistors using an ultrathin ZrO_x dielectric. *ACS Appl. Mater. Interfaces* 6 (20): 17364–17369.

18 Carlos, E., Branquinho, R., Kiazadeh, A. et al. (2017). Boosting electrical performance of high-κ nanomultilayer dielectrics and electronic devices by combining solution combustion synthesis and UV irradiation. *ACS Appl. Mater. Interfaces* 9 (46): 40428–40437.

19 Ma, P., Du, L., Wang, Y. et al. (2018). Low voltage operation of IGZO thin film transistors enabled by ultrathin Al_2O_3 gate dielectric. *Appl. Phys. Lett.* 112 (2): 4.

20 Chiu, C., Chang, S., and Chang, S. (2010). High-performance a-IGZO thin-film transistor using Ta_2O_5 gate dielectric. *IEEE Electron Device Lett.* 31 (11): 1245–1247.

21 Liu, A., Liu, G., Zhu, H. et al. (2015). Water-induced scandium oxide dielectric for low-operating voltage n- and p-type metal-oxide thin-film transistors. *Adv. Funct. Mater.* 25 (46): 7180–7188.

22 Cai, W., Wilson, J., Zhang, J. et al. (2019). Low-voltage, flexible InGaZnO thin-film transistors gated with solution-processed, ultra-thin Al_xO_y. *IEEE Electron Device Lett.* 40 (1): 36–39.

23 Cai, W., Wilson, J., and Song, A. (2021). Present status of electric-double-layer thin-film transistors and their applications. *Flex. Print. Electron.* 6 (4): 043001.

24 Du, L., He, D., Liu, Y. et al. (2018). Low-voltage, flexible IGZO transistors gated by PSSNa electrolyte. *IEEE Electron Device Lett.* 39 (9): 1334–1337.

25 Yuan, H., Shimotani, H., Tsukazaki, A. et al. (2009). High-density carrier accumulation in ZnO field-effect transistors gated by electric double layers of ionic liquids. *Adv. Funct. Mater.* 19 (7): 1046–1053.

26 Zare, B.F., Tang, B., Ma, R. et al. (2020). Sub-3 v ZnO electrolyte-gated transistors and circuits with screen-printed and photo-crosslinked ion gel gate dielectrics: new routes to improved performance. *Adv. Funct. Mater.* 30 (20): 1902028.

27 Lu, A., Sun, J., Jiang, J. et al. (2009). Microporous SiO_2 with huge electric-double-layer capacitance for low-voltage indium tin oxide thin-film transistors. *Appl. Phys. Lett.* 95 (22): 222905.

28 Lu, A., Sun, J., Jiang, J. et al. (2010). Low-voltage transparent electric-double-layer ZnO-based thin-film transistors for portable transparent electronics. *Appl. Phys. Lett.* 96 (4): 043114.

29 Dou, W., Jiang, J., Sun, J. et al. (2011). Low-voltage electric-double-layer TFTs on SiO_2-covered paper substrates. *IEEE Electron Device Lett.* 32 (11): 1543–1545.

30 Jiang, J., Wan, Q., Sun, J. et al. (2009). Ultralow-voltage transparent electric-double-layer thin-film transistors processed at room-temperature. *Appl. Phys. Lett.* 95 (15): 152114.

31 Dou, W. and Tan, Y. (2019). Low-voltage self-assembled indium tin oxide thin-film transistors gated by microporous SiO_2 treated by H_3PO_4. *RSC Adv.* 9 (53): 30715–30719.

32 Dou, W. and Tan, Y. (2019). Effects of NaCl treatment on the performance and environmental stability of microporous SiO_2-based thin film transistors. *RSC Adv.* 9 (64): 37436–37439.

33 Dou, W. and Tan, Y. (2020). Dual-gate low-voltage transparent electric-double-layer thin-film transistors with a top gate for threshold voltage modulation. *RSC Adv.* 10 (14): 8093–8096.

34 Ma, X., Zhang, J., Cai, W. et al. (2017). A sputtered silicon oxide electrolyte for high-performance thin-film transistors. *Sci. Rep.* 7 (1): 809.

35 Cai, W., Zhang, J., Wilson, J. et al. (2017). Oxide-based electric-double-layer thin-film transistors on a flexible substrate. *IEEE Electron Device Lett.* 38 (12): 1680–1683.

36 Lee, S. and Nathan, A. (2016). Subthreshold schottky-barrier thin-film transistors with ultralow power and high intrinsic gain. *Science* 354 (6310): 302–304.

37 Li, S., Tian, M., Gao, Q. et al. (2019). Nanometre-thin indium tin oxide for advanced high-performance electronics. *Nat. Mater.* 18 (10): 1091–1097.

38 Cai, W., Zhang, J., Wilson, J. et al. (2020). Significant performance improvement of oxide thin-film transistors by a self-assembled monolayer treatment. *Adv. Electron. Mater.* 6 (5): 1901421.

39 Jeong, K., Won, H., Jeong, J. et al. (2008). Origin of threshold voltage instability in indium-gallium-zinc oxide thin film transistors. *Appl. Phys. Lett.* 93 (12): 123508.

40 Cai, W., Wilson, J., Zhang, J. et al. (2020). Significant performance enhancement of very thin InGaZnO thin-film transistors by a self-assembled monolayer treatment. *ACS Appl. Electron. Mater.* 2 (1): 301–308.

41 Xiao, P., Lan, L., Dong, T. et al. (2015). Ingazno thin-film transistors modified by self-assembled monolayer with different alkyl chain length. *IEEE Electron Device Lett.* 36 (7): 687–689.

42 Zhong, W., Li, G., Lan, L. et al. (2018). Insnzno thin-film transistors with vapor-phase self-assembled monolayer as passivation layer. *IEEE Electron Device Lett.* 39 (11): 1680–1683.

43 Kim, M., Cho, S., Shin, Y. et al. (2020). Improving electrical stability of a-InGaZnO thin-film transistors with thermally deposited self-assembled monolayers. *Electron. Mater. Lett.* 16 (5): 451–456.

44 Ho, D., Jeong, H., Choi, S. et al. (2020). Organic materials as a passivation layer for metal oxide semiconductors. *J. Mater. Chem. C* 8 (43): 14983–14995.

45 Hong, S., Park, S., Kim, Y. et al. (2017). Low-temperature fabrication of an HfO_2 passivation layer for amorphous indium–gallium–zinc oxide thin film transistors using a solution process. *Sci. Rep.* 7 (1): 16265.

46 Choi, S. and Han, M. (2012). Effect of deposition temperature of SiO_x passivation layer on the electrical performance of a-IGZO TFTs. *IEEE Electron Device Lett.* 33 (3): 396–398.

47 Ji, K., Kim, J., Jung, H. et al. (2011). Effect of high-pressure oxygen annealing on negative bias illumination stress-induced instability of InGaZnO thin film transistors. *Appl. Phys. Lett.* 98 (10): 103509.

48 Ide, K., Kikuchi, Y., Nomura, K. et al. (2011). Effects of excess oxygen on operation characteristics of amorphous In-Ga-Zn-O thin-film transistors. *Appl. Phys. Lett.* 99 (9): 093507.

49 Kim, J., Bang, J., Nakamura, N. et al. (2018). Ultra-wide bandgap amorphous oxide semiconductors for NBIS-free thin-film transistors. *APL Mater.* 7 (2): 022501.

50 Chang, Y., Lee, D., Herman, G. et al. (2007). High-performance, spin-coated zinc tin oxide thin-film transistors. *Electrochem. Solid State Lett.* 10 (5): H135.

51 Street, R., Ng, T.N., Lujan, R. et al. (2014). Sol–gel solution-deposited InGaZnO thin film transistors. *ACS Appl. Mater. Interfaces* 6 (6): 4428–4437.

52 Cai, W., Li, H., and Zang, Z. (2021). One-volt, solution-processed InZnO thin-film transistors. *IEEE Electron Device Lett.* 42 (4): 525–528.

53 Kim, Y., Heo, J., Kim, T. et al. (2012). Flexible metal-oxide devices made by room-temperature photochemical activation of sol–gel films. *Nature* 489 (7414): 128–132.

54 Kim, M., Kanatzidis, M., Facchetti, A. et al. (2011). Low-temperature fabrication of high-performance metal oxide thin-film electronics via combustion processing. *Nat. Mater.* 10 (5): 382–388.

55 Song, K., Young, C., Jun, T. et al. (2011). Low-temperature soluble InZnO thin film transistors by microwave annealing. *J. Cryst. Growth* 326 (1): 23–27.

56 Jun, T., Song, K., Jeong, Y. et al. (2011). High-performance low-temperature solution-processable ZnO thin film transistors by microwave-assisted annealing. *J. Mater. Chem.* 21 (4): 1102–1108.

57 Lin, Y., Faber, H., Zhao, K. et al. (2013). High-performance ZnO transistors processed via an aqueous carbon-free metal oxide precursor route at temperatures between 80–180 °C. *Adv. Mater.* 25 (31): 4340–4346.

58 Scheideler, W., Kumar, R., Zeumault, A. et al. (2017). Low-temperature-processed printed metal oxide transistors based on pure aqueous inks. *Adv. Funct. Mater.* 27 (14): 1606062.

59 Hwang, Y., Seo, J., Yun, M. et al. (2013). An 'aqueous route' for the fabrication of low-temperature-processable oxide flexible transparent thin-film transistors on plastic substrates. *NPG Asia Mater.* 5 (4): e45–e45.

60 Li, X., Cheng, J., Chen, Y. et al. (2019). Low-temperature aqueous route processed indium oxide thin-film transistors by NH_3 plasma-assisted treatment. *IEEE Trans. Electron Devices* 66 (3): 1302–1307.

61 Jeong, Y., Pearson, C., Lee, Y. et al. (2014). Effects of hydrogen plasma treatment on the electrical behavior of solution-processed ZnO transistors. *J Appl. Phys.* 116 (7): 074509.

62 Miyakawa, M., Nakata, M., Tsuji, H. et al. (2018). Highly stable low-temperature aqueous solution-processed oxide thin-film transistors by the hydrogen injection and oxidation method. *Flex. Print. Electron.* 3 (2): 025005.

63 Park, S., Kim, K., Jo, J. et al. (2015). In-depth studies on rapid photochemical activation of various sol–gel metal oxide films for flexible transparent electronics. *Adv. Funct. Mater.* 25 (19): 2807–2815.

64 Heo, J., Jo, J., Kang, J. et al. (2016). Water-mediated photochemical treatments for low-temperature passivation of metal-oxide thin-film transistors. *ACS Appl. Mater. Interfaces* 8 (16): 10403–10412.

65 Hennek, J., Kim, M., Kanatzidis, M. et al. (2012). Exploratory combustion synthesis: amorphous indium yttrium oxide for thin-film transistors. *J. Am. Chem. Soc.* 134 (23): 9593–9596.

66 Wang, B., Yu, X., Guo, P. et al. (2016). Solution-processed all-oxide transparent high-performance transistors fabricated by spray-combustion synthesis. *Adv. Electron. Mater.* 2 (4): 1500427.

67 Kang, Y., Jeong, S., Ko, J. et al. (2014). Two-component solution processing of oxide semiconductors for thin-film transistors via self-combustion reaction. *J. Mater. Chem. C* 2 (21): 4247–4256.

68 Moreira, M., Carlos, E., Dias, C. et al. (2019). Tailoring IGZO composition for enhanced fully solution-based thin film transistors. *Nanomaterials* 9 (9): 1273.

69 Kim, D., Kim, D., Rim, Y. et al. (2012). Improved electrical performance of an oxide thin-film transistor having multistacked active layers using a solution process. *ACS Appl. Mater. Interfaces* 4 (8): 4001–4005.

70 Jeong, W., Kim, K., Kim, D. et al. (2012). The effects of dual-active-layer modulation on a low-temperature solution-processed oxide thin-film transistor. *IEEE Trans. Electron Devices* 59 (8): 2149–2152.

71 Rim, Y.S., Chen, H., Kou, X. et al. (2014). Boost up mobility of solution-processed metal oxide thin-film transistors via confining structure on electron pathways. *Adv. Mater.* 26 (25): 4273–4278.

72 Tetzner, K., Lin, Y., Regoutz, A. et al. (2017). Sub-second photonic processing of solution-deposited single layer and heterojunction metal oxide thin-film transistors using a high-power xenon flash lamp. *J. Mater. Chem. C* 5 (45): 11724–11732.

73 Faber, H., Das, S., Lin, Y. et al. (2017). Heterojunction oxide thin-film transistors with unprecedented electron mobility grown from solution. *Sci. Adv.* 3 (3): e1602640.

74 Khim, D., Lin, Y., and Anthopoulos, T. (2019). Impact of layer configuration and doping on electron transport and bias stability in heterojunction and superlattice metal oxide transistors. *Adv. Funct. Mater.* 29 (38): 1902591.

75 Yarali, E., Faber, H., Yengel, E. et al. (2020). Low-voltage heterojunction metal oxide transistors via rapid photonic processing. *Adv. Electron. Mater.* 6 (6): 2000028.

76 Matsuzaki, K., Nomura, K., Yanagi, H. et al. (2008). Epitaxial growth of high mobility Cu_2O thin films and application to p-channel thin film transistor. *Appl. Phys. Lett.* 93 (20): 202107.

77 Fortunato, E., Figueiredo, V., Barquinha, P. et al. (2010). Thin-film transistors based on p-type Cu_2O thin films produced at room temperature. *Appl. Phys. Lett.* 96 (19): 192102.

78 Zou, X., Fang, G., Yuan, L. et al. (2010). Top-gate low-threshold voltage *p*-type Cu_2O thin-film transistor grown on SiO_2/Si substrate using a high-k HfON gate dielectric. *IEEE Electron Device Lett.* 31 (8): 827–829.

79 Bae, J., Lee, J., Park, S. et al. (2020). Gallium doping effects for improving switching performance of p-type copper(i) oxide thin-film transistors. *ACS Appl. Mater. Interfaces* 12 (34): 38350–38356.

80 Kim, S., Ahn, C., Lee, J. et al. (2013). P-channel oxide thin film transistors using solution-processed copper oxide. *ACS Appl. Mater. Interfaces* 5 (7): 2417–2421.

81 Pattanasattayavong, P., Thomas, S., Anthopoulos, G. et al. (2013). P-channel thin-film transistors based on spray-coated Cu_2O films. *Appl. Phys. Lett.* 102 (16): 163505.

82 Jo, J., Lenef, J., Mashooq, K. et al. (2020). Causes of the difference between hall mobility and field-effect mobility for p-type RF sputtered Cu_2O thin-film transistors. *IEEE Trans. Electron Devices* 67 (12): 5557–5563.

83 Ogo, Y., Hiramatsu, H., Nomura, K. et al. (2009). Tin monoxide as an s-orbital-based p-type oxide semiconductor: electronic structures and TFT application. *Phys. Status Solidi A* 206 (9): 2187–2191.

84 Ogo, Y., Hiramatsu, H., Nomura, K. et al. (2008). P-channel thin-film transistor using p-type oxide semiconductor, SnO. *Appl. Phys. Lett.* 93 (3): 032113.

85 Fortunato, E., Barros, R., Barquinha, P. et al. (2010). Transparent p-type SnO_x thin film transistors produced by reactive RF magnetron sputtering followed by low temperature annealing. *Appl. Phys. Lett.* 97 (5): 052105.

86 Yabuta, H., Kaji, N., Hayashi, R. et al. (2010). Sputtering formation of p-type SnO thin-film transistors on glass toward oxide complimentary circuits. *Appl. Phys. Lett* 97 (7): 072111.

87 Martins, R., Nathan, A., Barros, R. et al. (2011). Complementary metal oxide semiconductor technology with and on paper. *Adv. Mater.* 23 (39): 4491–4496.

88 Fortunato, E. and Martins, R. (2011). Where science fiction meets reality? With oxide semiconductors! *Phys. Status Solidi (RRL)* 5 (9): 336–339.

89 Martins, R., Ahnood, A., Correia, N. et al. (2013). Recyclable, flexible, low-power oxide electronics. *Adv. Funct. Mater.* 23 (17): 2153–2161.

90 Hsu, P., Chen, W., Tsai, Y. et al. (2013). Fabrication of p-type SnO thin-film transistors by sputtering with practical metal electrodes. *Jpn. J. Appl. Phys.* 52 (5S1): 05DC07.

91 Chiu, I. and Cheng, I. (2014). Gate-bias stress stability of p-type SnO thin-film transistors fabricated by RF-sputtering. *IEEE Electron Device Lett.* 35 (1): 90–92.

92 Cho, I., Myeonghun, U., Song, S. et al. (2014). Effects of air-annealing on the electrical properties of p-type tin monoxide thin-film transistors. *Semicond. Sci. Technol.* 29 (4): 045001.

93 Hsu, P., Hsu, C., Chang, C. et al. (2014). Sputtering deposition of p-type SnO films with SnO_2 target in hydrogen-containing atmosphere. *ACS Appl. Mater. Interfaces* 6 (16): 13724–13729.

94 Luo, H., Liang, L., Liu, Q. et al. (2014). Magnetron-sputtered sno thin films for p-type and ambipolar TFT applications. *ECS J. Solid State Sci. Technol.* 3 (9): Q3091–Q3094.

95 Chiu, I., Li, Y., Tu, M. et al. (2014). Complementary oxide–semiconductor-based circuits with n-channel ZnO and p-channel SnO thin-film transistors. *IEEE Electron Device Lett.* 35 (12): 1263–1265.

96 Han, Y., Choi, Y., Cho, I. et al. (2014). Improvement of long-term durability and bias stress stability in p-type SnO thin-film transistors using a SU-8 passivation layer. *IEEE Electron Device Lett.* 35 (12): 1260–1262.

97 Han, Y., Choi, Y., Jeong, C. et al. (2015). Environment-dependent bias stress stability of p-type SnO thin-film transistors. *IEEE Electron Device Lett.* 36 (5): 466–468.

98 Jeong, C., Lee, D., Han, Y. et al. (2015). Subgap states in p-channel tin monoxide thin-film transistors from temperature-dependent field-effect characteristics. *Semicond. Sci. Technol.* 30 (8): 085004.

99 Hsu, P., Tsai, S., Chang, C. et al. (2015). Preparation of p-type SnO thin films and transistors by sputtering with robust Sn/SnO_2 mixed target in hydrogen-containing atmosphere. *Thin Solid Films* 585: 50–56.

100 Lee, J., Choi, Y., Jeong, C. et al. (2016). Electrical instability of p-channel SnO thin-film transistors under light illumination. *IEEE Electron Device Lett.* 37 (3): 295–298.

101 Lee, J., Choi, Y., Jeong, C. et al. (2016). Temperature-dependent electrical instability of p-type SnO thin-film transistors. *J. Vac. Sci. Technol. B* 34 (4): 041210.

102 Zhong, C., Lin, H., Tsai, J. et al. (2016). Impact of gate dielectrics and oxygen annealing on tin-oxide thin-film transistors. *Jpn. J. Appl. Phys.* 55 (4S): 04EG02.

103 Li, Y., He, J., Hsu, S. et al. (2016). Flexible complementary oxide–semiconductor-based circuits employing n-channel ZnO and p-channel SnO thin-film transistors. *IEEE Electron Device Lett.* 37 (1): 46–49.

104 Chen, P., Wu, Y., Zheng, Z. et al. (2016). Bipolar conduction in tin-oxide semiconductorchannel treated by oxygen plasma for low-power thin-film transistorapplication. *J. Display Technol.* 12 (3): 224–227.

105 Zhang, J., Kong, X., Yang, J. et al. (2016). Analysis of carrier transport and band tail states in p-type tin monoxide thin-film transistors by temperature dependent characteristics. *Appl. Phys. Lett.* 108 (26): 263503.

106 Zhang, J., Yang, J., Li, Y. et al. (2017). High performance complementary circuits based on p-SnO and n-IGZO thin-film transistors. *Materials* 10 (3): 319.

107 Li, Y., Yang, J., Wang, Y. et al. (2018). Complementary integrated circuits based on p-type SnO and n-type IGZO thin-film transistors. *IEEE Electron Device Lett.* 39 (2): 208–211.

108 Qu, Y., Yang, J., Li, Y. et al. (2018). Organic and inorganic passivation of p-type SnO thin-film transistors with different active layer thicknesses. *Semicond. Sci. Technol.* 33 (7): 075001.

109 Guzmán-Caballero, D., Quevedo-López, M., De la Cruz, W. et al. (2018). Fully patterned p-channel SnO TFTs using transparent Al_2O_3 gate insulator and ITO as source and drain contacts. *Semicond. Sci. Technol.* 33 (3): 035010.

110 Yang, J., Wang, Y., Li, Y. et al. (2018). Highly optimized complementary inverters based on p-SnO and n-InGaZnO with high uniformity. *IEEE Electron Device Lett.* 39 (4): 516–519.

111 Jeong, H., Park, M., Kwon, S. et al. (2019). Highly sensitive and selective room-temperature NO_2 gas-sensing characteristics of SnO_x-based p-type thin-film transistor. *Sens. Actuators B Chem.* 288: 625–633.

112 Yim, S., Kim, T., Yoo, B. et al. (2019). Lanthanum doping enabling high drain current modulation in a p-type tin monoxide thin-film transistor. *ACS Appl. Mater. Interfaces* 11 (50): 47025–47036.

113 Li, Y., Zhang, J., Yang, J. et al. (2019). Complementary integrated circuits based on n-type and p-type oxide semiconductors for applications beyond flat-panel displays. *IEEE Trans. Electron Devices* 66 (2): 950–956.

114 Rajshekar, K., Hsu, H., Kumar, K. et al. (2019). Effect of plasma fluorination in p-type SnO TFTs: experiments, modeling, and simulation. *IEEE Trans. Electron Devices* 66 (3): 1314–1321.

115 Rajshekar, K., Hsu, H., Kumar, K. et al. (2020). Physical modeling of *p*-type fluorinated Al-doped tin-oxide thin film transistors. *IEEE J. Electron Devices Soc.* 8: 948–958.

116 Shin, M., Bae, K., Jeong, H. et al. (2020). Effects of capping layers with different metals on electrical performance and stability of p-channel SnO thin-film transistors. *Micromachines* 11 (10): 917.

117 Ahn, S., Jang, S., Song, A. et al. (2021). Performance enhancement of p-type SnO semiconductors via SiO_x passivation. *Mater. Today Commun.* 26: 101747.

118 Yen, T., Chin, A., and Gritsenko, V. (2021). Exceedingly high performance top-gate p-type SnO thin film transistor with a nanometer scale channel layer. *Nanomaterials* 11 (1): 92.

119 Mohanty, S., Wu, C., Chang, S. et al. (2021). Microwave and furnace annealing in oxygen ambient for performance enhancement of p-type SnO thin-film transistors. *Semicond. Sci. Technol.* 36 (12): 125024.

120 Caraveo-Frescas, J., Nayak, P., Al-Jawhari, H. et al. (2013). Record mobility in transparent p-type tin monoxide films and devices by phase engineering. *ACS Nano* 7 (6): 5160–5167.

121 Liang, L., Liu, Z., Cao, H. et al. (2010). Phase and optical characterizations of annealed SnO thin films and their p-type TFT application. *J. Electrochem. Soc.* 157 (6): H598.

122 Kim, H., Choi, S., Jeong, H. et al. (2021). Highly dense and stable p-type thin-film transistor based on atomic layer deposition SnO fabricated by two-step crystallization. *ACS Appl. Mater. Interfaces* 13 (26): 30818–30825.

123 Kim, S., Baek, I., Kim, D. et al. (2017). Fabrication of high-performance p-type thin film transistors using atomic-layer-deposited sno films. *J. Mater. Chem. C* 5 (12): 3139–3145.

124 Chae, M., Han, S., Park, B. et al. (2021). Atomic-layer-deposited SnO film using novel $Sn(dmamb)_2$ precursor for p-channel thin film transistor. *Appl. Surf. Sci.* 547: 148758.

125 Okamura, K., Nasr, B., Brand, R. et al. (2012). Solution-processed oxide semiconductor SnO in p-channel thin-film transistors. *J. Mater. Chem.* 22 (11): 4607–4610.

126 Liang, L., Cao, H., Chen, X. et al. (2012). Ambipolar inverters using SnO thin-film transistors with balanced electron and hole mobilities. *Appl. Phys. Lett.* 100 (26): 263502.

127 Nomura, K., Kamiya, T., and Hosono, H. (2011). Ambipolar oxide thin-film transistor. *Adv. Mater.* 23 (30): 3431–3434.

128 Li, Y., Yang, J., Qu, Y. et al. (2018). Ambipolar SnO_x thin-film transistors achieved at high sputtering power. *Appl. Phys. Lett.* 112 (18): 182102.

129 Shimotani, H., Suzuki, H., Ueno, K. et al. (2008). P-type field-effect transistor of nio with electric double-layer gating. *Appl. Phys. Lett.* 92 (24): 242107.

130 Jiang, J., Wang, X., Zhang, Q. et al. (2013). Thermal oxidation of Ni films for p-type thin-film transistors. *Phys. Chem. Chem. Phys.* 15 (18): 6875–6878.

131 Liu, A., Liu, G., Zhu, H. et al. (2016). Hole mobility modulation of solution-processed nickel oxide thin-film transistor based on high-k dielectric. *Appl. Phys. Lett.* 108 (23): 233506.

132 Napari, M., Huq, T., Hoye, R. et al. (2021). Nickel oxide thin films grown by chemical deposition techniques: potential and challenges in next-generation rigid and flexible device applications. *Info Mat.* 3 (5): 536–576.

133 Shan, F., Liu, A., Zhu, H. et al. (2016). High-mobility p-type NiO_x thin-film transistors processed at low temperatures with Al_2O_3 high-k dielectric. *J. Mater. Chem. C* 4 (40): 9438–9444.

134 Liu, A., Zhu, H., Guo, Z. et al. (2017). Solution combustion synthesis: Low-temperature processing for p-type Cu:NiO thin films for transparent electronics. *Adv. Mater.* 29 (34): 1701599.

135 Yang, J., Wang, B., Zhang, Y. et al. (2018). Low-temperature combustion synthesis and UV treatment processed p-type Li:NiO_x active semiconductors for high-performance electronics. *J. Mater. Chem. C* 6 (46): 12584–12591.

136 Lin, T., Li, X., and Jang, J. (2016). High performance p-type NiO_x thin-film transistor by Sn doping. *Appl. Phys. Lett.* 108 (23): 233503.

137 Hu, H., Zhu, J., Chen, M. et al. (2018). Inkjet-printed p-type nickel oxide thin-film transistor. *Appl. Surf. Sci.* 441: 295–302.

138 Presley, R., Hong, D., Chiang, H. et al. (2006). Transparent ring oscillator based on indium gallium oxide thin-film transistors. *Solid-State Electron.* 50 (3): 500–503.

139 Ofuji, M., Abe, K., Shimizu, H. et al. (2007). Fast thin-film transistor circuits based on amorphous oxide semiconductor. *IEEE Electron Device Lett.* 28 (4): 273–275.

140 Lee, K., Kim, K., Lee, K. et al. (2008). ZnO-based low voltage inverter with low-k/high-k double polymer dielectric layer. *Appl. Phys. Lett.* 93 (19): 193514.

141 Cha, S., Oh, M.S., Lee, K. et al. (2008). ZnO-based low-voltage inverter with quantum-well-structured nanohybrid dielectric. *IEEE Electron Device Lett.* 29 (10): 1145–1147.

142 Lee, J., Cho, I., Lee, J. et al. (2009). Full-swing InGaZnO thin film transistor inverter with depletion load. *Jpn. J. Appl. Phys.* 48 (10): 100202.

143 Heineck, D., McFarlane, B., and Wager, J. (2009). Zinc tin oxide thin-film-transistor enhancement/depletion inverter. *IEEE Electron Device Lett.* 30 (5): 514–516.

144 Yin, H., Kim, S., Jaechul, P. et al. (2009). High performance low voltage amorphous oxide tft enhancement/depletion inverter through uni-/bi-layer channel hybrid integration. In *2009 IEEE International Electron Devices Meeting (IEDM)*, IEEE, IEDM, Baltimore, MD, USA, pp. 1–4.

145 Seok, M., Choi, M., Mativenga, M. et al. (2011). A full-swing a-IGZO TFT-based inverter with a top-gate-bias-induced depletion load. *IEEE Electron Device Lett.* 32 (8): 1089–1091.

146 Debnath, P. and Lee, S. (2012). Full swing logic inverter with amorphous SiInZnO and GaInZnO thin film transistors. *Appl. Phys. Lett.* 101 (9): 092103.

147 Li, Y., Ramirez, J., Sun, K. et al. (2013). Low-voltage double-gate ZnO thin-film transistor circuits. *IEEE Electron Device Lett.* 34 (7): 891–893.

148 Yu, S., Kim, B., Kang, M. et al. (2013). In/Ga-free, inkjet-printed charge transfer doping for solution-processed ZnO. *ACS Appl. Mater. Interfaces* 5 (19): 9765–9769.

149 Huang, X., Wu, C., Lu, H. et al. (2014). Large-swing a-IGZO inverter with a depletion load induced by laser annealing. *IEEE Electron Device Lett.* 35 (10): 1034–1036.

150 Han, S. and Lee, S. (2015). High performance of full swing logic inverter using all n-types amorphous ZnSnO and SiZnSnO thin film transistors. *Appl. Phys. Lett.* 106 (21): 212104.

151 Leppäniemi, J., Eiroma, K., Majumdar, H. et al. (2016). In_2O_3 thin-film transistors via inkjet printing for depletion-load nmos inverters. *IEEE Electron Device Lett.* 37 (4): 445–448.

152 Nayak, P., Wang, Z., and Alshareef, H. (2016). Indium-free fully transparent electronics deposited entirely by atomic layer deposition. *Adv. Mater.* 28 (35): 7736–7744.

153 Lyu, R., Lin, H., Li, P. et al. (2016). A film-profile-engineered 3-D InGaZnO inverter technology with systematically tunable threshold voltage. *IEEE Trans. Electron Devices* 63 (9): 3533–3539.

154 Wang, M., Liang, L., Luo, H. et al. (2016). Threshold voltage tuning in a-IGZO TFTs with ultrathin SnO_x capping layer and application to depletion-load inverter. *IEEE Electron Device Lett.* 37 (4): 422–425.

155 Chi, L., Yu, M., Chang, Y. et al. (2016). 1-v full-swing depletion-load a-In–Ga–Zn–O inverters for back-end-of-line compatible 3D integration. *IEEE Electron Device Lett.* 37 (4): 441–444.

156 Xiao, H., Zhu, L., Wan, C. et al. (2016). Low-voltage depletion-load inverter using solid-state electrolyte gated oxide transistors. *IEEE Electron Device Lett.* 37 (5): 591–594.

157 Han, S. and Lee, S. (2017). Full swing depletion-load inverter with amorphous SiZnSnO thin film transistors. *Phys. Status Solidi A* 214 (2): 1600469.

158 Xu, H., Ye, Z., Liu, N. et al. (2017). Low-power transparent RFID circuits using enhancement/depletion logic gates based on deuterium-treated ZnO TFTs. *IEEE Electron Device Lett.* 38 (10): 1383–1386.

159 Alshammari, F., Hota, M., Wang, Z. et al. (2017). Atomic-layer-deposited SnO_2 as gate electrode for indium-free transparent electronics. *Adv. Electron. Mater.* 3 (9): 1700155.

160 Fernandes, C., Santa, A., Santos, Â. et al. (2018). A sustainable approach to flexible electronics with zinc-tin oxide thin-film transistors. *Adv. Electron. Mater.* 4 (7): 1800032.

161 Feng, Z., Lu, L., Wang, S. et al. (2018). Fluorination-enabled monolithic integration of enhancement- and depletion-mode indium-gallium-zinc oxide tfts. *IEEE Electron Device Lett.* 39 (5): 692–695.

162 Lee, B., Kim, S., and Lee, S. (2019). Investigation on dependency mechanism of inverter voltage gain on current level of photo stressed depletion mode thin-film transistors. *Solid-State Electron.* 156: 5–11.

163 Cai, W., Zhang, J., Wilson, J. et al. (2019). Low-voltage, full-swing ingazno-based inverters enabled by solution-processed, ultra-thin Al_xO_y. *IEEE Electron Device Lett.* 40 (8): 1285–1288.

164 Hong, S., Na, J., Lee, I. et al. (2020). Simultaneously defined semiconducting channel layer using electrohydrodynamic jet printing of a passivation layer for oxide thin-film transistors. *ACS Appl. Mater. Interfaces* 12 (35): 39705–39712.

165 Liang, K., Ren, H., Li, D. et al. (2021). Fully-printed flexible n-type tin oxide thin-film transistors and logic circuits. *J. Mater. Chem. C* 9 (35): 11662–11668.

166 Shao, S., Liang, K., Li, X. et al. (2021). Large-area (64 × 64 array) inkjet-printed high-performance metal oxide bilayer heterojunction thin film transistors and n-metal-oxide-semiconductor (nmos) inverters. *J Mater. Sci. Technol.* 81: 26–35.

167 Seok, M., Mativenga, M., Geng, D. et al. (2013). Achieving high performance oxide TFT-based inverters by use of dual-gate configurations with floating and biased secondary gates. *IEEE Trans. Electron Devices* 60 (11): 3787–3793.

168 Cho, I., Lee, J., Park, J. et al. (2012). Full-swing a-IGZO inverter with a depletion load using negative bias instability under light illumination. *IEEE Electron Device Lett.* 33 (12): 1726–1728.

169 Luo, H., Wellenius, P., Lunardi, L. et al. (2012). Transparent IGZO-based logic gates. *IEEE Electron Device Lett.* 33 (5): 673–675.

170 Lee, B., Sohn, A., Kim, S. et al. (2019). Mechanism of carrier controllability with metal capping layer on amorphous oxide SiZnSnO semiconductor. *Sci. Rep.* 9 (1): 886.

171 Rockelé, M., Vasseur, K., Mityashin, A. et al. (2018). Integrated tin monoxide p-channel thin-film transistors for digital circuit applications. *IEEE Trans. Electron Devices* 65 (2): 514–519.

172 Hsu, S., Su, D., Tsai, F. et al. (2021). Flexible complementary oxide thin-film transistor-based inverter with high gain. *IEEE Trans. Electron Devices* 68 (3): 1070–1074.

5

Metal Oxide Semiconductors for Sensors

5.1 Introduction

With the rapid development of Internet of Things (IoT), sensors have been playing an increasingly important role in the fields of industrial and agricultural production, environment monitoring, medical diagnosis, and so on. In the last few decades, metal oxide semiconductors (MOS) have been widely investigated for integration in sensors due to their high specific surface area, remarkable surface reaction activity, strong adsorption capacity, and excellent stability [1]. According to the semiconducting properties, the representative MOS materials can be classified into *n*-type MOS such as SnO_2, ZnO, TiO_2, WO_3, and In_2O_3, as well as *p*-type ones such as NiO, CuO, Co_3O_4, and Cr_2O_3 [2].

Among various sensors based on MOS, gas sensors are the most popular sensing device that has been studied. In addition, pressure sensors and pH sensors based on MOS are frequently utilized in terms of electronic skins, clinical diagnosis, and pollution monitoring among other areas. This chapter mainly reviews these three types of MOS-based sensors. In particular, this includes the detailed discussion on their working mechanism and representative devices reported recently.

5.2 Metal Oxide-Based Gas Sensors

Ever since the initiative utilization of MOS in gas detection in 1950s, MOS gas sensors have attracted extensive attention [3, 4]. The first commercial MOS gas sensor was fabricated by Taguchi in 1970s (Taguchi-type sensors) [5]. Nowadays, there are many commercial companies to produce various MOS gas sensors such as Figaro, FIS, MICS, UST, CityTech, and Applied Sensors. To meet the growing demand, much effort has been recently devoted to promoting the sensor performance including response, sensitivity, response/recovery speed, repeatability, selectivity, operating temperature, and long-term stability.

In this section, we initially introduce the operation mechanisms of MOS gas sensors and the strategies including morphological design, noble-metal doping, heterojunction formation, and light illumination to enhance the sensor performance.

Metal Oxide Semiconductors: Synthesis, Properties, and Devices,
First Edition. Zhigang Zang, Wensi Cai, and Yong Zhou.

Then, typical gas-sensing devices and their evaluation criteria are discussed. Furthermore, we investigate a variety of MOS gas sensors according to the species of target gases, such as volatile organic compounds (VOCs), environmental pollutants, humidity, explosives, and chemical warfare agents. Eventually, some novel gas sensors, i.e. flexible and self-powered ones based on MOS are discussed.

5.2.1 Mechanisms of Gas Sensors

As for chemiresistive MOS gas sensors, the electrical conductivity or resistance of the sensing layer varies on exposure to target gas and recovers to the initial state after air/nitrogen purification. The preadsorbed oxygen on the surface of MOS always plays a significant role during the sensing performance. First, oxygen adsorption is accompanied with the formation of electronic core-shell structure (Figure 5.1). Specifically, oxygen molecules are adsorbed in the forms of molecules O_2 (ads) and anions such as O_2^-, O^-, and O^{2-}. These ionic species are closely related to the operating temperature as revealed in Eqs. (5.1–5.4) [6]:

$$O_2\,(\text{gas}) \rightarrow O_2\,(\text{ads}) \tag{5.1}$$

$$O_2\,(\text{ads}) + e^- \rightarrow O_2^-\,(\text{ads})\ T < 100\,^\circ\text{C} \tag{5.2}$$

$$O_2\,(\text{ads}) + e^- \rightarrow 2O^-\,(\text{ads})\ 100\,^\circ\text{C} < T < 300^\circ\text{C} \tag{5.3}$$

$$O^-\,(\text{ads}) + e^- \rightarrow O^{2-}\,(\text{ads})\ 300\,^\circ\text{C} < T \tag{5.4}$$

The adsorbed oxygen on the MOS surface plays a crucial role in capturing electrons and forming oxygen anions, particularly under different operating temperatures. In the case of *n*-type MOS, where electrons are the majority carriers, the presence of oxygen anions leads to the formation of an electron depletion layer (EDL) near the surface (see Figure 5.1a). Conversely, for *p*-type MOS, a hole accumulation layer

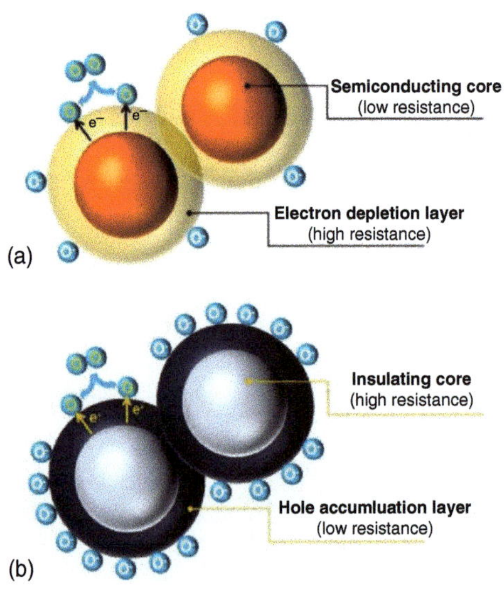

Figure 5.1 Formation of electronic core-shell structures in (a) *n*-type and (b) *p*-type oxide semiconductors. Source: Reproduced from Kim and Lee [2]/with permission of Elsevier.

(HAL) is formed near the surface (see Figure 5.1b). Let us focus on *n*-type MOS as an example. The formation of the EDL has a direct impact on the density of the majority carriers (electrons) and creates a region of high resistance, as depicted in Figure 5.1a. When exposed to reducing gases like NH_3, H_2S, H_2, and acetone, the electrons are drawn toward the EDL. Consequently, the EDL becomes narrower, leading to a simultaneous decrease in sensor resistance. Conversely, when the MOS is exposed to oxidizing gas species such as NO_2, SO_2, and O_3, electrons are extracted from the sensing materials, causing the EDL to broaden. Consequently, the sensor resistance increases as a result of this process.

To illustrate the sensing mechanism, we take *n*-type In_2O_3 as an example. As reported in Ref. [7], hierarchical In_2O_3 was prepared via a simple one-step hydrothermal method on the surface of clinoptilolite substrates and then utilized for highly sensitive NO_2 detection. The as-prepared sensor showed a remarkable response of 19.6 toward 5 ppm NO_2 at 125 °C and a fast response/recovery speed of ~3.5/4.2 minutes. When In_2O_3 is exposed to air atmosphere, adsorbed oxygen existed on the surface of In_2O_3 nanoparticles, in the form of oxygen species O_2^- at room temperature. This leads to the formation of an EDL near the surface, causing the generation of a potential barrier. The reactions involved in this process are depicted in Eq. (5.1)–(5.4). Upon exposure to NO_2 gas, NO_2 molecules interact with the oxygen species on the surface of In_2O_3, capturing electrons from the material, as indicated in Eq. (5.5).

$$NO_2 + O^- + 3e^- \leftrightarrow NO + 2O^{2-} \tag{5.5}$$

At a temperature of 125 °C, NO_2 molecules can chemically react with the chemisorbed oxygen (O^-) presented on the surface of In_2O_3 microflowers. This reaction consumes free electrons within the conduction band and increases the potential barrier, consequently leading to a significant increase in sensor resistance. The reverse reaction occurs during the desorption process, resulting in a decrease in sensor resistance.

Figure 5.2a shows the circuitry for simultaneous resistance and work function measurement of gas sensors. The model consists of partial crystallites which are connected to their neighbors by necks. The carrier transport path is located between two electrodes across various metal oxide grains. Hence, the grain size and depth of EDL or HAL impose an important effect on the transport path. For *n*-type MOS, the depth of EDL is limited (around several nanometers), which is defined as Debye length (L) calculated by Eq. (5.6) [8]:

$$L = \sqrt{\frac{\varepsilon k_B T}{q^2 N_d}}, \tag{5.6}$$

where ε is the dielectric constant, k_B is the Boltzmann's constant, T is the operating temperature (K), q is the elementary charge, and N_d is the density of dopants. For large grain, i.e. when the diameter of grain (D) is much larger than two times of EDL depth (2L), most of the crystallites are unaffected by the surface interactions with the target gases, as shown in Figure 5.2b. In this case, the predominant effect of the ambient atmosphere on the sensor resistance is introduced via grain boundaries

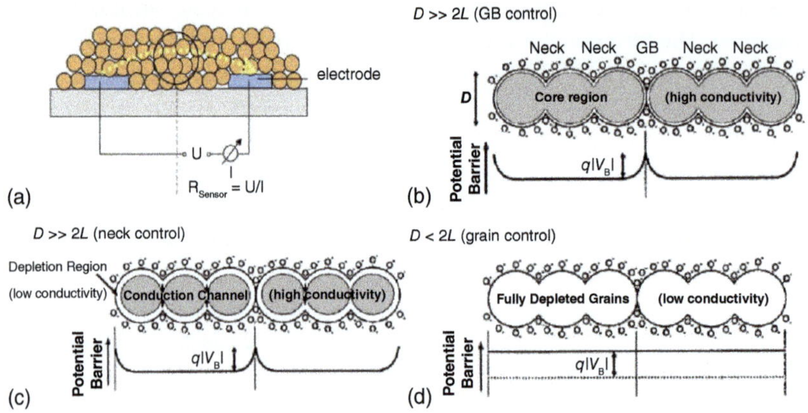

Figure 5.2 (a) Circuitry for simultaneous resistance and work function measurement of gas sensor. Source: Reproduced from Barsan et al. [5]/with permission of Elsevier. Schematic model of the effect of the crystallite size on the sensitivity of metal- oxide gas sensor. (b) D≫2 L (GB control), (c) D≥2 L (neck control), (d) D<2 L (grain control). Source: Reproduced from Rothschild and Komem [11]/with permission of AIP Publishing.

(GB) for intracrystallite charge transport from one grain to another. The electrical conductivity σ depends exponentially on the barrier height (qV_B) as explained in Eq. (5.7),

$$\sigma \approx \exp(-qV_B/k_BT) \tag{5.7}$$

where V_B is the GB potential. Therefore, for large grains ($D \gg 2L$) the sensing mechanism is controlled by the GB barriers. When D approaches but still larger than 2L ($D \geq 2L$), the depletion region that surrounded each neck forms a constricted conduction channel within each grain, as exhibited in Figure 5.2c. The conductivity depends not only on the GB barriers but also on the cross-section area of those channels. When D is smaller than 2L, the depletion region extends throughout the whole grain and the mobile charge carriers within crystallites are nearly depleted (Figure 5.2d). Hence, the conductivity sharply decreases when the conduction channels between the grains are vanished. The energy band is nearly flat throughout the whole structure of the interconnected grains, and the conductivity is controlled by the intracrystallite conductivity (grain control). The conductivity depends on the difference between Fermi level (E_F) and the bottom of conduction band (E_C), as shown in Eq. (5.8). Empirically, the highest gas sensitivity is obtained in this case ($D < 2L$) [8, 9]. The response of In_2O_3 sensors toward 1 ppm NO_2 increases with the decreasing of In_2O_3 particle size [10].

$$\sigma \approx \exp(-[E_C - E_F]/k_BT) \tag{5.8}$$

To improve the sensing performance of MOS gas sensors, various strategies have been investigated such as morphological design, noble metal modification, heterojunctions, heteroatom doping, and light illumination.

Morphological design of MOS nanostructures has been widely investigated to promote the performance of MOS gas sensors. Various geometrical structures,

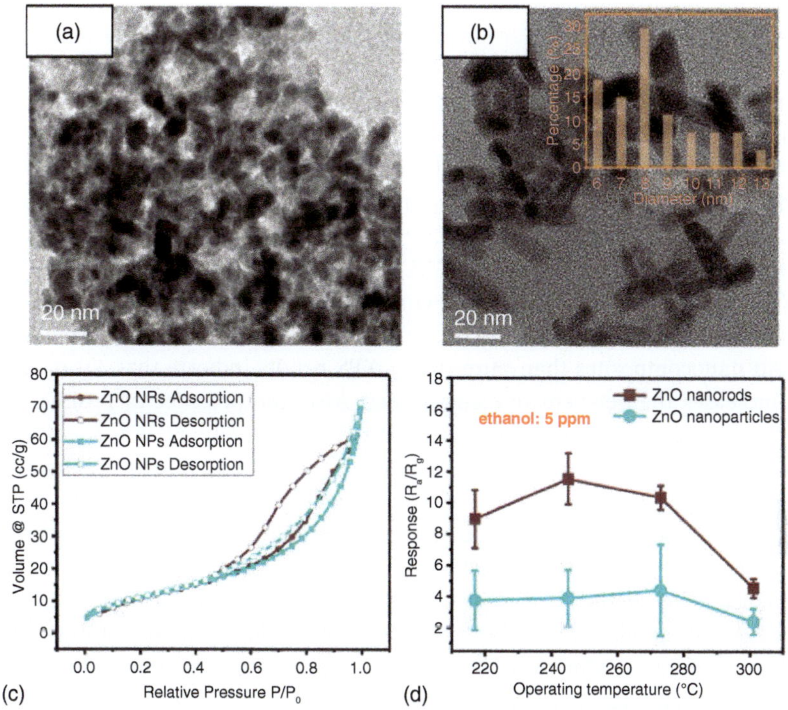

Figure 5.3 TEM images of (a) ZnO nanoparticles (NPs) and (b) ZnO nanorods (NRs). (c) N_2-adsorption/desorption isotherms of ZnO NRs and NPs. (d) Response-operation temperature relationship for ZnO NRs and NPs toward 5 ppm ethanol. Source: Reproduced from Wang et al. [12]/with permission of John Wiley & Sons.

including nanowires, nanofibers, nanoparticles, nanorods, nanosheets, and hollow spheres, are conducive to enhance the sensing properties primarily due to the larger specific surface area compared with conventional MOS, offering more adsorption sites. Moreover, these porous structures could facilitate the diffusion of target gas molecules and promote the response/recovery speed. Our group prepared MEMS ethanol gas sensors based on ZnO nanorods (NRs) and nanoparticles (NPs) for trace ethanol detection [12]. The TEM images of ZnO NPs and NRs are shown in Figure 5.3a,b, respectively. In contrast to the severe NPs agglomeration in ZnO NPs, ZnO NRs exhibit homogeneous rod-like morphology (length: 15–50 nm, diameter: 6–13 nm) with less aggregation, and larger specific surface area (42.11 vs. 39.48 $m^2 g^{-1}$), as shown in Figure 5.3c, indicating a more porous structure. Thus, ZnO NRs exhibit a better ethanol-sensing performance than ZnO NPs, including a higher sensitivity and a lower optimal operation temperature (Figure 5.3d).

Noble metals (e.g. Pd, Pt, Au, and Ag) have been extensively employed to functionalize MOS materials via the electronic and chemical sensitization effect [13]. Yang et al. [14] investigated a novel amine sensor based on Au decorated In_2O_3 hollow nanospheres prepared via the calcination of solid organic precursors and *in situ* reduction of Au nanoparticles on the In_2O_3 hollow nanospheres. The prepared

In_2O_3 hollow nanospheres possessed a diameter of 200 nm and a shell thickness of 30 nm with Au nanoparticles of around 10 nm in diameter. The resulting composite sensor exhibited better performance toward amine than pure In_2O_3 counterpart in terms of a reduced optimized operating temperature (340 °C vs. 420 °C), a higher response (26.3 vs. 2.9, toward 100 ppm amine), and a shorter response/recovery time (71 s/180 s vs. 105 s/310 s). The performance enhancement could be attributed to electronic and chemical sensitization. The electronic sensitization resulted from a Schottky junction form at the interface between Au nanoparticles and In_2O_3 surface, and the work function of Au (5.1 eV) is much higher than that of In_2O_3 (4.3 eV) [15]. Thus, electrons from the conduction band of In_2O_3 transfer to Au nanoparticles, resulting in an interface dipole layer, together higher resistance of In_2O_3-Au nanocomposites than pure In_2O_3. XPS results indicate that Au modification increases the deficient oxygen vacancy. Also, the presence of positive Au ions trap electrons, which may facilitate electronic sensitizations. When exposing the nanocomposites to air, more oxygen molecules are ionized due to the spillover effect. During the adsorption process of 1-butylamine, the reaction between oxygen species and amine molecules would release electrons to In_2O_3 and impoverish the width of EDL. Therefore, the resistance decreases, thus improving the sensing performance. The chemical sensitization, generated from the excellent catalytic properties of Au nanoparticles, could reduce the reaction potential and lower the working temperature. Moreover, Au modification benefits a better diffusion of oxygen into surface vacancies and accelerates the process of oxygen ionization, thereby shortening the response time.

Heterojunction construction is a feasible method to improve sensing behaviors [16]. The heterojunctions can be classified into three types, namely, *p–n*, *n–n*, and *p–p* heterojunctions. Here *p–n* heterojunction is adopted as one typical example to reveal this sensitization effect.

ZnO is a chemically and thermally stable *n*-type semiconductor with a bandgap of 3.37 eV at room temperature while NiO is a *p*-type semiconductor with the bandgap of 4.2 eV [17, 18]. Ju et al. [19] prepared trimethylamine (TEA) gas sensors based on NiO-ZnO *p–n* heterojunctions. The ZnO nanosheets were directly grown on Al_2O_3 ceramic tubes by introducing a seed layer with a simple and cost-effective hydrothermal method. By employing pulsed laser deposition (PLD) method, the construction of NiO-ZnO *p–n* heterojunction was highly controllable and reproducible. The composite sensor showed an *n*-type sensing behavior toward reducing TEA gas, in consistent with pure ZnO one. Compared with pure ZnO sensor, a larger response (185.1 vs. 78.4) was found in the composite counterpart toward 100 ppm TEA gas at 320 °C. As for composites, the majority carrier of *n*- and *p*-type semiconductors are electrons and holes, respectively. When NiO nanoparticles implant onto the surface of ZnO nanosheet, the electrons in ZnO and holes in NiO diffuse to the opposite side due to the existing concentration gradient. Then, an internal built-in electric field is induced at the ZnO–NiO interface, and the carrier diffusion is finally balanced. As a result, the energy band bends within the depletion layer until the system reaches a uniform Fermi level (E_F). As mentioned above, when exposing ZnO nanosheets to air, EDL is produced near the surface of ZnO, leading to a high-resistance region.

Moreover, the formation of NiO-ZnO *p–n* heterojunction brings a new depletion layer at the interface, which further increases the sensor resistance in air. When exposing the composites to reducing TEA gas, TEA molecules react with oxygen ions on the surface of ZnO nanosheets and release electrons back to ZnO, resulting in a reduced sensor resistance. Simultaneously, TEA releases electrons into *p*-type NiO and the electron–hole recombination leads to a decrease of hole concentration. According to the law of mass action ($n_0 \times p_0 = n_i^2$) for semiconductors, the decrease of holes in NiO results in an increase of electrons and reduces the concentration gradient of the same carrier within both sides of *p–n* heterojunction. Consequently, the carrier diffusion is weakened and the depletion layer at the interface becomes thinner, causing a further decreasing of NiO-ZnO sensor resistance in TEA. In brief, compared with ZnO sensor, the formation of *p–n* heterojunction greatly increases the resistance in air and decreases the resistance in TEA gas, resulting in a greatly enhanced sensor response since $S = R_a/R_g$.

Heteroatom doping is another effective strategy to boost the performance of MOS gas sensors. The common doping elements include nonmetallic elements (N, C) [20, 21], metallic elements (Cr, Co, Fe, Cu, Ni, Al) [22–27], and rare earth elements (Ce, La) [28, 29]. The element doping could influence the size of grain, crystal cell, conductivity, and the oxygen adsorption within MOS, all of which might affect the gas-sensing performance [30]. Du et al. [20] synthesized N-doped SnO_2 microspheres by calcining the pristine SnO_2 in NH_3 atmosphere. The N-doped SnO_2 sensors showed a high response of 155 toward 5 ppm NO_2 at 80 °C and a fast response/recovery time of 184 s/125 s, superior to those using pure counterparts. According to the characterization results, both N-doping and oxygen vacancies occur on the surface of SnO_2 after NH_3 treatment, resulting in rich free electrons and a narrow energy band. The density functional theory (DFT) calculation results reveal that the improved sensing performance is attributed to the formation of N-containing active sites and effective electron transfer, promoting the adsorption of NO_2 molecules. Zhou et al. reported a highly sensitive carbon monoxide gas sensor based on Ni-doped SnO_2 nanomaterials [24]. Such Ni-SnO_2 composite sensors showed a better sensing performance than pure SnO_2 toward 100 ppm CO, including a lower optimal operation temperature (280 °C vs. 310 °C), a faster response/recovery time (15 s/16 s vs. 16 s/28 s), and a higher response (25.086 vs. 7.2978). The promotion was demonstrated to originate from the small crystallite size, high surface area, and the catalytic effect of doped metal ions. Liu et al. [28] proposed Ce-doped In_2O_3 nanostructure for glycol sensing. After Ce doping, the response of the composite sensor was found to enhance with a reduction of response/recovery time. Moreover, the optimal operating temperature lowered (240 °C vs. 260 °C) while retaining good stability compared with pure In_2O_3. The EDL is initially formed near the surface of In_2O_3 and then narrowed down when exposing to reducing glycol. As for Ce-doped In_2O_3, the EDL is thicker when exposed to air, with more ionized oxygen species and reactive sites being generated. Once glycol is introduced, more electrons are released back, leading to a thinner EDL. Additionally, when Ce^{4+} is incorporated into In_2O_3 lattices by occupying In^{3+} sites, the electron-donor defects are formed which are conductive to more oxygen adsorption.

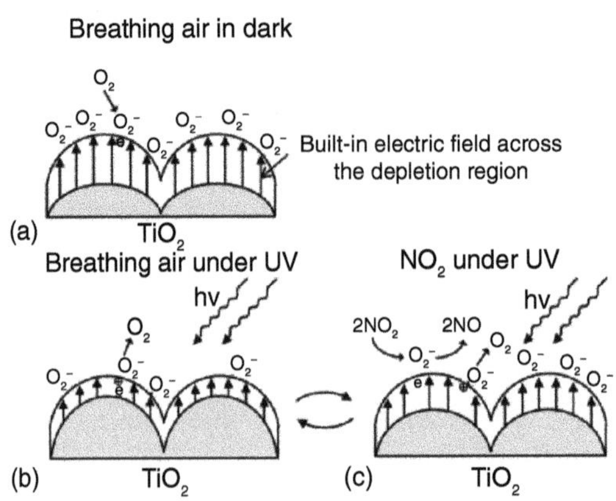

Figure 5.4 Schematic of proposed NO_2 gas-sensing mechanism of the TiO_2 sensor under UV illumination: (a) In dark environment with air. (b) Under UV illumination in air. (c) Under UV illumination with NO_2. Source: Reproduced from Xie et al. [33]/with permission of Elsevier.

Light illumination is also feasible to improve the sensing performance of MOS gas sensors [31, 32]. Xie et al. [33] proposed UV-assisted room temperature chemiresistive NO_2 sensors based on TiO_2 thin films. The prepared TiO_2 sensors showed a larger response to 500 ppm NO_2 under UV illumination than that in dark environment (2.4% vs. 1.6%). The schematic shown in Figure 5.4a reveals that O_2^- is adsorbed on the polycrystalline TiO_2 surface in dark that induces the built-in electric field across the depletion region and thus a high-resistance domain. UV–Vis absorption spectroscopy and Tauc plots show an optical bandgap of 3.26 eV for the prepared TiO_2. Under UV activation with the energy of UV light (365 nm, 3.4 eV) larger than the bandgap of TiO_2, massive electron–hole pairs could be generated, as described in Eq. (5.8). In this case, photogenerated electrons transfer from valence band to conduction band, further lowering the barrier height. When photogenerated electrons are driven into the bulk, photogenerated holes migrate to the surface and recombine with the adsorbed oxygen species. Both processes decrease the width of the depletion region, resulting in an increase of current roughly 5.5 times larger than the dark case. Then the surface adsorption and desorption of oxygen reach equilibrium (Figure 5.4b). On exposure to NO_2 molecules (Figure 5.4c), NO_2 acts as a scavenger for photogenerated electrons, as depicted in Eq. (5.10), leading to a decrease of photocurrent. Meanwhile, this reaction restores the concentration of surface-adsorbed oxygen species, which broadens the depletion region and further reduces the current.

$$hv \leftrightarrow h^+\ (hv) + e^-\ (hv) \tag{5.9}$$

$$2NO_2 + e^-\ (hv) \leftrightarrow 2NO\ (g) + O_2^-\ (hv) \tag{5.10}$$

Currently, three typical MOS-based devices including sintered-block type, MEMS type with micro-heater chips, and thick- or thin-film type (independent heating source) have been developed for molecular gas detection, as shown in Figure 5.5.

To evaluate the sensor performance, response, sensitivity, selectivity, stability, response/recovery time, and repeatability are fundamental parameters. Among these, response displays the degree of resistance change of gas sensors when reacting with target gases and is one of the foremost indicators to directly exhibit

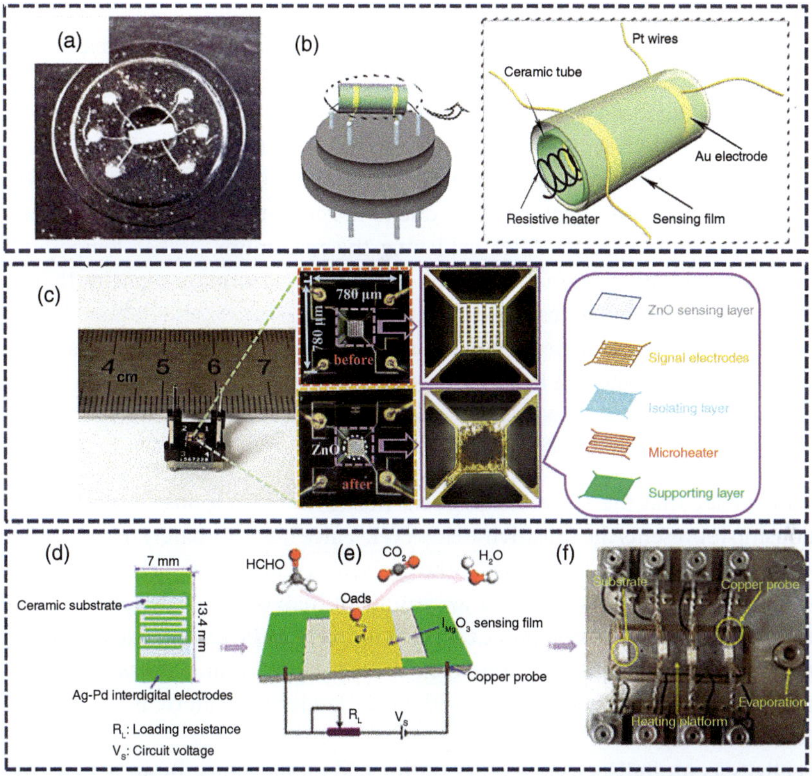

Figure 5.5 Sintered block type gas sensors: (a) a digital photo. Source: Reproduced from Jia et al. [34]/with permission of Elsevier. (b) A schematic diagram. Source: Reproduced from Wang et al. [35]/with permission of Elsevier. MEMS sensors: (c) the optical photograph of the as-prepared ethanol-sensing device, MEMS sensors before and after sensing material deposition, and a schematic diagram of the working area from the exploded view. Source: Reproduced from Wang et al. [12]/with permission of John Wiley & Sons. Thick- or thin-film type gas sensors: (d) a schematic diagram of the sensor, (e) a schematic diagram of gas-sensing mechanism, and (f) a photograph of the test system. Source: Reproduced from Wang et al. [25]/with permission of American Chemical Society.

the initial output signal. Typically, there are three definitions of sensor response, as displayed in Eqs. (5.11)–(5.13):

$$\text{Response} = R_a/R_g \tag{5.11}$$

$$\text{Response} = R_g/R_a \tag{5.12}$$

$$\text{Response} = (R_a - R_g)/R_g = \Delta R/R_a \tag{5.13}$$

where R_a is the steady sensor resistance in pure carrier gas (dry air or nitrogen), R_g is the sensor resistance in target gas of a certain concentration. For n-type MOS gas sensors, Eqs. (5.11) and (5.12) usually describe the response toward reducing and oxidizing gases, respectively. Meanwhile, Eq. (5.13) stresses the variation degree of gas sensors. In most cases, the operating temperature poses a great effect on the sensor response. To provide enough activation energy for the surface reaction, typically the gas sensors need to be heated to a temperature between 100 and 500 °C. Generally speaking, the response increases first and then decreases with the operating temperature, resembling a bell-shaped curve. The ethanol-sensing performance of SnO_2 sensor probed by Liu et al. clearly illustrates this phenomenon [36]. This relationship is attributed to the adsorbed oxygen species and the dynamic equilibrium of gas adsorption/desorption. Initially, the response is boosted with the increasing of operating temperature due to the enhanced activation energy and more active oxygen species. After achieving the maximum, higher temperature favors the desorption of gas molecules and reduces the adsorption capacity, thereby resulting in the descending response [37].

The sensitivity describes the change extent of response within the investigated gas concentration range, and can be defined as Eq. (5.14):

$$\text{Sensitivity} = \Delta\text{Response}/\Delta\text{Concentration} \tag{5.14}$$

This is to say that the slope of the linear-fitting line of response–concentration relationship represents the sensitivity. However, a nice linearity barely occurs within a wide concentration range. Actually, the gas sensors show a higher sensitivity at low-concentration regions and tend to saturate at high ones. The phenomenon is closely related to the diffusion process within the sensing layer. At a fixed operating temperature, the gas concentration within the sensitive film decreases with the diffusion depth [38]. Hence, the sensitivity is laddering pattern in a wide gas concentration range.

Selectivity is an indispensable criterion for real applications. In an actual environment, various interfering gases are always mixed together. Hence, the ideal gas sensor should show specific response to a target gas and no signal toward the interfering one, just like "antigen–antibody reaction." Nevertheless, MOS gas sensors are readily plagued by limited selectivity as various gas molecules can react with the adsorption sites within MOS. In this regard, sensor arrays combined with algorithms is a popular strategy to improve selectivity [39, 40]. Alternatively, the intentional modulation of the operating temperature is also feasible to achieve this goal [41, 42].

The response and recovery speed also play a vital role in the application of gas sensors. In most cases, the response and recovery times are separately defined as the time taken for the sensor to reach 90% of maximum response in target gas and 10% after carrier gas purification. During the adsorption/desorption kinetics, the operating temperature, gas diffusion rate, and mutual reaction between sensing film and gas molecules jointly decide the speed. Generally speaking, the elevated temperature is helpful for accelerating the gas diffusion within the sensing film and improving the response/recovery speed. Besides, porous film morphology is conducive to providing more gas diffusion channels, which facilitates the adsorption/desorption process.

Repeatability is utilized to evaluate the response consistency over several sensing cycles while long-term stability is obtained by monitoring the sensing performance of gas sensors over several weeks or longer. MOS are more stable than two-dimensional materials such as graphene, molybdenum disulfide, and black phosphorus. The biggest challenge so far is the recrystallization of MOS at a high operating temperature, which readily results in the baseline drift [43]. Hence, MOSs with inherent high crystallinity always perform a decent long-term stability [44].

5.2.2 VOCs Detection

The World Health Organization (WHO) defines volatile organic compounds (VOCs) as "organic chemical with a low melting point below room temperature and a boiling temperature between 50–260 °C" [45]. It is well documented that VOCs in the ambient environment mainly result from human activities such as painting houses, furniture manufacture, driving automobiles, and chemical production [46]. Some VOCs have been verified as hazardous gases that could lead to irreversible damage and severe diseases such as leukemia, lung cancer, and rectal cancer [47]. Therefore, the sensitive monitoring of VOCs is essential for human health and production safety. Ethanol, formaldehyde, acetone, and toluene are common chemicals in our daily life. Hereafter, MOS gas sensors for the detection of these VOCs are successively discussed in detail in this subsection.

Ethanol is a colorless and volatile liquid with a melting point of −114 °C and boiling point of 78.3 °C. Ethanol is widely used in fields of food industry, industrial and agricultural production, as well as health services. Generally, ethanol vapor does not bring serious damage to human health, but may cause dizziness, headache, fatigue, and difficulty in breath [48]. Sufficient ethanol mixing with air within confined space would form explosive mixture. Besides, ethanol vapor is a typical marker in exhaled air of drunk drivers. Hence, it is necessary to monitor the concentration of ethanol vapor in the fields of industrial production and transportation. Several ethanol gas sensors based on MOS are summarized in Table 5.1. Thereinto, ZnO is one typical *n*-type MOS for ethanol detection. Wan et al. prepared ethanol gas sensors based on ZnO nanowires (diameter of about 25 nm) by using MEMS technology, which showed high sensitivity and fast response toward ethanol gas at 300 °C [49]. Wang et al. synthesized ZnO nanorods (length: 1.7–2.1 μm, diameter: 90–200 nm) via a simple low-temperature hydrothermal process to recognize ethanol vapor [50]. The sensor response increased with temperature from 40 to 220 °C, and then decreased with temperature above 220 °C. In addition, the sensor showed a high, reversible, and fast response. Bhooloka Rao et al. [51] employed ZnO prepared by the hydrolysis method to detect 1000 ppm ethanol in air with a maximum response of 85%. Saito et al. [52] found that the pyramid-shaped ZnO particles based gas sensor showed a high response (more than 10 000) toward 50 ppm ethanol vapor, mainly attributed to the pyramidal shape with active planes for gas adsorption.

Analogous to ZnO, SnO_2 is also a widely investigated *n*-type MOS for ethanol sensing. Zhang et al. [53] synthesized SnO_2 nanofibers with an average diameter of ~100 nm by electronspinning to detect 0.01–5000 ppm ethanol at 330 °C. Large response (~4.5) and fast response/recovery time (~13 s/~13.9 s) were achieved toward 10 ppm ethanol. Such a high performance was ascribed into the porous microstructure of nanofiber mesh that provided a high surface-to-volume ratio and a larger surface accessibility. Liu et al. [36] prepared novel sea urchin-like hollow core-shell SnO_2 superstructure by a facile template-free hydrothermal method.

Table 5.1 Gas-sensing performance of MOS sensors toward ethanol.

Materials	Structure	$T_{operation}$ (°C)	Gas concentration (ppm)	Response	Response/ recovery time (s)	Ref.
ZnO	Nanowire	300	200	47^A	–/–	[49]
ZnO	Nanorod	320	50	$\sim21^C$	60/47	[50]
ZnO	Nanoparticle	200	1000	$85\%^B$	–/–	[51]
ZnO	Pyramid nanoparticle	350	50	$10\,000^A$	–/–	[52]
SnO_2	Nanofiber	330	10	$\sim4.5^A$	~13/~13.9	[53]
SnO_2	Hollow core-shell	260	50	60.5^A	9/20	[36]
SnO_2	Nanotube	200	300	$\sim16^A$	<5/<5	[54]
TiO_2	Nanotube array	250	300	$68\,000\%^B$	110/19	[55]
TiO_2	3D hierarchical flower-like	RT	200	$\sim2.8^A$	–/–	[56]
TiO_2-Ag	Nanobelt	200	20	$\sim6.59^A$	1–2/1–2	[57]
SnO_2-Pd	Nanoparticle	300	50	$98\%^B$	1.5/18	[58]
In_2O_3-Au	Nanofiber	140	500	14^A	12/24	[59]

A: (R_g/R_a) or (R_a/R_g), B: $(R_g - R_a)/R_a$, %, C: (V_g/V_a) or (I_g/I_a).

This superstructure displayed a larger response and quicker response/recovery time than common hollow SnO_2 spheres for ethanol detection, since the spikes with a diameter of 5–15 nm on the sea urchin-like structure provided more adsorption sites. Wang et al. [54] got SnO_2 nanotubes by a sol–gel template technique. The as-obtained nanotubes possessed an outer diameter of 200 nm, an inner diameter of about 150 nm, and a length of tens of micrometers. It was found that the gas sensor based on SnO_2 nanotubes showed a bigger response toward ethanol vapor (10–1000 ppm) than the counterpart based on SnO_2 nanoparticles due to the larger specific surface area ($45.6\,m^2\,g^{-1}$ vs. $15.2\,m^2\,g^{-1}$).

Alternatively, TiO_2 is another typical ethanol-sensitive material. Kwon et al. [55] prepared TiO_2 nanotube arrays and unveiled a high response of 680 toward 300 ppm ethanol at 250 °C. This is because TiO_2 nanotubes prepared here possessed a high surface area with a large inner and outer surface compared to common nanoparticles only with outer surfaces. Wang et al. [56] prepared a novel room-temperature ethanol gas sensor based on 3D hierarchical flower-like TiO_2 microstructures (Figure 5.6a). The sensing mechanism was ascribed to the reactions between adsorbed oxygen species and target gas on the surface of the sensitive material, as illustrated in Figure 5.6b. The flower-like microstructures composed of thin nanosheets were synthesized using a combination of the hydrothermal process and the calcination method. The excellent sensing performance toward ethanol was shown in Figure 5.6c,d. Additionally, the larger response toward ethanol than toluene, benzene, acetone, and methanol indicated an excellent selectivity.

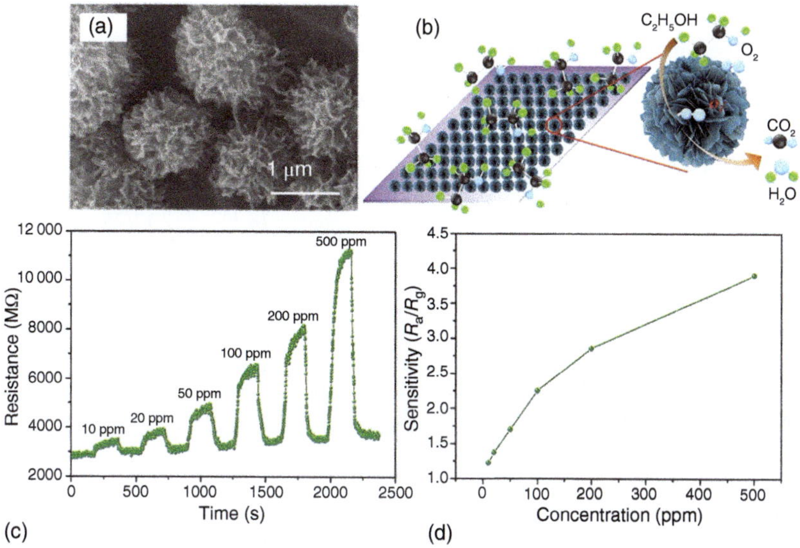

Figure 5.6 (a) An FESEM image of TiO_2 flower-like microstructures. (b) A schematic illustration of ethanol-sensing mechanism for TiO_2 flower-like microstructures. (c) The dynamic response and recovery curve of TiO_2 sensors to different ethanol concentrations varying from 10 to 500 ppm (d) Relationship between the response and ethanol concentration measured at room temperature. Source: Reproduced from [56]/with permission of Elsevier.

To further improve the sensing performance, incorporating noble metals with MOS has been verified as a feasible strategy. Hu et al. [57] reported an ethanol sensor based on TiO_2-Ag composites. During the *in-situ* synthesis process, Ag nanoparticles were formed on the surface of TiO_2 nanobelts by a photo-reduction method. The TiO_2-Ag sensors showed a larger sensitivity, a better selectivity, shorter response/recovery time, and a lower optimal working temperature compared with pure TiO_2 one, which might be attributed to the chemical sensitization mechanism of noble metal nanoparticles. Li et al. [58] discussed a porous SnO_2-Pd architecture for MEMS low-power ethanol detection. The rich catalytic Pd active sites, proper mesopore size distribution, and large specific surface area could account for the promoted sensing performance of SnO_2-Pd composites. Moreover, Au-loaded In_2O_3 nanofibers were successfully obtained for ethanol sensing by Xu et al. [59]. Different binary composites with varying Au concentrations were investigated and the results showed that 0.2 wt% Au-In_2O_3 based sensors exhibited a superior performance, including a high response (13.8), short response/recovery time (12 s/24 s), a good selectivity, and a low power consumption (222.5 mW) over the rest counterparts.

Formaldehyde (HCHO) is a stimulating, colorless, and reducing gas with a strong water solubility. In our daily life, HCHO usually emits from furniture, antiseptics, glues, and plastics [60]. Low-concentration HCHO would lead to dizziness, chest distress, sore eyes, and nausea while high-level HCHO might induce breathing difficulties and lung edema [61]. Besides, HCHO has been identified as a carcinogen.

Table 5.2 Gas-sensing performance of MOS sensors toward formaldehyde.

Materials	Structure	$T_{operation}$ (°C)	Gas concentration (ppm)	Response	Response/recovery time (s)	Ref.
In_2O_3	Nanoribbon	300	100	4.214[A]	16/15	[63]
SnO_2	Microsphere	200	100	38.3[A]	17/25	[64]
TiO_2	Nanotube	22	50	~36%[B]	−/−	[65]
In_2O_3-Co	Nanorod	130	10	23.2[A]	60/120	[25]
WO_3-Ag	Nanosheet	300	100	20.83[A]	5/5	[66]
SnO_2-NiO	Powder	300	22.5	~9[A]	11/60	[67]

A: (R_g/R_a) or (R_a/R_g), B: $(R_g - R_a)/R_a$, %.

The WHO has established a standard of 0.08 ppm averaged over 30 minutes [62]. Inspiringly, HCHO sensors based on MOS have shown an excellent performance. Table 5.2 summarizes the sensing performance of MOS gas sensors toward HCHO detection. Previously, In_2O_3 nanofibers and nanoribbons were prepared by an electrospinning method for HCHO sensing [63]. The nanoribbons-based sensors showed shorter response/recovery time, a higher response, and a lower optimal working temperature than nanofibers-based ones when detecting HCHO from 50 to 250 ppm Li et al. [64] employed SnO_2 microspheres with diameters of ~250 nm to selectively detect HCHO, which exhibited a high response (38.3) and fast response/recovery speed (17 s/25 s) toward 100 ppm HCHO. A selective room-temperature HCHO gas sensor based on TiO_2 nanotube arrays was proposed, which demonstrated a good response to HCHO (10 to 50 ppm) and an excellent selectivity over other reducing gases [65]. Wang et al. [25] found that 1 wt% Co-doped In_2O_3 nanorods exhibited a high response of 23.2 toward 10 ppm HCHO at 130 °C and a low detection limit of 1 ppm due to abundant adsorption sites provided by the rod-like In_2O_3 materials with a large specific surface area. The incorporation of Co served as a donor by replacing In^{3+} ion, which could increase the number of oxygen vacancies and boost the sensing performance. Yu et al. [66] discovered that 2 at% Ag-WO_3 nanosheets demonstrated the best performance in terms of a high response (20.83) and rapid response/recovery time (5 s/5 s) toward 100 ppm HCHO at 300 °C. Lv et al. [67] utilized a SnO_2-NiO sensitive film to detect HCHO with a wide concentration ranging from 0.06 to 22.5 ppm at 300 °C, and achieved a high response of ~9 and a fast response/recovery speed (11 s/60 s) toward 22.5 ppm HCHO at 300 °C.

Acetone is a transparent and colorless liquid with a pungent odor at room temperature. As one reagent, acetone is extensively used as a diluting, cleaning, extraction agent in laboratories. Moreover, acetone is widely used as an industrial solvent in paints and agglomerants. Long-time acetone exposure could lead to bronchitis, headache, and coma [68]. Besides, acetone is a biomarker in breath analysis to screen preliminarily for some diseases such as diabetes [69]. Therefore, monitoring acetone is significant for human health. Table 5.3 concludes the sensing

Table 5.3 Gas-sensing performance of MOS sensors toward acetone.

Materials	Structure	$T_{operation}$ (°C)	Gas concentration (ppm)	Response	Response/ recovery time (s)	Ref.
TiO_2	Nanoporous	370	500	25.97^A	13/8	[70]
SnO_2	Powder	RT	8	$\sim 18^A$	<20/<60	[71]
ZnO	Hierarchical	230	100	33^A	3/–	[34]
SnO_2-Rh	Nanofiber	200	50	60.6^A	2/64	[72]
ZnO-NiO	Hollow nanosphere	275	100	$\sim 29.8^A$	1/20	[73]
In_2O_3-Au	Nanosphere	320	10	53.08^A	4/9	[74]

A: (R_g/R_a) or (R_a/R_g).

performance of recently reported acetone gas sensors based on MOS. Chen et al. [70] synthesized nanoporous TiO_2 with a small grain size (diameter: ~12 nm) and a high surface area (147.17 $m^2\ g^{-1}$) via a facile hydrothermal method. The prepared TiO_2 sensors displayed a decent performance toward 20–1500 ppm acetone at 370 °C, including short response/recovery time of 13 s/8 s and a high response of 25.97 toward 500 ppm acetone. Zhao et al. [71] utilized SnO_2 films composed of different grain sizes to detect acetone and found that the reduction of grain size could strongly increase the response. Jia et al. [34] prepared hierarchically nanostructured ZnO dandelion-like spheres for acetone detection, reaching a low detection limit of 0.25 ppm, fast response time of 3 s, and a high response of 33 toward 100 ppm acetone at 230 °C, as a result of the porous structure and large polar facet. Kou et al. [72] obtained Rh-doped SnO_2 nanofibers by electrospinning and found that 0.5 mol% Rh-doped SnO_2 nanofiber showed the best performance, including a high response of 60.6 to 50 ppm acetone, which was 9.6 times than that of pure SnO_2. In addition, the Rh doping improved the cross selectivity for acetone, originated from the wider space charge layer in Rh-SnO_2 nanofibers than pure SnO_2 ones. Liu et al. [73] presented an acetone sensor based on NiO-ZnO composites with a fast response and recovery, and a low detection limit in ppb level. In detail, this includes a faster response/recovery time of 1 s/20 s and a higher response of 29.8 toward 100 ppm acetone at 275 °C in NiO incorporated composites than those with pure ZnO (response/recovery time of 6 s/130 s and a response of 9.2). Such an improvement was found mainly attributed to the electrons tuning and high oxidative catalytic activity of NiO during the adsorption/desorption process of acetone. Zhang et al. [74] prepared mesoporous In_2O_3 nanospheres decorated with Au nanoparticles. The porous structure, large specific surface area, and catalytic effect of Au nanoparticles enabled an excellent sensing performance toward 10 ppm acetone in terms of a high response of 53.08 and fast response/recovery time of 4 s/9 s at 320 °C.

Toluene is a colorless volatile liquid at room temperature and widely used in manufacture of adhesives, fingernail polish, pesticides, and medicine [75]. Because of the inflammability, toluene vapor mixed with air could form explosive mixture

Table 5.4 Gas-sensing performance of MOS sensors toward toluene.

Materials	Structure	$T_{operation}$ (°C)	Gas concentration (ppm)	Response	Response/recovery time (s)	Ref.
SnO_2	Hierarchical	400	50	20.2[A]	52/74	[77]
TiO_2	Nanowire	25	40	~6.5[A]	120/100	[78]
WO_3	Nanoparticle	225	100	132[A]	2/6	[35]
ZnO	Flower like	350	50	32[A]	53/151	[81]
ZnO-TiO_2	Flower like	290	100	17.1[A]	8/20	[79]
SnO_2-Fe_2O_3	Nanotube	260	1	~3.8[A]	5/11	[80]

A: (R_g/R_a) or (R_a/R_g).

even under a relative low concentration. Besides, toluene also threatens human health, such as irritant to the mucous membrane, skin, and anesthetic on the central nervous system. The Occupational Safety and Health Administration (OSHA) has provided the exposure limit for toluene as 100 ppm for 8 hours [76]. Furthermore, toluene is also a cancer biomarker in the exhaled breath of patients with lung cancer. Therefore, fabricating advanced toluene gas sensors is of great importance in fields of safety production and medical care. Table 5.4 summarizes the recent MOS gas sensors toward toluene detection. Hermawan et al. investigated the toluene sensing of micro-/mesoporous SnO_2 spheres [77]. The SnO_2 spheres with a particle size of 0.9 μm and a pore size of 1.4 nm showed a high response of 20.2 and fast response/recovery time of 52 s/74 s toward 50 ppm toluene. Tshabalala et al. [78] prepared TiO_2 sensors with a high response of 80 toward 40 ppm toluene and robust stability over eight months. The as-prepared sensors displayed a temperature-dependent dual-functionality performance to selectively detect toluene and pxylene vapors at 25 and 150 °C, respectively. Wang et al. [35] introduced porous WO_3 nanoparticles to detect toluene at 225 °C, showing a high response of 132.0 and short response/recovery time of 2 s/6 s. Zeng et al. [79] designed a TiO_2-doped flowerlike ZnO nanostructure as the active material for toluene sensing. After the incorporation of TiO_2, the nanocomposite sensor exhibited a higher response toward 1–100 ppm toluene and a lower optimal working temperature than pure ZnO counterpart. Similarly, Shan et al. [80] synthesized sensors based on SnO_2-Fe_2O_3 interconnected nanotubes (Figure 5.7a,b), which displayed an outstanding capability of trace toluene detection, including the lowest detection limit of 50 ppb and fast response/recovery time of 5 s/11 s to 1 ppm toluene at an optimal working temperature. Moreover, the SnO_2-Fe_2O_3 sensors exhibited good repeatability, long-term stability, and selectivity, as shown in Figure 5.7c,d.

Although MOS-based VOCs detection has been extensively explored, the sensor selectivity remains challenging, thus necessitating the development of new metal oxide materials with a higher specificity. In addition, VOCs sensors based on MOS nanostructures always need a high temperature to realize ppb-level detection limit,

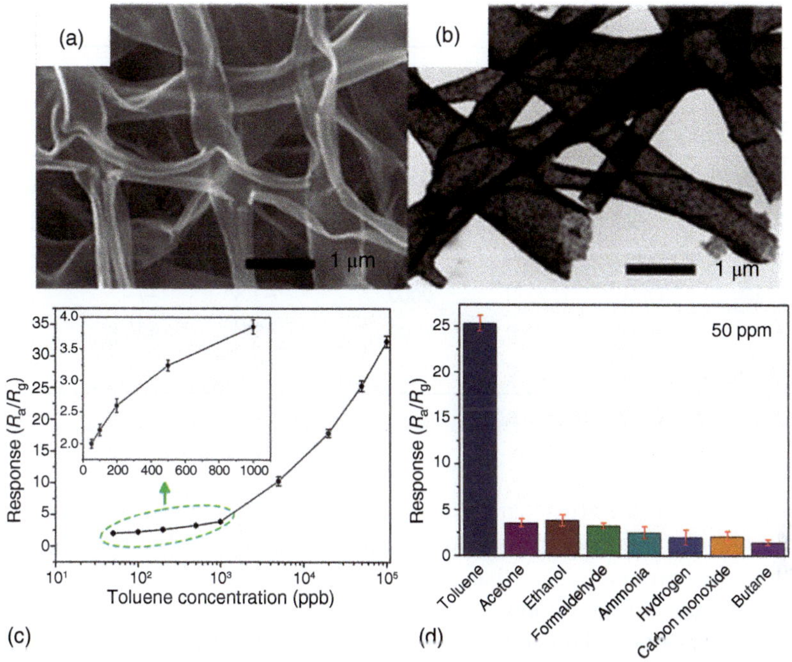

Figure 5.7 (a) SEM and (b) TEM images of SnO_2-Fe_2O_3 interconnected nanotubes. (c) The response of SnO_2-Fe_2O_3 interconnected nanotube gas sensors to different concentrations of toluene in the range of 50 to 100 ppm at 260 °C. (d) Responses of prepared sensors to 50 ppm of different gases at 260 °C. Source: Reproduced from Shan et al. [80]/with permission of American Chemical Society.

readily leading to explosives and other safety issues. Further efforts could be focused on methods to lower the operation temperature, which is also beneficial to reduce the power consumption.

5.2.3 Environmental Pollution Gas Detection

With the rapid expansion of industry production, the emission of environmental pollution gases is becoming increasingly fierce. Common environmental hazardous gases include greenhouse gases such as CO_2, toxic gases such as H_2S, NH_3, and gases with both features, such as NO_2, NO, and SO_2. These pollution gases have brought critical disasters to human, such as terrible smog in London and Los Angeles in the twentieth century. Hence, it is of great significance to monitor these pollution gases for human health. Compared with optical spectroscopy and gas chromatography/mass spectrometry, chemiresistive sensors based on MOS have attracted more attention for pollution monitoring owing to the unique advantages of low cost, miniaturization, robustness, and easy operation.

CO_2, a colorless and odorless gas, is the main contributor for global warming with far more contributions than other greenhouse gases [82]. In addition, CO_2 is identified as the indispensable raw material for the photosynthesis of plants and the

Table 5.5 Gas-sensing performance of MOS sensors toward CO_2.

Materials	Structure	$T_{operation}$ (°C)	Gas concentration (ppm)	Response	Response/recovery time (s)	Ref.
La_2O_3-Pd	Nanorod	250	400	64%[B]	80/50	[83]
TiO_2-Cr	Powder	65	1000	~170%[B]	81/63	[26]
ZnO	Powder	300	1000	1.13%[B]	20/20	[84]
In_2O_3- CaO	Powder	230	2000	~1.8[A]	–/–	[85]
ZnO-La	Powder	400	5000	65%[B]	90/38	[29]
SnO_2-Au-La_2O_3	Nanofiber	300	100	10.1%[B]	–/–	[86]

A: (R_g/R_a) or (R_a/R_g), B: $(R_g - R_a)/R_a$, %.

normal recycling in the ecosystem. However, excess CO_2 is emitted into air from fossil fuel burning with the expansion of human activities since 1750. Besides, CO_2 gas with a concentration higher than the threshold limit of 5000 ppm indoor can lead to an oxygen deficiency and then asphyxiation/suffocation. MOS-based CO_2 sensors that were reported previously are summarized in Table 5.5. Yadav et al. [83] prepared pure La_2O_3 and La_2O_3-Pd nanoparticles, as shown in Figure 5.8a,b. Pure La_2O_3 sensors displayed a response of 25% toward 400 ppm CO_2 at 450 °C. After the incorporation of Pd nanoparticles, La_2O_3-Pd sensors showed a higher response of 64% and a lower working temperature of 250 °C (Figure 5.8c,d). In addition, short response/recovery time of 80 s/50 s toward 400 ppm, a nice selectivity, and a good long-term stability were reached. Figure 5.8e,f schematically illustrate the CO_2-sensing mechanism for La_2O_3-Pd sensors. Mardare et al. [26] prepared Cr-doped TiO_2 composite sensors, which presented a decent response of ~170% to 1000 ppm CO_2 and fast response/recovery time (81 s/63 s) at 65 °C. Kannan et al. [84] prepared CO_2 gas sensors based on ZnO thin films of different thicknesses deposited on glass substrates via DC reactive magnetron sputtering. The sensor with 40-nm film thickness implied a better performance than its counterparts at 300 °C. Prim et al. [85] reported a novel CO_2 sensor based on mesoporous CaO-loaded In_2O_3. The 6.0 at% CaO/In_2O_3 composite sensor showed a wide detection concentration from 300 to 5000 ppm at 230 °C. Jeong et al. [29] prepared La-loaded ZnO composites and found that 50 at% La-loaded ZnO displayed a maximum response of 65% and fast response/recovery time of 90 s/38 s to 5000 ppm CO_2 at 400 °C. The enhanced sensing performance for the composite sensor compared to pure ZnO ones could be attributed to the increase of oxygen vacancies and other defects introduced by La loading. Hsu et al. [86] prepared Au-5 at% La_2O_3-loaded SnO_2 nanofibers. Benefiting from the spillover and catalytic effects of Au nanoparticles, the ternary composite sensor showed a higher response of 10.1% toward 100 ppm CO_2 than that (5.1%) of 5 at% La_2O_3-SnO_2 ones at 300 °C.

Ammonia (NH_3) is a colorless gas with a pungent odor. The ever-increasing development of agricultural and animal husbandry production has brought about

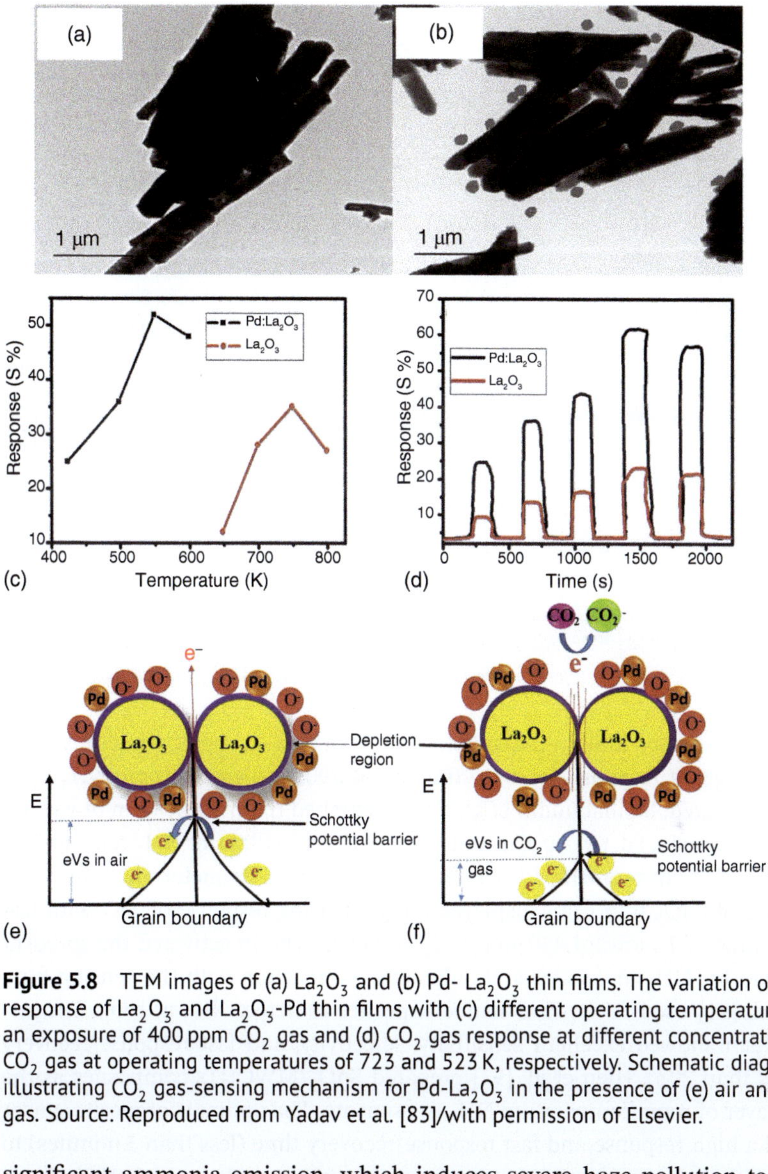

Figure 5.8 TEM images of (a) La_2O_3 and (b) Pd- La_2O_3 thin films. The variation of CO_2 gas response of La_2O_3 and La_2O_3-Pd thin films with (c) different operating temperatures under an exposure of 400 ppm CO_2 gas and (d) CO_2 gas response at different concentrations of CO_2 gas at operating temperatures of 723 and 523 K, respectively. Schematic diagrams illustrating CO_2 gas-sensing mechanism for Pd-La_2O_3 in the presence of (e) air and (f) CO_2 gas. Source: Reproduced from Yadav et al. [83]/with permission of Elsevier.

significant ammonia emission, which induces severe haze pollution to ambient environment. Moreover, ammonia is caustic and hazardous to human eyes and respiratory system. The US OSHA has established an exposure limit of 35 ppm for 15 minutes, and 25 ppm for 8 hours [87]. Due to the unique merits of MOS, ammonia gas sensors based on MOS have also attracted extensive focus. Table 5.6 lists the sensing performance of MOS gas sensors for NH_3 detection. Wagh et al. discovered that 0.37 at% RuO_2-doped ZnO thick-film sensors showed a high response of 386 to 1000 ppm NH_3 at 250 °C and an excellent selectivity over various gases including liquefied petroleum gas – H_2, CO_2, and H_2S [88]. Dhivya et al. [89] realized a room-temperature NH_3 detection based on nanostructure TiO_2 films.

Table 5.6 Gas-sensing performance of MOS sensors toward NH_3.

Materials	$T_{operation}$ (°C)	Gas concentration (ppm)	Response	Response/ recovery time (s)	Ref.
$ZnO\text{-}RuO_2$	250	1000	38 600%[B]	18/960	[88]
TiO_2	RT	100	7857[A]	34/90	[89]
SnO_2-Pt	230	450	25.7[A]	1/59	[90]
$ZnO\text{-}\alpha Fe_2O_3$	RT	0.4	10000[A]	20/20	[91]
SnO_2/MWCNT	RT	200	~27[A]	593/854	[92]
TiO_2-RGO-Au	RT	2	8.9%[B]	–/–	[87]
WO_3	300	0.5	2.3[A]	–/–	[93]
WO_3-Pt	125	200	13.61[A]	43/272	[94]
rGO-Cu_2O	60	50	6.5%[B]	–/–	[95]
$MoS_2\text{-}Cu_2O$	75	100	~872%[B]	269/–	[96]
BP-TiO_2	18 ± 2	5	19.7%[B]	412/690	[97]
$Ti_3C_2T_x\text{-}TiO_2$	20	0.2	7.3%[B]	–/–	[98]

A: (R_g/R_a) or (R_a/R_g), B: $(R_g - R_a)/R_a$, %.

Besides a wide concentration-range detection from 5 to 100 ppm, a high response of 7857 and quick response/recovery time of 34 s/90 s toward 100 ppm NH_3 were also demonstrated. Shahabuddin et al. [90] researched the modification of various metallic clusters (Pt, Pd, Cr, Au, Cu, and In) on SnO_2 thin films for NH_3 sensing. The integration of Pt clusters with a thickness of ~8 nm and a diameter of 200 μm onto the surface of SnO_2 thin films displayed a high sensing response of 25.7 with fast response time of 1 s toward 450 ppm NH_3 at 230 °C. The Pt activated the spillover mechanism for NH_3 molecules and enhanced the reaction with the sensing film. Tang et al. [91] used Fe_2O_3-ZnO nanocomposites to detect NH_3 at room temperature, which showed excellent sensitivity, selectivity, and fast response/recovery time (less than 20 s). Hieu et al. [92] prepared SnO_2/MWCHTs composites as the sensing layer of room-temperature NH_3 gas sensors. The optimal composite sensor disclosed a high response and fast response/recovery time (less than 5 minutes) to NH_3 gas with the concentration varying from 60 to 800 ppm Dao et al. [93] disclosed novel WO_3 nanoparticles based NH_3 sensors with the lowest detection limit of 500 ppb and a response of 2.3 at 300 °C. Wang et al. [94] studied the NH_3 sensing of Pt-loaded mesoporous WO_3 and discovered that the 1.0%Pt-WO_3 sensor possessed an excellent sensing performance to 50–1500 ppm NH_3 (Figure 5.9a,b), including a high response (13.61), fast response/recovery time (43 s/272 s) toward 200 ppm NH_3 at 125 °C, and a good selectivity. The relevant sensing mechanism is shown in Figure 5.9c,d.

Our group also dedicated some efforts to NH_3 sensing based on MOS. NH_3 sensing is always plagued by sluggish recovery. To improve this situation, we took reduced

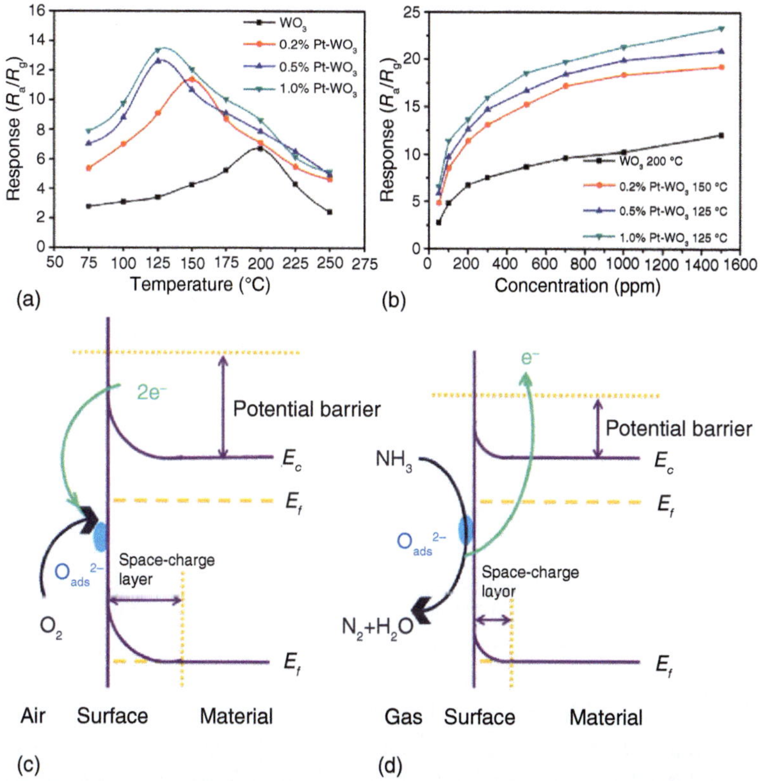

Figure 5.9 (a) The sensitivity of obtained materials to 200 ppm ammonia as a function of operating temperature. (b) The dot-line gas response patterns of the obtained sample to NH_3 with different concentrations under each optimal operating temperature. Schematic drawing of the reaction mechanism for MOS-based sensors in (c) air and (d) ammonia. Source: Reproduced from Wang et al. [94]/with permission of Elsevier.

graphene oxide (rGO)/Cu_2O composites as the sensing layer and introduced a 70 °C heat pulse with a duration period of 5 s upon the beginning of NH_3 desorption [95]. The as-prepared composite sensors showed a full and swift recovery within 26 s, which was considerably improved in comparison to a partial and sluggish one (77% recovery within 10 minutes) without this treatment. Moreover, the sensor response achieved 6.5% toward 50 ppm at 60 °C. Besides, we employed MoS_2-Cu_2O composites to successfully detect 20–100 ppm NH_3 [96]. The sensing response of the MoS_2-Cu_2O sensor (~872%) with an optimized composition for 100 ppm NH_3 was increased more than 8 times compared with pristine Cu_2O (~103%) at 75 °C. In another work, TiO_2 decorated black phosphorus (BP) nanosheets were prepared via a hydrothermal method [97]. The BP-TiO_2 composites were utilized to selectively discern 0.5 to 30 ppm NH_3 at room temperature (18 ± 2 °C) for the first time. The constituent ratio optimized BP-TiO_2 sensors exhibited a higher response, a less baseline drift, shorter response/recovery time, and a stronger long-term stability

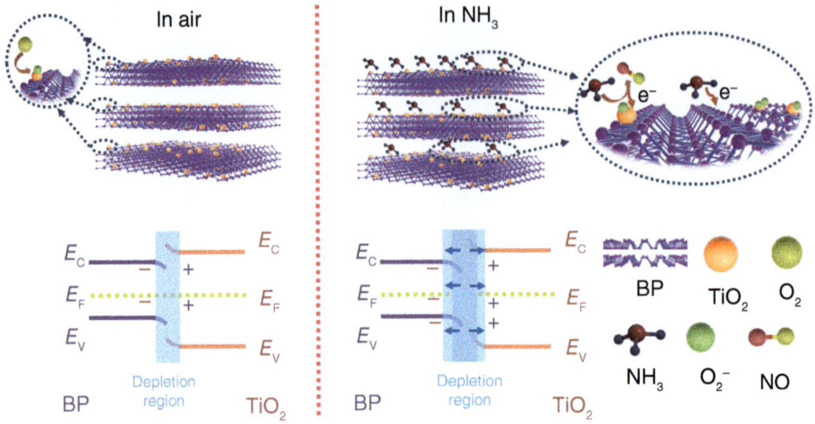

Figure 5.10 Schematic illustrations of NH_3-sensing mechanism of BP-TiO_2 sensors. Source: Reproduced from Wang et al. [97]/with permission of Elsevier.

compared with pure BP counterparts. Enlarged specific surface area, numerous *p-n* BP-TiO_2 heterojunctions, and TiO_2 nanoparticles involved passivation were primarily responsible for these improvements. The sensing mechanism is depicted in Figure 5.10. Lately, we prepared MXene $Ti_3C_2T_x$ derived nitrogen functionalized heterophase TiO_2 homojunctions (N-MXene) via a urea-involved solvothermal treatment with varied reaction time as the sensing layer to detect trace NH_3 gas at 20 °C [98]. Compared with no signal of pristine MXene counterpart, 18 h-treated sensors (N-MXene-18) achieved a detection limit of 200 ppb with an inspiring response of 7.3%, which was also superior to existing MXene-involved reports so far. Also, decent repeatability, stability, and selectivity were demonstrated. Furthermore, a commercial waterproof polytetrafluoroethylene (PTFE) membrane was anchored onto the N-MXene layer to enable a humidity-independent NH_3-sensing performance. The detailed interaction between NH_3 and the N-MXene sensing layer is schematically demonstrated in Figure 5.11. We also prepared rGO-TiO_2-Au sensors which unveiled a sensing response of 8.9% toward 2 ppm NH_3, a high

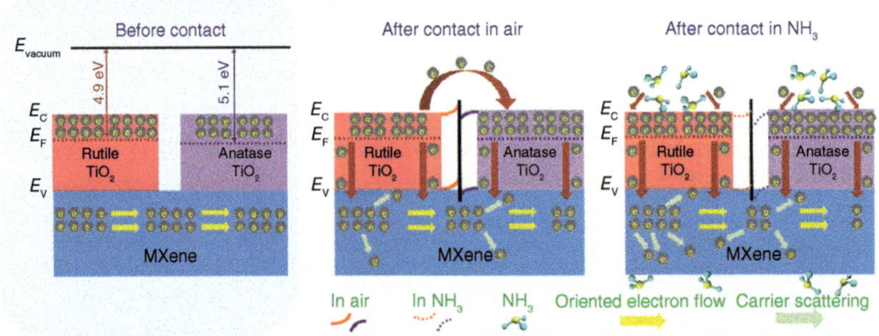

Figure 5.11 The diagram of sensing mechanism for N-MXene-18 sensors. Source: Reproduced from Zhou et al. [98]/with permission of American Chemical Society.

sensitivity of 1.43%/ppm over the test concentration range of 2–10 ppm, a nice selectivity, and a robust long-term stability at room temperature [87].

NO_2 is a reddish-brown gas with high toxicity and irritancy, and is mainly emitted by automobiles and combustion chemical plants. NO_2 works as the major contributor for production of photochemical smog, haze, and acid rain. The maximum threshold limit of NO_2 is 3 ppm for humans [82, 99]. Hence, the NO_2-emission monitoring is of crucial significance for environmental protection and human health. Table 5.7 summarizes the performance of MOS gas sensors for NO_2 exploration. Zhou et al. [100] reported a room-temperature trace NO_2 sensor based on ZnO nanowire-integrated films under UV illumination. The prepared sensor showed an excellent sensing performance toward ppb-level NO_2 at 25 °C, including a high response of 15 700% toward 50 ppb NO_2, and a good sensitivity of 7.8%/ppb. Chen et al. [101] compared the NO_2-sensing performance of flower-like and tube-like ZnO nanomaterials. The results showed that the tube-like ZnO sensors showed faster response/recovery time (15 s/36 s) and a higher response (350) toward 10 ppm NO_2 at 190 °C than flower-like counterparts. XPS analysis demonstrated that the enhancement was attributed to the higher capability of O_2 adsorption in tube-like ZnO than flower-like ones. Sonker et al. [102] synthesized ZnO nanopetals for NO_2 sensing. The as-prepared sensor displayed a high response of 119 toward 20 ppm NO_2 gas at room temperature with mean response and recovery time of 1.42 and 1.71 mintues, respectively. Shendage et al. [103] prepared 2D WO_3 nanoplates that showed a separate response of ~10 and 131.75 toward 5 and 100 ppm NO_2 at 100 °C, respectively. Zhu et al. [104] synthesized TiO_2 nanowires for rapid NO_2 detection.

Table 5.7 Gas-sensing performance of MOS sensors toward NO_2.

Materials	Structure	$T_{operation}$ (°C)	Gas concentration (ppm)	Response	Response/recovery time (s)	Ref.
ZnO	Nanowire	25	0.05	15 700%^{B}	~1200/–	[100]
ZnO	Nanotube	190	10	350^{A}	15/36	[101]
ZnO	Nanopetal	RT	20	~119^{A}	~85.2/~102.6	[102]
WO_3	Nanoplate	100	100	131.75^{A}	–/–	[103]
TiO_2	Nanowire	RT	100	3.1^{A}	10/19	[104]
WO_3-SnO_2	Nanocluster	100	10	54000^{A}	67/1023	[105]
ZnO-Sn	Nanoparticle	150	1.5	11^{A}	–/–	[106]
TiO_2-Cr	Nanoparticle	500	50	1.35^{A}	–/–	[107]
In_2O_3-Fe	Nanoparticle	150	1	71^{A}	276/150	[27]
TeO_2-SnO_2	Microdiscs	90	10	~22600^{A}	75/900.6	[108]
In_2O_3	Nanobrick	50	0.5	402^{A}	114/49	[109]
MoS_2-rGO-Cu_2O	Powder	RT	0.5	14.88%^{B}	–/–	[110]
ZnO-BP	Nanowire	RT	0.05	74%^{B}	–/–	[111]

A: (R_g/R_a) or (R_a/R_g), B: $(R_g - R_a)/R_a$, %.

The prepared sensor showed rapid response and recovery time of 10 and 19 s toward 100 ppm NO_2 at room temperature, respectively. Sharma et al. [105] proposed WO_3 nanoslusters-SnO_2 composite sensors for NO_2 detection, which exhibited a high response of 54 000 toward 10 ppm NO_2 gas with a fast response of 67 s at 100 °C. Shishiyanu et al. [106] prepared 5 at% Sn-ZnO composites that showed a response of 1.5 toward 1.5 ppm NO_2 at 25 °C. Ruiz et al. [107] proposed 10 at.% Cr-TiO_2 sensors, which manifested an excellent sensing performance toward 4–50 ppm NO_2 at 500 °C. Zhao et al. [27] discovered that ordered mesoporous Fe-doped In_2O_3 sensors showed a high response of 116 to 5 ppm NO_2, which was ~3 times higher than pure In_2O_3 at 150 °C. Moreover, the composite sensor displayed an ultralow detection concentration of 40 ppb NO_2 and a good selectivity. Sharma et al. [108] utilized TeO_2-SnO_2 *p–n* heterojunctions to detect trace-level NO_2 (1–10 ppm) at 90 °C. Han et al. [109] prepared brick-like In_2O_3 nanomaterials by an oil bath precipitation and a subsequent calcination. The In_2O_3 sensor exhibited a high response (402) toward 500 ppb NO_2 with fast response/recovery time (114 s/49 s) at 50 °C and a nice linearity over the low concentration range of 100 to 500 ppb.

Our group prepared hollow Cu_2O nanospheres decorated with MoS_2-rGO ternary composites by combining hydrothermal and soft-template methods and the composites provided a boosted ppb-level NO_2 sensing characteristics [110]. The optimum composite sensor exhibited the highest response to 500 ppb NO_2 at room temperature, which was 11 and 5 times higher than pure MoS_2 and binary MoS_2-rGO, respectively. We also proposed a ppb-level NO_2 gas sensor based on ZnO nanowires-black phosphorus nanosheets composite films [111], as shown in Figure 5.12a. The composite sensor showed excellent properties toward 25–100 ppb NO_2 at room temperature, including a response of 74% to 50 ppb NO_2 with a fast response, humidity independence, and a nice selectivity. The detailed sensing mechanism is demonstrated in Figure 5.12b.

NO_2 and NH_3, as representative oxidizing and reducing gases, are the most popular pollution gases researched so far. However, until now, there were still insufficient reports referring to MOS-based CO_2 and SO_2 sensors. Of these gas species, CO_2 is chemically inert and reluctant to exchange electrons with MOS, leading to a limited

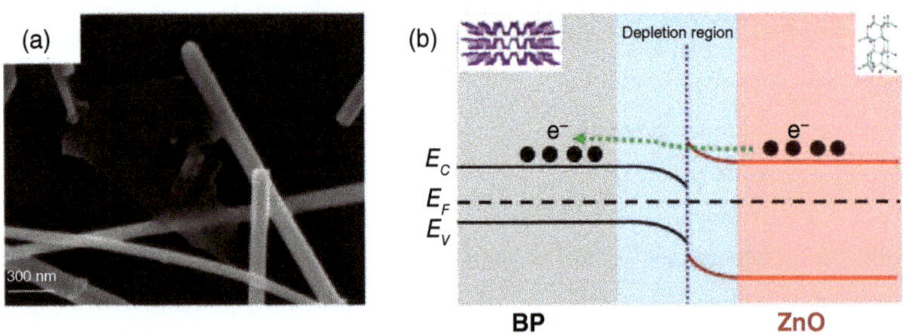

Figure 5.12 (a) A TEM image of BP nanosheets-ZnO nanowires composites. (b) A schematic illustration of NO_2 gas-sensing mechanism of BP-ZnO *p–n* heterojunctions. Source: Reproduced from Wang et al. [111]/with permission of American Chemical Society.

sensitivity. SO_2 molecules simultaneously possess oxidizing and reducing features, and the response polarity of MOS sensors readily undergoes abnormal transition over temperature. Therefore, achieving sensitive detection of CO_2 and SO_2 with a smaller detection limit, a higher selectivity, and a lower operation temperature is expected research directions.

5.2.4 Humidity Detection

Humidity defines the amount of water vapor present in air. Absolute humidity presents the mass of water in a unit volume of gas. To make the humidity simple and perceptible for people, relative humidity (RH) is introduced. RH is defined as the ratio of the partial pressure of water vapor to the saturated one at a particular temperature, quoted as a percentage. Humidity monitoring is very significant for industrial production, agriculture, storage service, and human health. Humidity sensors based on MOS have been widely applied due to low cost, easy fabrication, and high accuracy compared with other humidity measurements including hair hygrometer, wet and dry bulb hygrometer, Regnaults hygrometer, and chilled mirror hygrometer [112]. Unlike other gases, the adsorption of water molecules within the sensing materials can be classified into chemisorption and physisorption, as shown in Figure 5.13. The chemisorbed water layer is initially formed. As RH increases, the physisorbed layer is generated on the chemisorbed one. H_3O^+ is produced due to the ionization of water molecules under an electrostatic field. As the adsorption of water molecules continues, the multilayer physisorption exhibits a liquid-like behavior while the protons hopping-transport occurs via ionic conductivity, as depicted in the equation $H_2O + H_3O^+ = H_3O^+ + H_2O$ [113].

Besides humidity sensors based on direct current (DC) signals (resistance, current, and voltage), alternating current (AC) signals including impedance (Im) and capacitance are also used in humidity sensing. Table 5.8 summarizes the sensing performance of MOS gas sensors toward humidity. Tsai et al. [114] prepared laterally grown ZnO nanosheets for humidity sensing. The optical microscopy image of the as-prepared device is schematically exhibited in Figure 5.14a. At room temperature, a high response of 220 ($R_{12\%RH}/R_{96\%RH}$), a good linearity over the range of 12–96% RH, fast response/recovery time (3 s/600 s), a good selectivity, and a long-term stability were achieved (Figure 5.14b,c). Qi et al. [115] prepared flower-like ZnO nanorods-based humidity sensors in AC pattern. The sensors showed a high humidity response at 100 Hz, fast response/recovery time (5 s/11 s), low hysteresis,

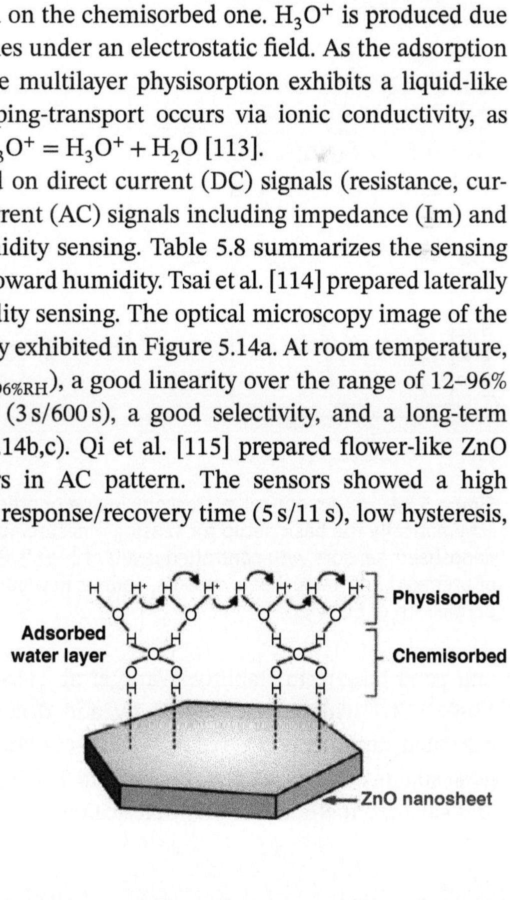

Figure 5.13 The adsorption of water molecules on ZnO nanosheets. Source: Reproduced from Tsai and Wang [114]/with permission of Elsevier.

Table 5.8 Gas-sensing performance of MOS sensors toward humidity.

Materials	type	$T_{operation}$ (°C)	Relative humidity (RH)	Response	Response/recovery time (s)	Ref.
ZnO nanosheet	DC	RT	12% → 96%	220[C]	3/600	[114]
ZnO nanorod	AC	RT	11% → 95%	5 orders[A]	5/11	[115]
TiO_2 nanoparticle	AC	RT	11% → 95%	4 orders[A]	5/8	[116]
TiO_2 nanotube	DC	25	11% → 95%	2 orders[C]	100/190	[117]
N-TiO_2	DC	25	11.3% → 93.6%	131.25[C]	–/–	[118]
TiO_2 nanowire	DC	RT	12% → 97%	3 orders[C]	<120/<120	[119]
SnO_2 nanowire	DC	30	5% → 85%	32[C]	170/60	[120]
SnO_2 nanosheet	DC	RT	30% → 40%	2.5%[B]	0.052/0.14	[121]
NiO-SnO_2 nanofiber	DC	RT	0% → 83%	86.6%[B]	18.4/37.2	[122]
KCl-SnO_2 nanofibers	AC	RT	11% → 95%	5 orders[A]	5/6	[123]
ZnO- In_2O_3	AC	RT	11% → 95%	4 orders[A]	15/40	[124]

A: (Im_g/Im_a) or (Im_a/Im_g), B: $(R_g - R_a)/R_a$, %, C: (R_g/R_a) or (R_a/R_g).

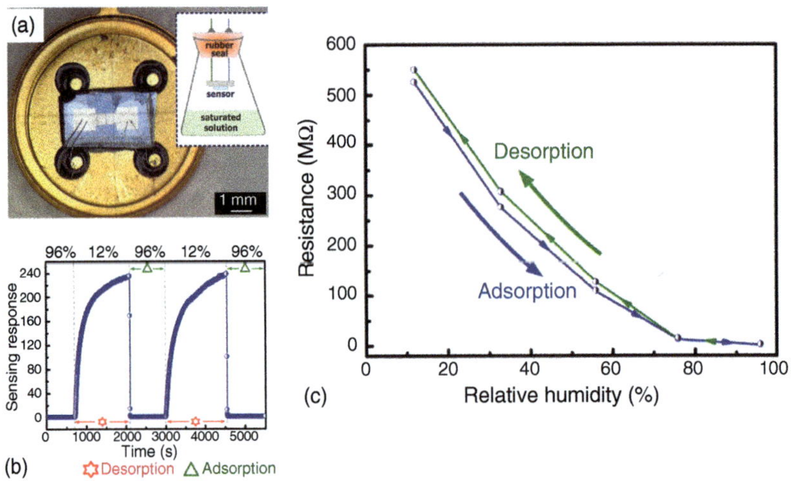

Figure 5.14 (a) An optical microscopy image of prepared sensing device. The inset shows schematically the basic setup for sensing measurements. (b) Transient response of ZnO nanosheets sensors with controlled levels: 12–96% RH. (c) Typical hysteresis characteristics of prepared ZnO nanosheets sensors. Source: Reproduced from Tsai and Wang [114]/with permission of Elsevier.

and good long-term stability. Wang et al. [116] employed porous TiO_2 to detect 11–95% RH with a high sensitivity and decent linearity. Moreover, the sensor exhibited rapid response/recovery time (5 s/8 s), little hysteresis of <1% RH, and good stability within 30 days. Zhang et al. [117] prepared TiO_2 nanotubes by anodic and calcined methods. The samples calcined at 600 °C displayed a high sensitivity with nearly two orders of magnitude change in the sensor resistance and short response/recovery time (100 s/190 s) toward a RH range of 11%–95%. Wang et al. [118] reported resistive and capacitive humidity sensors based on nitrogen-doped TiO_2 nanotubes. As RH increased from 11.3% to 93.6%, the sensor capacitance increased from 0.6 to 420 nF, and the sensor resistance decreased from 105 to 0.8 kΩ. Wu et al. [119] employed TiO_2 nanowires/nafion composites as a humidity-sensitive material. Kuang et al. [120] realized a sensitive humidity detection based on a single SnO_2 nanowire. The as-prepared sensor showed a high response of 32 to 85% RH at 30 °C, and fast response/recovery time of 120–170 s/20–60 s when detecting 30% to 85% RH. Zhong et al. [121] synthesized SnO_2 with rich oxygen vacancies for ultrafast humidity detections. The prepared sensor with a stoichiometric ratio of Sn/O = 1 : 1.55 revealed ultrafast response/recovery time of 52 ms/140 ms, hence being very competitive in further applications. In another work, NiO-SnO_2 nanofibers were utilized for humidity sensing. Pascariu et al. [122] proposed NiO-SnO_2 nanofibers for humidity sensing. The sensor displayed excellent sensing properties toward 0–100% RH and short response/recovery time (18.4 s/37.2 s). Song et al. [123] reported a humidity sensor based on KCl-doped SnO_2 nanofibers. The sensor impedance decreased by more than five orders of magnitude with the increase of RH from 11% to 95%, together with fast response/recovery time (5 s/6 s). Liang et al. [124] found that ZnO-In_2O_3 thin films displayed a large impedance change by more than four orders of magnitude over the whole humidity range (11–95% RH), and fast response/recovery time (15 s/40 s).

Although vast research has been conducted on humidity detection, MOS-based trace humidity sensing particularly at ppm level or below was scarcely reported. Therefore, incorporating hydrophilic polymer, alkali metal salts, or ionic liquids into MOS matrix is very prospective in this extreme scenario.

5.2.5 Explosives Detection

Methane (CH_4) is a colorless and odorless gas, and is a major component of coalmine gas, natural gas, and marsh gas. As one of the simplest hydrocarbons, CH_4 is widely used as fuel and chemical raw material [125]. However, CH_4 would lead to a dangerous explosion when the volume concentration exceeds 4.9% because of its flammable and volatile property. Table 5.9 summarizes previous work about CH_4-sensing performance. Typically, 20 mol% Pt-SnO_2 nanofibers exhibited a maximum response of 4.4825 toward 1000 ppm methane at 350 °C, as shown in Figure 5.15a,b [126]. The detailed sensing mechanism is shown in Figure 5.15c,d. In another report, the nanocrystalline ZnO thin films showed a high response of 87.3% and fast response/recovery time of 8.3 s/17.8 s to 1 vol% CH_4 at 250 °C [127]. Moreover, 0.5 vol% CH_4 was detectable with a low power of 120 mW. Alternatively, three-dimensional hierarchical WO_3-SnO_2 composites were utilized to detect CH_4 ranging from 10 to 5000 ppm at 110 °C, together with a high response of 2.85 toward 500 ppm CH_4 [128]. Also, NiO decorated In_2O_3 nanospheres were synthesized to probe CH_4 at a low temperature [129]. The composites displayed a response nearly 10.3 times higher than pure In_2O_3 toward 500 ppm CH_4 at 120 °C.

Table 5.9 Gas-sensing performance of MOS sensors toward CH_4.

Materials	Structure	$T_{operation}$ (°C)	Gas concentration (ppm)	Response	Response/ recovery time (s)	Ref.
ZnO	Powder	250	10 000	87.3%^{B}	8.3/17.8	[126]
SnO_2-Pt	Nanofiber	350	1000	4.4825^A	24.3/141.1	[127]
SnO_2-WO_3	Nanoflower	110	500	2.85^A	–/–	[128]
In_2O_3-NiO	Nanosphere	120	500	13.057^A	–/–	[129]
SnO_2-Cr	Nanoparticle	350	10 000	1268^A	3.9/–	[130]
In_2O_3-Ni_2O_3	Nanoparticle	350	200	52%^{B}	–/–	[131]
In_2O_3-Pd	Powder	300	500	4.1^A	–/–	[132]
WO_3-Rh	Powder	350	1	20^A	~32/~30	[133]
ZnO-Ag	Microsphere	200	5000	20^A	118/119	[134]
ZnO-Zn_2SnO_4	Microflower	250	5000	27.2^A	7/30	[135]

A: (R_g/R_a) or (R_a/R_g), B: $(R_g - R_a)/R_a$, %.

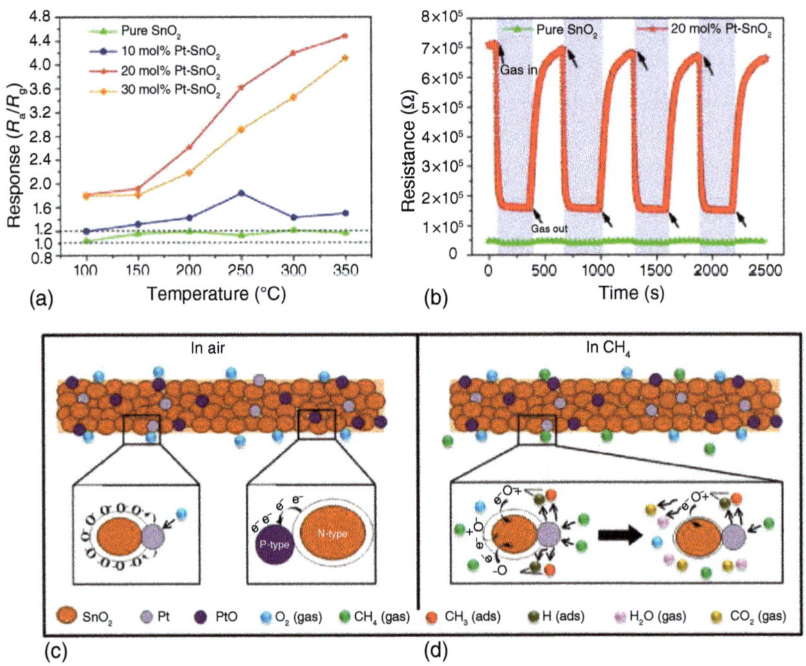

Figure 5.15 (a) Variations of response to 1000 ppm CH_4 with temperature for all samples with different Pt concentration ratios. (b) The response characteristic curves of 20 mol% Pt-SnO_2 nanofibers and pure SnO_2 nanofibers. (c-d) The schematic diagram of CH_4-sensing mechanisms for Pt-SnO_2. Source: Reproduced from Lu et al. [126]/with permission of Elsevier.

Besides, the composites could detect CH_4 with a wide concentration ranging from 50 to 5000 ppm, together with an excellent selectivity, and good repeatability and stability. Bunpang et al. [130] proposed a highly selective and sensitive CH_4 sensor based on flame-spray-made Cr-doped SnO_2 particulate films. The sensor with 0.5 wt% Cr-SnO_2 composites showed a remarkable response of 1268 and response time of 3.9 s toward 1 vol% CH_4 at 350 °C. Vuong et al. [131] used Ni_2O_3-SnO_2 composites to detect CH_4 at 350 °C, which showed a high response of 52% to 200 ppm CH_4. In addition, Pd-In_2O_3 sensors could selectively detect CH_4 ranging from 100 to 3000 ppm [132]. It was found that Rh-doped WO_3 sensors were capable of linearly detecting CH_4 from 1 to 5 ppm at 350 °C. The response/recovery time was evaluated as 32 s and 30 s toward 100 ppm CH_4, respectively [133]. Wang et al. [134] proposed a gas sensor based on Ag-modified ZnO flower-like microspheres for CH_4 sensing with a concentration ranging from 10 to 5000 ppm at 200 °C. Li et al. [135] synthesized porous nanosheets-assembled ZnO microflowers to detect CH_4 as low as 1.48 ppm at 250 °C.

Hydrogen (H_2) is the least dense gas with colorless, odorless, and flammable features. As a clean energy source, H_2 has been widely used in many fields, such as fuel cells, automobiles, spaceflight, and chemical production. However, when H_2 concentration in a gas mixture achieves the volume fraction between 4.0% and 75.6%, the mixture would be explosive once contacting with fire source [136]. Therefore, it is very important to sensitively detect hydrogen leakage in various dangerous spaces. Table 5.10 summarizes sensing performance of MOS gas sensors toward hydrogen. Li et al. [137] proposed a quick responding and ppb-level detectable H_2 gas sensor based on mesoporous In_2O_3. Mondal et al. [138] synthesized ZnO-SnO_2 composites for selective H_2 sensing at 150 °C. Lupan et al. [139] investigated H_2 sensing of a single ZnO nanorod, which showed a high response of ~4% to 200 ppm H_2, quick response/recovery time (30–40 s/50–90 s), and a nice selectivity. Hassan

Table 5.10 Gas-sensing performance of MOS sensors toward H_2.

Materials	Structure	$T_{operation}$ (°C)	Gas concentration (ppm)	Response	Response/recovery time (s)	Ref.
In_2O_3	Nanoparticle	260	500	18.0^A	1.7/1.5	[137]
ZnO-SnO_2	Powder	150	10 000	$90\%^B$	60/75	[138]
ZnO	Nanorod	RT	200	$\sim4\%^B$	30–40/50–90	[139]
ZnO	Nanorod	200	1000	$500\%^B$	176/116	[140]
WO_3-Pt	Powder	200	10 000	1.41×10^{6B}	201/26	[141]
WO_3-Pt	Microparticle	70	10 000	1600^C	180/360	[142]
WO_3-Pd	Nanoparticle	25	5000	11.78^A	80/10	[143]
In_2O_3-CeO_2	Hollow microsphere	160	50	20.7^A	1/9	[144]
In_2O_3-Pt	Nanocube	25	30 000	$\sim30^A$	~90/~72	[145]

A: (R_g/R_a) or (R_a/R_g), B: $(R_g - R_a)/R_a$, %, C: (V_g/V_a) or (I_g/I_a).

et al. [140] proposed a highly sensitive room-temperature H_2 gas sensor based on oblique and vertical ZnO nanorod arrays. Chang et al. [141] synthesized WO_3 thin films decorated with Pt nanoparticles for H_2 sensing at 200 °C. Ippolito et al. [142] compared the sensing performance of Pt-WO_3 and Au-WO_3 composites toward 0.125–1% H_2 at a wide temperature range of 30 to 300 °C. The Pt-WO_3 composites displayed a lower optimal temperature (70 °C) than both Au-WO_3 and pure WO_3, and a high response of 1600 toward 1 vol% H_2. Wu et al. [143] prepared Pd-doped mesoporous WO_3 to detect H_2 at room temperature. Hu et al. [144] prepared CeO_2-loaded In_2O_3 hollow spheres for ppb-level H_2 detection (Figure 5.16a). The 2 at% CeO_2-loaded In_2O_3 composites (Ce_2In) sensor showed a high response of 20.7 toward 50 ppm H_2 at 160 °C, fast response/recovery time (1 s/9 s), an ultra-low detection limit of 10 ppb, an excellent reversibility (Figure 5.16b), and good long-term stability. Figure 5.16c,d show the electron transportation and energy band variation of CeO_2-In_2O_3 composites during the sensing process. Wang et al. [145] realized room-temperature H_2 detection by using Pt-decorated In_2O_3 nanocubes.

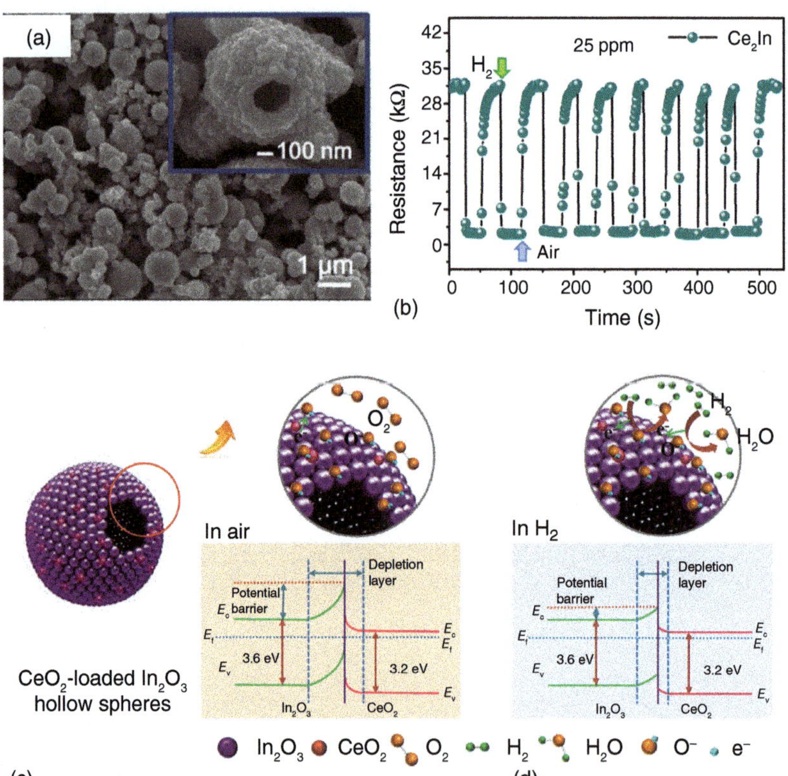

Figure 5.16 (a) An SEM image of different contents of CeO_2-loaded In_2O_3 hollow spheres after annealing (Ce_2In). (b) Reversibility of Ce_2In gas sensor at 160 °C. Schematic presentations of electron transportation and energy band diagrams of CeO_2-loaded In_2O_3 hollow spheres to (c) air and (d) hydrogen gas ambient. Source: Reproduced from Hu et al. [144]/with permission of Elsevier.

Table 5.11 Gas-sensing performance of MOS sensors toward LPG.

Materials	Structure	$T_{operation}$ (°C)	Gas concentration (ppm)	Response	Response/ recovery time (s)	Ref.
TiO_2-Pd	Nanorod	325	5200	49%^{B}	100/200	[146]
TiO_2-Ppy	Powder	RT	1040	55%^{B}	154/161	[147]
SnO_2-Pd	Nanoparticle	100	1000	95^B	–/–	[148]
SnO_2-CuO	Nanoparticle	25	5000	4100^A	180/200	[149]
TiO_2	Powder	400	2600	29%^{B}	–/–	[150]
TiO_2-PANI	Nanoparticle	RT	1000	63%^{B}	140/180	[151]
SnO_2-Pt	Powder	260	100	750^A	~100/~120	[152]
TiO_2	Web-like	425	800	35.8%^{B}	165/240	[153]
WO_3	Nanoparticle	250	830	~16.8%^{B}	–/–	[154]

A: (R_g/R_a) or (R_a/R_g), B: $(R_g - R_a)/R_a$, %.

Liquefied petroleum gas (LPG) is mainly composed of propane, butane, and other alkanes. The LPG leakage would induce explosions in kitchen and chemical plants. The sensing performance of MOS-based LPG sensors is summarized in Table 5.11. TiO_2 nanorods decorated with Pd nanoparticles were utilized for LPG sensing [146]. Especially, the composites showed a wide detection concentration from 1300 to 4200 ppm at 598 K, a high response of 49% toward 5200 ppm, fast response/recovery time of 100 s/200 s, and a lower optimal operating temperature than pure TiO_2 (Figure 5.17a,b). Figure 5.17c illustrates the adsorption process and the variation of energy band of TiO_2 in air and LPG. Bulakhe et al. [147] proposed polypyrrole (Ppy)-TiO_2 heterojunctions for room-temperature LPG sensing. Vaishampayan et al. [148] reported that Pd-doped SnO_2 composites with 1.5 mol% Pd exhibited excellent LPG-sensing behaviors at 100 °C. Singh et al. [149] synthesized CuO-SnO_2 nanocomposite for LPG detection at room temperature (25 °C). Dhawale et al. [150] prepared electron-beam irradiated TiO_2 thin films for LPG sensing at 400 °C. Dhawale et al. [151] studied the room-temperature LPG sensing based on p-polyaniline/n-TiO_2 heterojunctions. Haridas et al. [152] proposed a novel LPG sensor based on SnO_2 thin films loaded with nanoscale Pt catalysts at 260 °C. More et al. [153] investigated LPG sensors based on interconnected web-like TiO_2 films at 425 °C. Singh et al. [154] synthesized WO_3 nanoparticles to detect 125–830 ppm LPG at 250 °C, which showed fast response and recovery time as well as good repeatability.

As explosives detection was always executed under elevated temperature within oxygen-enriched atmospheres, safety hazards inclined to occur. The future work should be stressed more on their sensitive detection in terms of smaller detection limit, higher selectivity, and lower operation temperature. To achieve these, subtle nanostructure modulation and adjuvant light illumination are promising strategies to be exploited during the sensing process.

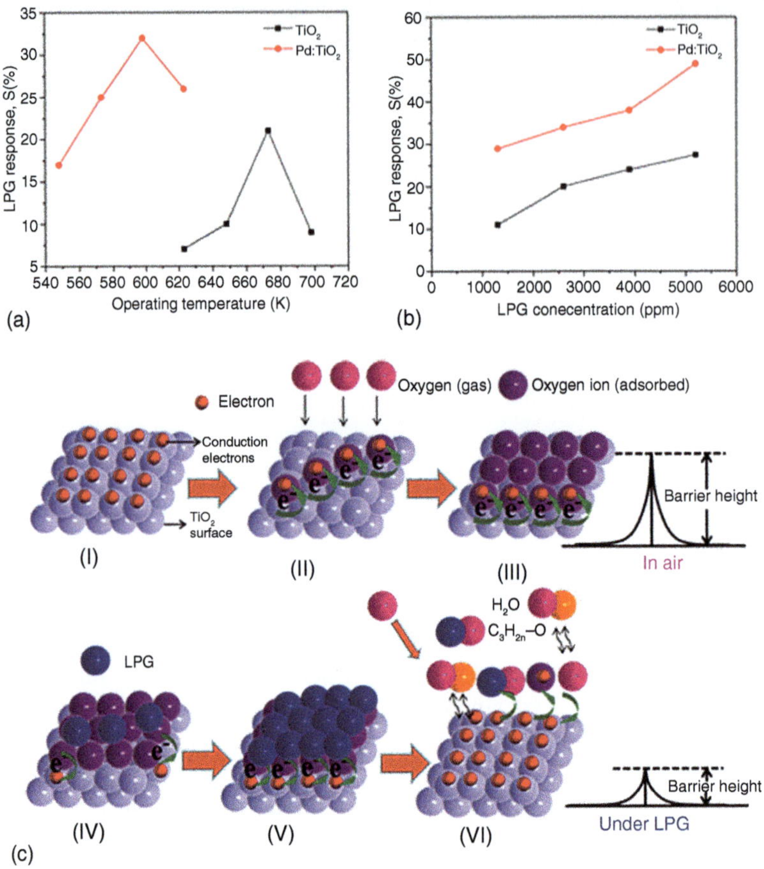

Figure 5.17 (a) LPG response upon exposure of 2600 ppm LPG as a function operating temperature. (b) Variation of LPG response with LPG concentration at 673 K and 598 K of pristine TiO_2 and TiO_2-Pd nanorods. (c) Schematic illustration of the sensing mechanism for TiO_2 in air and under LPG exposure. Source: Reproduced from Dhawale et al. [146]/with permission of American Chemical Society.

5.2.6 Chemical Warfare Agent Detection

Chemical warfare agents (CWAs) are a variety of chemicals used in war with highly toxic for humans, animals, and plants. Hence, sensitive and fast detection of CWAs is very important, especially in battleground. For example, sarin gas ($C_4H_{10}FO_2P$) is an organophosphorus compound that can cause death within 1–10 minutes by suffocation from lung muscle paralysis [155]. Due to the severe perniciousness of CWAs, the simulants of CWAs are investigated widely as target gases, such as dimethyl methylphosphonate (DMMP, $C_3H_9O_3P$), acetonitrile, 1,5-dichloropentane (DCP), dipropylene glycol methyl ether (DPGME), and dichoromethane (DCM) [156, 157]. Table 5.12 summarizes the sensing performance of relevant MOS gas sensors. Tomchenko et al. [157] synthesized various nanocrystalline MOS, such

Table 5.12 Gas-sensing performance of MOS sensors toward warfare agent.

Materials	Target gas	$T_{operation}$ (°C)	Gas concentration (ppm)	Response	Response/recovery time (s)	Ref.
ZnO-Al	DMMP	350	10	4347^A	2/96	[155]
SnO_2-Al_2O_3	Acetonitrile	350	10	$50\%^B$	80/200	[158]
SnO_2	DCM	350	0.5	$60\%^B$	–/–	[156]
TiO_2-SnO_2	DCM	250	20	$30\%^B$	–/–	[159]
SnO_2-Ni-Sb-Mo	DMMP	350	0.5	$70\%^B$	–/–	[160]
ZnO-Pt	DMMP	300	2	240^A	15/45	[161]
WO_3-CuO	DMMP	25	10	19.3^A	9.2/15.5	[162]

A: (R_g/R_a) or (R_a/R_g), B: $(R_g - R_a)/R_a$, %.

as SnO_2, WO_3, In_2O_3, CuO, and Y_2O_3 for detecting CWAs simulants. The sensors showed an excellent sensing performance in terms of sensitivity, stability, and response speed toward simulants. And the sensors based on WO_3, In_2O_3, and CuO were tested with CWAs at certified live agent facility, showing a reliable detection ability toward sarin, soman, and mustard gas at 10 ppb and higher concentrations. Yoo et al. [155] prepared Al-doped ZnO nanoparticles for DMMP detection. The composites significantly responded (4347) toward 10 ppm DMMP at 350 °C, showing fast response speed (2 s) and a low detection limit of 100 ppb. Choi et al. [158] employed a MEMS-structure sensor for the detection of CWA simulant acetonitrile (CH_3CN). The Al_2O_3-SnO_2 sensor exhibited an excellent sensing performance toward 0.5–30 ppm CH_3CN at 350 °C. Lee et al. [156] proposed a SnO_2-based gas sensor to detect dichloromethane (DCM) as low as 0.1 ppm Similarly, Park et al. [159] also proposed a SnO_2-based gas sensor to detect DCM as low as 0.1 ppm Lee et al. [160] proposed SnO_2-based composites to sense 0.5 ppm DMMP at 350 °C. Patil et al. [161] proposed DMMP gas sensors based on 0.12 wt% Pt-ZnO composites at 300 °C. Alali et al. [162] proposed *n–p* WO_3/CuO nanofibers as the sensing material to recognize sarin nerve agents at 25 °C. The composites achieved an excellent sensing performance to DMMP at room temperature compared with pure CuO and WO_3, showing a high response of 19.3 to 10 ppm DMMP with fast response/recovery time of 9.2 s/15.5 s, as well as a remarkable response of 6.2 to 0.5 ppm DMMP, good long-term stability and repeatability, and a high reliability. Figure 5.18 illustrates the detailed sensing mechanism for DMMP detection.

Most of the current laboratory experiments aim at the detection of simulants. Although the molecular structure is very similar between simulants and CWA, the electronic properties probably differentiate to a large extent that poses a significant effect on the sensor response. To verify the effectiveness of the gas-sensing material system, further on-site experiments under real environments are meaningful under the guidelines of the professional institutes.

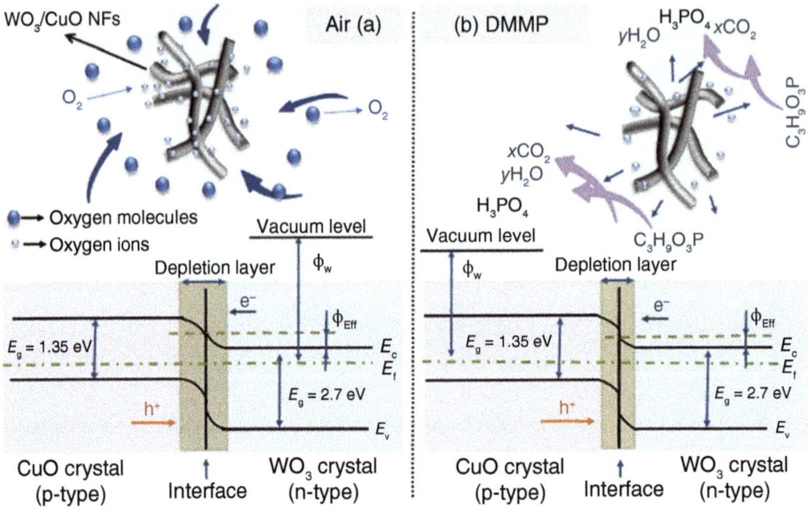

Figure 5.18 Schematic diagrams showing the reaction of WO_3/CuO NFs with oxygen and DMMP molecules in (a) air and (b) DMMP atmosphere. Illustration diagrams of the energy band gap in each case are shown in the figure down. Φ_{eff} is the potential junction barrier on the surface of crystals, E_c is the lower level of the conduction band gap, E_f is the Fermi level, Φ_w is work function, and E_v is the higher level of the conduction band gap. Source: Reproduced from Alali et al. [162]/with permission of Elsevier.

5.3 Metal Oxide-Based Pressure Sensors

Pressure sensors are playing an ever-increasingly important role in various application scenarios, such as electronic skins, health monitoring, tactile displays, and wearable systems [163–165]. Recently, pressure sensors based on MOS have attracted growing attention [166].

5.3.1 Working Mechanisms and Performance Characterizations

According to the difference of transduction mechanisms, pressure sensors based on metal oxides could be mainly classified into piezoresistive, capacitive, and piezoelectric pressure sensors, as shown in Figure 5.19.

For piezoresistive pressure sensors based on metal oxides, the basic mechanism is piezoresistive effect [168, 169]. Namely, the applied strain would lead to energy band variation of metal oxide, leading to the change of resistivity [170]. The piezoresistive pressure sensors convert the external mechanical stimuli into electrical signals based on the piezoresistive effect, as presented in Figure 5.19a. The sensor resistance variation ratios ($[R_m - R_0]/R_0 = \Delta R/R_0$) are used to calculate the piezoresistive response of the device, where R_0 and R_m indicate steady resistance of sensor at presence and absence of strain, respectively. The sensitivity is calculated by $S = \Delta(\Delta R/R_0)/\Delta p$, where p is the pressure applied on sensor and Δp represents

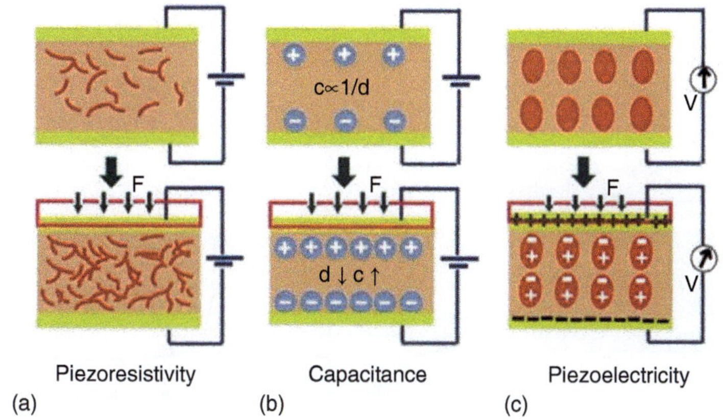

Figure 5.19 Schematic illustrations of three common transduction mechanisms and representative devices. (a) Piezoresistivity, (b) capacitance, and (c) piezoelectricity. Source: Reproduced with permission from Xu et al. [167]. Copyright 2018 MDPI/Public Domain CC BY 4.0.

the variation of pressure. Yang et al. [168] proposed a flexible piezoresistive strain sensor based on single Sb-doped ZnO nanobelt. The device used a two-end bonded Sb-doped ZnO nanobelt on a flexible polystyrene substrate (Figure 5.20a–c). Such sensors showed a decrease of resistance with the increase of compressed strains in the nanobelt (Figure 5.20d,f). The longitudinal piezoresistance coefficient of the Sb-doped ZnO nanobelt is about 350 (Figure 5.20e), available for detecting the strains in a signature pen.

As for capacitive pressure sensors, the typical structure contains a dielectric material sandwiched between the two parallel metal conducting plates [171]. When applying an external force, the distance between plate capacitors would be altered, leading to capacitance changing, as exhibited in Figure 5.19b. The performance of capacitive sensor is highly dependent on the dielectric material. The pressure response is defined as the capacitance variation ratio (($[C_m - C_0]/C_0 = \Delta C/C_0$), where C_m and C_0 are the capacitance under presence and absence of pressure, respectively). The sensitivity (S) is calculated from the capacitance variation ratios as a function of the applied pressure (p), namely, $S = \Delta(\Delta C/C_0)/\Delta p$. Kumaresan et al. [172] proposed soft capacitive pressure sensors with an enhanced sensitivity assisted by ZnO nanowires interlayers. Compared with pure polydimethylsiloxane (PDMS) pressure sensors, the performance of capacitive pressure sensors was significantly improved by the addition of ZnO nanowire interlayers. At a pressure range of 1–10 kPa, PDMS-ZnO sensors showed a sensitivity more than 7 times higher than pure PDMS sensors (5.6452% kPa^{-1} vs. 0.81% kPa^{-1}). This enhancement could be attributed to the piezoelectric effects introduced by the ZnO interlayer, which increased the dielectric constant of interlayer materials under applied pressures.

Regarding piezoelectric pressure sensors, when applying an external strain, metal oxides create a piezoelectric potential inside the crystal due to the noncentral symmetric crystal structure (Figure 5.19c). The inner-crystal piezopotential could

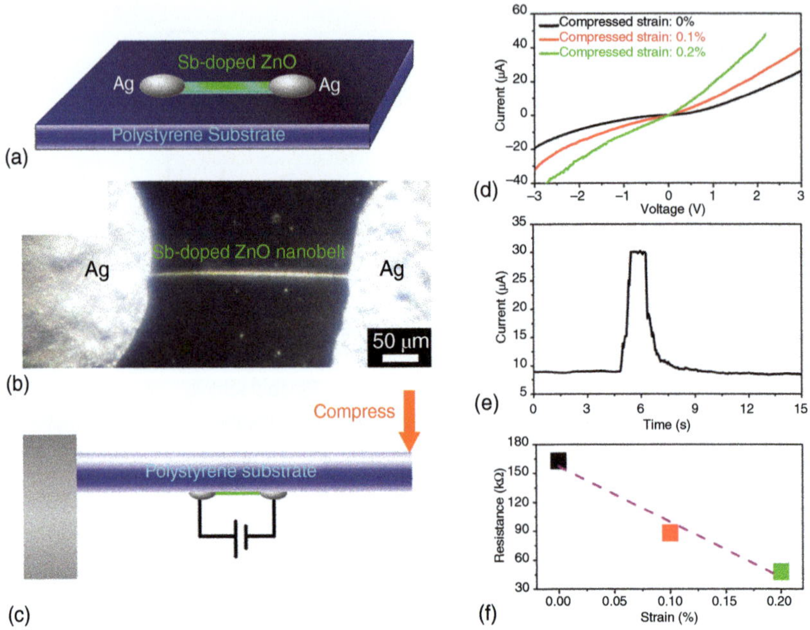

Figure 5.20 (a) A schematic and (b) an optimal image showing a single Sb-doped ZnO nanobelt fixed by silver paste on a PS substrate. (c) Schematic of the measurement system to characterize the electromechanical property of a single Sb-doped ZnO nanobelt. (d) Typical *I–V* curves of the Sb-doped ZnO nanobelt under different compressed stains. (e) Current response of the single Sb-doped ZnO nanobelt to the compressed stain (0.2%) under a voltage of 1.5 V. (f) A plot showing the relationship between the resistance of Sb-doped ZnO nanobelt and the compressed strains under a voltage of 1.5 V. The dashed curve is accomplished by fitting three data points. Source: Reproduced from Yang et al. [168]/with permission of AIP Publishing.

tune the charge generation, separation, transportation, and recombination, thus changing the electrical properties of metal oxide devices [173]. Generally, the pressure responses of the piezoelectric devices are defined from the current variation ratio (($[I_m - I_0]/I_0 = \Delta I/I_0$), where I_m and I_0 are the current under presence and absence of pressure, respectively). Moreover, the sensitivity (S) is calculated from the electronic current variation ratios as a function of the applied pressure (p), namely, $S = \Delta(\Delta I/I_0)/\Delta p$. Park et al. [174] proposed highly sensitive and flexible pressure sensors using position- and dimension-controlled ZnO nanotube arrays. The measurement setup is exhibited schematically in Figure 5.21a. Figure 5.21b shows the I–V curves of ZnO nanotube sensors under different applied pressures, showing a decrease of device current with the increase of applied pressure. This could be attributed to a decline in conductance, originating from a piezo-induced electric field that formed across the width of ZnO under applied pressure. The current variation under different applied pressures with a constant bias voltage of 0.4 V is shown in Figure 5.21c. Figure 5.21d shows the corresponding linear relationship between pressure response and applied pressure, from which a sensitivity of $-4.4\ \text{kPa}^{-1}$ can be obtained.

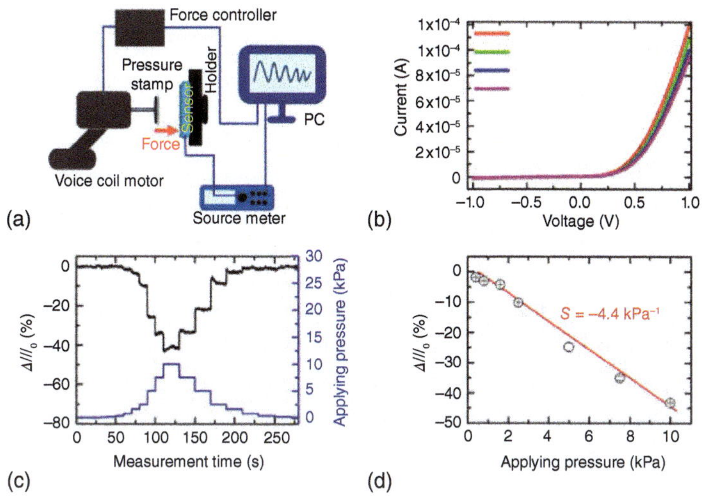

Figure 5.21 (a) A schematic diagram of the measurement system used to characterize the performance of the ZnO nanotube pressure sensors, with pressure being applied by a VCM. (b) Typical I–V characteristics of the sensors under different applied pressures. (c) Real-time current response of ZnO nanotube pressure sensors under a fixed bias of 0.4 V and with different applied pressures. (d) Current varying ratios of ZnO nanotube pressure sensors under different applied pressures. Source: Reproduced from Park et al. [174]/with permission of Springer Nature/Public Domain CC BY 4.0.

5.3.2 ZnO-Based Pressure Sensors

As a famous piezoelectric material, ZnO is widely investigated for the pressure sensors. Chey et al. [175] proposed a fast piezoresistive sensor based on Mn-doped ZnO nanorods. Under on/off states of a mass at 3 V forward bias, the piezoresistive sensor showed a good reliability with relatively fast response and recovery time of ~0.03 and ~0.27 s, respectively. The linear relationship between $\Delta R/R_0$ and applied pressure can be divided into two parts, including a linear range up to 20 kPa with a pressure sensitivity of 0.00617 kPa^{-1} and a linear range from 20 to 320 kPa with a pressure sensitivity of 0.000180 kPa^{-1}. Ferreirra et al. [169] reported pressure sensors based on nanostructured ZnO/Ag thin films, which showed a high piezoresistive response. Zigzag nanostructured silver (Ag, 23 at%) doped ZnO films exhibited a piezoelectric sensitivity of 0.25 with a z-deformation of 0.5 mm. Tan et al. [173] proposed high-performance textile piezoelectric pressure sensors with a novel structural hierarchy based on ZnO nanorod arrays for wearable applications. The sensor exhibited a low detection limit of 8.71 Pa, a superior mechanical stability, and a high sensitivity of 0.62 kPa^{-1} in the pressure range of 0–25 kPa, qualified to detect human movements such as bending/relaxation motion of the wrist, bending/stretching motion of each finger. Huo et al. [176] reported Sb-doped ZnO nanowire films for strain sensors. The sensor could not only detect the motion of each finger individually and accurately, but also evaluate change in gestures based on the output signal. Pan et al. [177] reported a high-resolution electroluminescent imaging of pressure distribution using a piezoelectric nanowire LED array. The device was based on a patterned array of *n*-ZnO nanowires grown on a *p*-GaN thin-film substrate. And the nanowire light-emitting diode-based pressure sensor array can map two-dimensional distributions of strain with an unprecedented spatial resolution of 2.7 mm, corresponding to a pixel density of 6350 dpi. Wang et al. [178] proposed a piezoelectric pressure sensor constituted by flexible ZnO nanorod (NR) arrays and a 2D indium selenide (InSe)-based field-effect transistor (FET). The ZnO NR arrays transformed stress into piezoelectric voltage, contributing to the gate potential. The piezoelectric voltage was effectively amplified by the InSe-based FET, which had high field-effect mobility and electrical stability. The loading pressure on ZnO was analyzed by measuring the current variation of the InSe-based FET. The pressure sensor could sense a minimum load of 0.1 g corresponding to a potential of 0.2 mV.

5.3.3 Other Pressure Sensors

In addition to ZnO, other metal oxides such as SnO_2 and TiO_2, also exhibited a high sensitivity to pressure [179]. For example, Korostynska et al. investigated the real-time pressure sensing performance of SnO_2, TiO_2, and TiO_2/ZnO_2 thick-film capacitors with interdigitated electrodes [180]. At an applied pressure of 5 kPa, SnO_2, TiO_2, and TiO_2/ZnO_2 thick-film capacitors showed short response time of 2.5 s, 5.6 s, and 4 s, respectively. All three sensors also exhibited less than 18.4% repeatability error by constant loading and unloading over 25 cycles [181]. Thereinto, PDMS/Ag composites were served as a flexible substrate and dandelion-like rutile TiO_2 nanorods were grown on the substrate by a hydrothermal method. The sensitivity of the piezoelectric TiO_2 pressure sensors was 2.56×10^{-5} V/N. Wang et al. [182] proposed a novel pressure sensor based on Fe_2O_3/SnO_2 heterostructures. The piezoresistive pressure sensor exhibited a high sensitivity of 680 kPa^{-1}, fast response time of 10 ms, a wide pressure range of 0–150 kPa, and a good reproducibility over 3500 cycles under a pressure of 110 kPa. They also reported highly sensitive pressure sensors based on laser-induced graphene and sea urchin-shaped Fe_2O_3 sandwiched structure [183]. Thereinto, the sea urchin-like Fe_2O_3 served as a space layer between two laser-induced graphene/polyimide films. The piezoresistive sensor exhibited a high sensitivity of 603 kPa^{-1} for a pressure range of 0–10 kPa. Liao et al. [184] proposed a solid-phase sintering and vapor–liquid–solid growth method for preparing quantum-size oxide materials, i.e. black phosphorus MgO quantum dots (QDs) crystals with a high piezoelectric response. MgO QDs can control the piezoelectric effect on the nanoscale and promote the compatibility between piezoelectric materials and nanodevices. The prepared BP@MgO QDs crystal films with a thickness of 7.5 μm exhibited clear output voltage signals when the sensors were applied with a dynamic finger pressure of 9–11 N.

Currently, the thickness of the sensing film for pressure sensing is usually located at the scale of hundreds of micrometers or more, which was very unfavorable for the device miniaturization and system portability. Moreover, pressure sensors relying on the deformation of conducting paths readily suffer from low sensitivity. It is still quite challenging to develop high-performance pressure sensors with a small device size and a high sensitivity. In this regard, nanomaterial engineering and/or novel device structure modification via MEMS or NEMS techniques are appealing research fields in the future.

5.4 Metal Oxide-Based pH Sensors

PH is a crucial parameter in various technologies, such as clinical diagnosis, pollution monitoring, and industrial and agricultural production [185, 186]. Hence, it is urgent to develop accurate, reliable, and highly sensitive pH sensors. Glass is a conventional sensitive electrode material for pH measurement. However, the mechanical fragility hinders its further applications [187]. Recently, metal oxides-based pH sensors have attracted considerable attention due to their excellent pH-sensing performance, such as high sensitivity, fast response time, nice long-term stability, easy miniaturization, and integration [188].

5.4.1 Working Mechanisms and Performance Characterizations

Metal oxides-based pH sensors could be mainly put into three categories according to their working mechanisms, including potentiometric, ion-sensitive field-effect transistor (ISPET), and chemiresistive/conductimetric pH sensors [186]. Among them, potentiometry is the most preferred strategy [189]. In this method, the potential difference between a sensitive electrode and a reference electrode is measured to detect the pH value of test solution. The reference electrode (RE) with Ag/AgCl is commonly used, which provides a stable constant reference potential. And thin or thick films of metal oxides are used as the sensitive electrode (SE) to respond to the change of pH. When both electrodes are immersed into the solution, the electrical potential difference between them is measured to determine the H^+ concentration in the solution, as shown in Figure 5.22a. Two possible mechanisms for potentiometric pH sensors are widely accepted: redox reactions and ion-selective permeation. In the former one, reversible chemical reactions will produce a potential difference by the free energy change if the metal oxide on sensitive electrode has redox reaction with H_3O^+. As for the later one, if the sensitive material serves as an ion-selective membrane, the concentration gradient of ions across the membrane would also generate a potential difference. The sensitivity of potentiometric-based sensors is determined by the potential difference between SE and RE immersed in a solution according to the Nernstian relation. If the sensitivity is close to the ideal Nernstian response (59.14 mV pH^{-1} at 25 °C derived from the Nernst equation), the sensor is anticipated to obtain excellent performance [190].

Chemiresistive/conductimetric pH sensor is one two-terminal device with the sensitive material deposited between two electrodes without RE, as shown in Figure 5.22b. The intrinsic resistance of sensing material would change with the change of H_3O^+ concentration in test solutions. The chemical reaction between H_3O^+/OH^- with sensing materials or the change induced by H_3O^+ in the depletion/accumulation of metal oxides could account for the resistance variation in sensors.

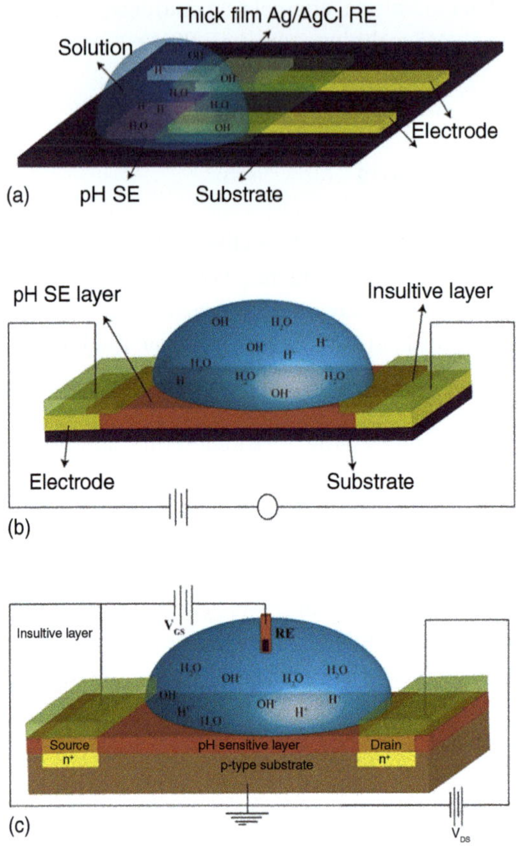

Figure 5.22 Schematic diagrams of three pH sensors. (a) Potentiometric pH sensor, (b) chemiresistive pH sensor, and (c) ISFET-based pH sensor. Source: Reproduced from Manjakkal et al. [187]/with permission of Elsevier.

Ion-sensitive field-effect transistor (ISFET)-type pH sensors are operated similarly to metal oxide field effect transistors (MOSFETs), as shown in Figure 5.22c. The gate is covered with an ion-sensitive layer and is placed between the source and drain electrodes. The drain current is controlled by the electrostatic field generated by the gate.

The performance of pH sensors is evaluated by their sensitivity, response time, drift rate, and hysteresis [188]. The sensitivity is determined by the linear slope of response in sensors as a function of pH. The response time is defined as the time required for electrical signal reaching 90% of an equilibrium value after immersing the sensor in test solutions. The potential drift is defined as the difference between the peak potential value and the 90% value of the saturated potential. And hysteresis is defined as the variation of potential of sensors in a test solution at the same pH level after several measurements.

5.4.2 Potentiometric PH Sensors

Potentiometric pH sensor is the most utilized device among pH sensors owing to the advantages of easy fabrication, low cost, miniaturization, and precise and reliable operation [191]. Thin or thick films of metal oxides have been widely utilized as sensitive electrodes. Tsai et al. prepared SnO_2 thin films by radio frequency sputtering for pH potentiometric sensors [192]. The sensitivity of SnO_2 pH sensors was 59 mV/pH in a PH range of 2–12. The hysteresis experiments were carried out under short- and long-loop circles, pH 7–4–7–10–7 and pH 7–2–7–12–6, respectively. The hysteresis value was less than 7 mV in short-loop circle and the hysteresis value of pH 7 was 9 mV in acid side (HA) and 11 mV in alkali side (HB). Manjakkal et al. [193] developed thick-film pH sensors based on the nanostructured RuO_2-SnO_2 composites. The potentiometric sensor showed a high sensitivity of 56.5 mV pH^{-1} in a pH range of 2–12, short response time (5 s/9 s) for acidic to basic and basic to acidic solutions), small hysteresis level of 2 and 7 mV in acidic solutions and basic solutions, respectively. Özdemir et al. [194] proposed a flexible wearable potentiometric sensor based on TiO_2-coated cotton for noninvasive pH monitoring. The sensor showed a linear potentiometric response in a pH range of 2 to 9 with a sensitivity of −23.9 mV pH^{-1}. Santos et al. [195] prepared WO_3 nanoparticle-based flexible pH sensors. The sensor exhibited a high sensitivity of −56.7 mV pH^{-1} in a wide pH range of 5 to 9 and low hysteresis values of 6–10 mV. The nanostructure of metal oxide has a significant effect on sensitivity. Fulati et al. [196] reported miniaturized potentiometric pH sensors based on ZnO nanotubes/nanorods. The sensitivity of ZnO nanotubes sensors was −45.9 mV pH^{-1}, higher than that of ZnO nanotubes sensors (−28.4 mV pH^{-1}). The promoted sensitivity could be attributed to the larger effective surface area with a higher surface-to-volume ratio of ZnO nanotubes than nanorods. Zaman et al. [197] synthesized well-crystallized flower-shaped cupric oxide (CuO) nanostructures composed of thin leaves by a simple low-temperature chemical bath method for pH sensors. The sensors exhibited an excellent sensing behavior, including a high sensitivity of 28 mV pH^{-1} with a good linearity in a wide pH range of 2 to 11.

5.4.3 Ion-Sensitive Field-Effect Transistor PH Sensors

ISFETs have attracted ever-growing attention in chemical and biomedical measurements with unique merits including miniaturization, fast response speed, and low cost [198]. Huang et al. [199] prepared TiO_2 nanowire (diameter: ~100 nm, length: ~1.8 μm) arrays by a hydrothermally grown method as the sensing electrode for an extended-gate-field effect-transistor PH sensor. The sensor exhibited a high sensitivity of 62 mV pH^{-1} in a wide pH range of 2–12. Lee et al. [200] proposed integrated ISFET pH sensors based on ZnO. The sensor showed a high performance including a dynamic response of 0.55 μA pH^{-1}, a sensitivity of 42.13 mV pH^{-1}, and response time of less than 12 s. Besides, the study verified that the photoelectrochemical (PEC) passivation could improve the sensing performance of ZnO-based ISFET pH sensors, exhibiting an enhanced dynamic response of 0.67 μA pH^{-1} and a promoted sensitivity of 48.27 mV pH^{-1}. Chin et al. [201] reported an integrated ISFET pH sensor based on SnO_2/Al, fabricated by a feasible standard CMOS IC process. Such sensors exhibited a high linearity of 58 mV pH^{-1} in a wide pH range of 2–10. Mishra et al. [202] proposed CuO nanowire-based extended gate FETs for PH sensing.

The CuO nanowires electrode-based sensors exhibited an excellent sensing performance including a high sensitivity of 48.34 mV pH^{-1}, a good linearity, a remarkable stability, a wide detection pH range of 2–12, and a nice reversibility with low hysteresis loss of 2.5 mV. Moreover, the sensor showed a significantly higher sensitivity toward H^+ than Na^+, K^+, Zn^{++}, and Mg^{++} ions, demonstrating a good selectivity. Khizir et al. [203] synthesized TiO_2 nanorods as sensing membrane for extended gate FET pH sensing. The TiO_2 nanorods were grown onto fluorine-doped tin oxide (FTO) coated glass substrates at a hydrothermal reaction temperature of 170 °C for 6 h. The TiO_2 nanorods sensors exhibited a high sensitivity of 78.25 mV pH^{-1}, a nice linearity of 99.27% in a wide sensing range of pH 2–12, a good stability, and a reliability with low hysteresis value of 9.1 mV. Bae et al. [204] proposed high-performance graphene FETs with a solution-processed Al_2O_3 membrane, offering a sensitivity of 36.8 mV pH^{-1} and a nice linearity of 97.32% in the pH range of 4–10. In addition, a small hysteresis of 17.3 mV and a tiny drift rate of 6.34 mV h^{-1} were achieved.

5.4.4 Chemiresistive/Conductimetric pH Sensors

The chemiresistive/conductimetric method is another attractive method to detect pH due to the unique merits such as facile operation and affordable cost. Arshak et al. proposed SnO_2 as sensing materials in conductimetric pH sensors, which exhibited a wide pH detection range of 2–11 [205]. Lee et al. [206] synthesized hierarchically structured suspended TiO_2 nanofibers for pH sensor devices. The TiO_2-based chemiresistive pH sensors showed a linear decrease of the conductance value with a resolution of 5.68 ± 0.28 nS pH^{-1} when increasing the pH from 2 to 12, Simić et al. [207] proposed an interdigitated conductimetric pH sensor using screen printing of TiO_2 thick films on an alumina substrate. The impedance spectroscopic studies of the fabricated thick-film sensors were carried out in a frequency range of 5–20 kHz for the test solutions at a pH range of 4–10. The relationship between impedance and pH exhibited a nice linearity. Manjakkal et al. [208] reported conductimetric interdigitated thick-film pH sensors based on RuO_2. Impedance spectroscopic studies in the frequency range of 10–2 MHz revealed that the response was distinctly dependent on the concentration of H^+ or OH^- ions adsorbed on the surface of the sensitive electrodes. Copal et al. [209] reported a resistance-based pH sensor using ZnO nanorods. The ZnO nanorods were grown on interdigitated finger electrodes via a low temperature chemical bath deposition, which showed a linear decrease of resistance with the increase of pH levels from 4 to 7. Chen et al. [210] reported a wide range of conductometric pH sensors based on TiO_2/MWCNTs/cellulose nanocomposites. The conductivity of such composites sensors linearly varied depending on the pH value with two parts: high sensitivity region between pH of 1–6 and low one between pH of 7–12.

Even though huge advances have been obtained in the MOS-based pH sensors, some challenges remain especially regarding the various application scenarios such as biomedical, food, and water quality monitoring [211]. The primary research efforts should focus on wearable, portable, and miniaturized sensors for online monitoring applications.

References

1 Nunes, D., Pimentel, A., Gonçalves, A. et al. (2019). Metal oxide nanostructures for sensor applications. *Semicond. Sci. Technol.* 34 (4): 043001.

2 Kim, H. and Lee, J. (2014). Highly sensitive and selective gas sensors using p-type oxide semiconductors: overview. *Sens. Actuators, B* 192: 607–627.

3 Moseley, P. (2017). Progress in the development of semiconducting metal oxide gas sensors: a review. *Meas. Sci. Technol.* 28 (8): 082001.

4 Zhang, J., Qin, Z., Zeng, D. et al. (2017). Metal-oxide-semiconductor based gas sensors: screening, preparation, and integration. *Phys. Chem. Chem. Phys.* 19 (9): 6313–6329.

5 Barsan, N., Koziej, D., and Weimar, U. (2007). Metal oxide-based gas sensor research: how to? *Sens. Actuators, B* 121 (1): 18–35.

6 Barsan, N. and Weimar, U. (2001). Conduction model of metal oxide gas sensors. *J. Electroceram.* 7: 143–167.

7 Zhou, P., Shen, Y., Lu, W. et al. (2020). Highly selective NO_2 chemiresistive gas sensor based on hierarchical In_2O_3 microflowers grown on clinoptilolite substrates. *J. Alloys Compd.* 828: 154395.

8 Xu, C., Tamaki, J., Miura, N. et al. (1991). Grain size effects on gas sensitivity of porous SnO_2-based elements. *Sens. Actuators, B* 3 (2): 147–155.

9 Tamaki, J., Zhang, Z., Fujimori, K. et al. (1994). Grain-size effects in tungsten oxide-based sensor for nitrogen oxides. *J. Electrochem. Soc.* 141 (8): 2207–2210.

10 Gurlo, A., Ivanovskaya, M., Bârsan, N. et al. (1997). Grain size control in nanocrystalline In_2O_3 semiconductor gas sensors. *Sens. Actuators, B* 44 (1–3): 327–333.

11 Rothschild, A. and Komem, Y. (2004). The effect of grain size on the sensitivity of nanocrystalline metal-oxide gas sensors. *J. Appl. Phys.* 95 (11): 6374–6380.

12 Wang, Y., Gao, R., Zhao, H. et al. (2023). Oxygen vacancy-rich ZnO nanorods-based MEMS sensors for swift trace ethanol recognition. *J. Am. Ceram. Soc.* 106 (2): 1050–1061.

13 Yamazoe, N., Sakai, G., and Shimanoe, K. (2003). Oxide semiconductor gas sensors. *Catal. Surv. Asia* 7: 63–75.

14 Yang, X., Fu, H., Tian, Y. et al. (2019). Au decorated In_2O_3 hollow nanospheres: a novel sensing material toward amine. *Sens. Actuators, B* 296: 126696.

15 Pan, C. and Ma, T. (1980). Work function of In_2O_3 film as determined from internal photoemission. *Appl. Phys. Lett.* 37 (8): 714–716.

16 Liu, Y., Xiao, S., and Du, K. (2021). Chemiresistive gas sensors based on hollow heterojunction: a review. *Adv. Mater. Interfaces* 8 (12): 2002122.

17 Liu, Y., Li, G., Mi, R. et al. (2014). An environment-benign method for the synthesis of p-NiO/n-ZnO heterostructure with excellent performance for gas sensing and photocatalysis. *Sens. Actuators, B* 191: 537–544.

18 Xu, H., Liu, X., Cui, D. et al. (2006). A novel method for improving the performance of ZnO gas sensors. *Sens. Actuators, B* 114 (1): 301–307.

19 Ju, D., Xu, H., Qiu, Z. et al. (2014). Highly sensitive and selective triethylamine-sensing properties of nanosheets directly grown on ceramic tube by forming NiO/ZnO PN heterojunction. *Sens. Actuators, B* 200: 288–296.

20 Du, W., Si, W., Du, W. et al. (2020). Unraveling the promoted nitrogen dioxide detection performance of N-doped SnO_2 microspheres at low temperature. *J. Alloys Compd.* 834: 155209.

21 Xiao, T., Wang, X., Zhao, Z. et al. (2014). Highly sensitive and selective acetone sensor based on C-doped WO_3 for potential diagnosis of diabetes mellitus. *Sens. Actuators, B* 199: 210–219.

22 Kumar, V., Sen, S., Muthe, K. et al. (2009). Copper doped SnO_2 nanowires as highly sensitive H_2S gas sensor. *Sens. Actuators, B* 138 (2): 587–590.

23 Shen, J., Li, F., Yin, B. et al. (2017). Enhanced ethyl acetate sensing performance of Al-doped In_2O_3 microcubes. *Sens. Actuators, B* 253: 461–469.

24 Zhou, Q., Chen, W., Xu, L. et al. (2018). Highly sensitive carbon monoxide (CO) gas sensors based on Ni and Zn doped SnO_2 nanomaterials. *Ceram. Int.* 44 (4): 4392–4399.

25 Wang, Z., Hou, C., De, Q. et al. (2018). One-step synthesis of Co-doped In_2O_3 nanorods for high response of formaldehyde sensor at low temperature. *ACS Sens.* 3 (2): 468–475.

26 Mardare, D., Cornei, N., Mita, C. et al. (2016). Low temperature TiO_2 based gas sensors for CO_2. *Ceram. Int.* 42 (6): 7353–7359.

27 Zhao, J., Yang, T., Liu, Y. et al. (2014). Enhancement of NO_2 gas sensing response based on ordered mesoporous Fe-doped In_2O_3. *Sens. Actuators, B* 191: 806–812.

28 Liu, X., Jiang, L., Jiang, X. et al. (2018). Synthesis of Ce-doped In_2O_3 nanostructure for gas sensor applications. *Appl. Surf. Sci.* 428: 478–484.

29 Jeong, Y., Balamurugan, C., and Lee, D. (2016). Enhanced CO_2 gas-sensing performance of ZnO nanopowder by La loaded during simple hydrothermal method. *Sens. Actuators, B* 229: 288–296.

30 Degler, D., Weimar, U., and Barsan, N. (2019). Current understanding of the fundamental mechanisms of doped and loaded semiconducting metal-oxide-based gas sensing materials. *ACS Sens.* 4 (9): 2228–2249.

31 Zhang, Q., Xie, G., Xu, M. et al. (2018). Visible light-assisted room temperature gas sensing with ZnO-Ag heterostructure nanoparticles. *Sens. Actuators, B* 259: 269–281.

32 Zhou, Y., Gao, C., and Guo, Y. (2018). UV assisted ultrasensitive trace NO_2 gas sensing based on few-layer MoS_2 nanosheet-ZnO nanowire heterojunctions at room temperature. *J. Mater. Chem. A* 6 (22): 10286–10296.

33 Xie, T., Sullivan, N., Steffens, K. et al. (2015). UV-assisted room-temperature chemiresistive NO_2 sensor based on TiO_2 thin film. *J. Alloys Compd.* 653: 255–259.

34 Jia, Q., Ji, H., Zhang, Y. et al. (2014). Rapid and selective detection of acetone using hierarchical ZnO gas sensor for hazardous odor markers application. *J. Hazard. Mater.* 276: 262–270.

35 Wang, X., Chen, F., Yang, M. et al. (2019). Dispersed WO_3 nanoparticles with porous nanostructure for ultrafast toluene sensing. *Sens. Actuators, B* 289: 195–206.

36 Liu, S., Xie, M., Li, Y. et al. (2010). Novel sea urchin-like hollow core-shell SnO_2 superstructures: Facile synthesis and excellent ethanol sensing performance. *Sens. Actuators, B* 151 (1): 229–235.

37 Gupta, C.S., Chatterjee, S., Ray, A. et al. (2015). Graphene-metal oxide nanohybrids for toxic gas sensor: a review. *Sens. Actuators, B* 221: 1170–1181.

38 Sakai, G., Matsunaga, N., Shimanoe, K. et al. (2001). Theory of gas-diffusion controlled sensitivity for thin film semiconductor gas sensor. *Sens. Actuators, B* 80 (2): 125–131.

39 Baschirotto, A., Capone, S., Damico, A. et al. (2008). A portable integrated wide-range gas sensing system with smart A/D front-end. *Sens. Actuators, B* 130 (1): 164–174.

40 Röck, F., Barsan, N., and Weimar, U. (2008). Electronic nose: current status and future trends. *Chem. Rev.* 108 (2): 705–725.

41 Chakraborty, S., Sen, A., and Maiti, H. (2006). Selective detection of methane and butane by temperature modulation in iron doped tin oxide sensors. *Sens. Actuators, B* 115 (2): 610–613.

42 Nakata, S., Okunishi, H., and Nakashima, Y. (2006). Distinction of gases with a semiconductor sensor under a cyclic temperature modulation with second-harmonic heating. *Sens. Actuators, B* 119 (2): 556–561.

43 Hoa, N., Duy, N., El-Safty, S. et al. (2015). Meso–/nanoporous semiconducting metal oxides for gas sensor applications. *J. Nanomater.* 16 (1): 1–14.

44 Tiemann, M. (2007). Porous metal oxides as gas sensors. *Chem. A Eur. J.* 13 (30): 8376–8388.

45 Lin, T., Lv, X., Hu, Z. et al. (2019). Semiconductor metal oxides as chemoresistive sensors for detecting volatile organic compounds. *Sensors* 19 (2): 233.

46 Wang, H., Nie, L., Li, J. et al. (2013). Characterization and assessment of volatile organic compounds (VOCs) emissions from typical industries. *Chin. Sci. Bull.* 58: 724–730.

47 Boeglin, M., Wessels, D., and Henshel, D. (2006). An investigation of the relationship between air emissions of volatile organic compounds and the incidence of cancer in Indiana counties. *Environ. Res.* 100 (2): 242–254.

48 Leonardi, S., Mirzaei, A., Bonavita, A. et al. (2016). A comparison of the ethanol sensing properties of alpha-iron oxide nanostructures prepared via the sol-gel and electrospinning techniques. *Nanotechnology* 27 (7): 075502.

49 Wan, Q., Li, Q., Chen, Y. et al. (2004). Fabrication and ethanol sensing characteristics of ZnO nanowire gas sensors. *Appl. Phys. Lett.* 84 (18): 3654–3656.

50 Wang, L., Kang, Y., Liu, X. et al. (2012). ZnO nanorod gas sensor for ethanol detection. *Sens. Actuators, B* 162 (1): 237–243.

51 Rao, B. (2000). Zinc oxide ceramic semi-conductor gas sensor for ethanol vapour. *Mater. Chem. Phys.* 64 (1): 62–65.

52 Saito, N., Watanabe, K., Haneda, H. et al. (2018). Highly sensitive ethanol gas sensor using pyramid-shaped ZnO particles with (0001) Basal Plane. *J. Phys. Chem. C* 122 (13): 7353–7360.

53 Zhang, Y., He, X., Li, J. et al. (2008). Fabrication and ethanol-sensing properties of micro gas sensor based on electrospun SnO_2 nanofibers. *Sens. Actuators, B* 132 (1): 67–73.

54 Wang, G., Park, J., Park, M. et al. (2008). Synthesis and high gas sensitivity of tin oxide nanotubes. *Sens. Actuators, B* 131 (1): 313–317.

55 Kwon, Y., Kim, H., Lee, S. et al. (2012). Enhanced ethanol sensing properties of TiO_2 nanotube sensors. *Sens. Actuators, B* 173: 441–446.

56 Wang, M., Zhu, Y., Meng, D. et al. (2020). A novel room temperature ethanol gas sensor based on 3D hierarchical flower-like TiO_2 microstructures. *Mater. Lett.* 277: 128372.

57 Hu, P., Du, G., Zhou, W. et al. (2010). Enhancement of ethanol vapor sensing of TiO_2 nanobelts by surface engineering. *ACS Appl. Mater. Interfaces* 2 (11): 3263–3269.

58 Xiao, L., Xu, S., Yu, G. et al. (2018). Efficient hierarchical mixed Pd/SnO_2 porous architecture deposited microheater for low power ethanol gas sensor. *Sens. Actuators, B* 255: 2002–2010.

59 Xu, X., Fan, H., Liu, Y. et al. (2011). Au-loaded In_2O_3 nanofibers-based ethanol micro gas sensor with low power consumption. *Sens. Actuators, B* 160 (1): 713–719.

60 Ge, W., Chang, Y., Natarajan, V. et al. (2018). In_2O_3-SnO_2 hybrid porous nanostructures delivering enhanced formaldehyde sensing performance. *J. Alloys Compd.* 746: 36–44.

61 Gu, C., Cui, Y., Wang, L. et al. (2017). Synthesis of the porous NiO/SnO_2 microspheres and microcubes and their enhanced formaldehyde gas sensing performance. *Sens. Actuators, B* 241: 298–307.

62 Kawamura, K., Kerman, K., Fujihara, M. et al. (2005). Development of a novel hand-held formaldehyde gas sensor for the rapid detection of sick building syndrome. *Sens. Actuators, B* 105 (2): 495–501.

63 Li, Z., Fan, Y., and Zhan, J. (2010). In_2O_3 nanofibers and nanoribbons: preparation by electrospinning and their formaldehyde gas-sensing properties. *Eur. J. Inorg. Chem.* 2010 (21): 3348–3353.

64 Li, Y., Chen, N., Deng, D. et al. (2017). Formaldehyde detection: SnO_2 microspheres for formaldehyde gas sensor with high sensitivity, fast response/recovery and good selectivity. *Sens. Actuators, B* 238: 264–723.

65 Lin, S., Li, D., Wu, J. et al. (2011). A selective room temperature formaldehyde gas sensor using TiO_2 nanotube arrays. *Sens. Actuators, B* 156 (2): 505–509.

66 Yu, H., Li, J., Li, Z. et al. (2019). Enhanced formaldehyde sensing performance based on Ag@WO_3 2D nanocomposite. *Powder Technol.* 343: 1–10.

67 Lv, P., Tang, Z., Yu, J. et al. (2008). Study on a micro-gas sensor with SnO_2–NiO sensitive film for indoor formaldehyde detection. *Sens. Actuators, B* 132 (1): 74–80.

68 Lian, X., Li, Y., Tong, X. et al. (2017). Synthesis of Ce-doped SnO_2 nanoparticles and their acetone gas sensing properties. *Appl. Surf. Sci.* 407: 447–455.

69 Wang, Z. and Wang, C. (2013). Is breath acetone a biomarker of diabetes? A historical review on breath acetone measurements. *J. Breath Res.* 7 (3): 037109.

70 Chen, N., Li, Y., Deng, D. et al. (2017). Acetone sensing performances based on nanoporous TiO_2 synthesized by a facile hydrothermal method. *Sens. Actuators, B* 238: 491–500.

71 Zhao, J., Huo, L., Gao, S. et al. (2006). Alcohols and acetone sensing properties of SnO_2 thin films deposited by dip-coating. *Sens. Actuators, B* 115 (1): 460–464.

72 Kou, X., Xie, N., Chen, F. et al. (2018). Superior acetone gas sensor based on electrospun SnO_2 nanofibers by Rh doping. *Sens. Actuators, B* 256: 861–869.

73 Liu, C., Zhao, L., Wang, B. et al. (2017). Acetone gas sensor based on NiO/ZnO hollow spheres: fast response and recovery, and low (ppb) detection limit. *J. Colloid Interface Sci.* 495: 207–215.

74 Zhang, S., Song, P., Zhang, J. et al. (2017). Highly sensitive detection of acetone using mesoporous In_2O_3 nanospheres decorated with Au nanoparticles. *Sens. Actuators, B* 242: 983–993.

75 Kawamura, K., Vestergaard, M., Ishiyama, M. et al. (2006). Development of a novel hand-held toluene gas sensor: possible use in the prevention and control of sick building syndrome. *Measurement* 39 (6): 490–496.

76 Sui, L., Zhang, X., Cheng, X. et al. (2017). Au-loaded hierarchical MoO_3 hollow spheres with enhanced gas-sensing performance for the detection of BTX (Benzene, Toluene, And Xylene) and the sensing mechanism. *ACS Appl. Mater. Interfaces* 9 (2): 1661–1670.

77 Hermawan, A., Asakura, Y., Inada, M. et al. (2019). One-step synthesis of micro–/mesoporous SnO_2 spheres by solvothermal method for toluene gas sensor. *Ceram. Int.* 45 (12): 15435–15444.

78 Tshabalala, Z., Mokoena, T., Jozela, M. et al. (2020). TiO_2 nanowires for humidity-stable gas sensors for toluene and xylene. *ACS Appl. Nano Mater.* 4 (1): 702–716.

79 Zeng, Y., Zhang, T., Wang, L. et al. (2009). Enhanced toluene sensing characteristics of TiO_2-doped flowerlike ZnO nanostructures. *Sens. Actuators, B* 140 (1): 73–78.

80 Shan, H., Liu, C., Liu, L. et al. (2013). Excellent toluene sensing properties of SnO_2-Fe_2O_3 interconnected nanotubes. *ACS Appl. Mater. Interfaces* 5 (13): 6376–6380.

81 Tang, W. and Wang, J. (2015). Mechanism for toluene detection of flower-like ZnO sensors prepared by hydrothermal approach: charge transfer. *Sens. Actuators, B* 207: 66–73.

82 Wetchakun, K., Samerjai, T., Tamaekong, N. et al. (2011). Semiconducting metal oxides as sensors for environmentally hazardous gases. *Sens. Actuators, B* 160 (1): 580–591.

83 Yadav, A., Lokhande, A., Kim, J. et al. (2017). Enhanced sensitivity and selectivity of CO_2 gas sensor based on modified La_2O_3 nanorods. *J. Alloys Compd.* 723: 880–886.

84 Kannan, P., Saraswathi, R., and Rayappan, J. (2014). CO_2 gas sensing properties of DC reactive magnetron sputtered ZnO thin film. *Ceram. Int.* 40 (8): 13115–13122.

85 Prim, A., Pellicer, E., Rossinyol, E. et al. (2007). A novel mesoporous CaO-loaded In_2O_3 material for CO_2 sensing. *Adv. Funct. Mater.* 17 (15): 2957–2963.

86 Hsu, K., Fang, T., Hsiao, Y. et al. (2020). Highly response CO_2 gas sensor based on Au-La_2O_3 doped SnO_2 nanofibers. *Mater. Lett.* 261: 127144.

87 Zhou, Y., Li, X., Wang, Y. et al. (2019). UV illumination-enhanced molecular ammonia detection based on a ternary-reduced graphene oxide-titanium dioxide-au composite film at room temperature. *Anal. Chem.* 91 (5): 3311–3318.

88 Wagh, M., Jain, G., Patil, D. et al. (2006). Modified zinc oxide thick film resistors as NH_3 gas sensor. *Sens. Actuators, B* 115 (1): 128–133.

89 Dhivya, P., Prasad, A., and Sridharan, M. (2014). Nanostructured TiO_2 films: enhanced NH_3 detection at room temperature. *Ceram. Int.* 40 (1): 409–415.

90 Shahabuddin, M., Sharma, A., Kumar, J. et al. (2014). Metal clusters activated SnO_2 thin film for low level detection of NH_3 gas. *Sens. Actuators, B* 194: 410–418.

91 Tang, H., Yan, M., Zhang, H. et al. (2006). A selective NH_3 gas sensor based on Fe_2O_3-ZnO nanocomposites at room temperature. *Sens. Actuators, B* 114 (2): 910–915.

92 Van Hieu, N., Thuy, L., and Chien, N. (2008). Highly sensitive thin film NH_3 gas sensor operating at room temperature based on SnO_2/MWCNTs composite. *Sens. Actuators, B* 129 (2): 888–895.

93 Dao, D., Shibuya, K., Bui, T. et al. (2011). Micromachined NH_3 gas sensor with ppb-level sensitivity based on WO_3 nanoparticles thinfilm. *Procedia Eng.* 25: 1149–1152.

94 Wang, Y., Liu, J., Cui, X. et al. (2017). NH_3 gas sensing performance enhanced by Pt-loaded on mesoporous WO_3. *Sens. Actuators, B* 238: 473–481.

95 Liu, G., Zhou, Y., Zou, C. et al. (2018). Heat-pulse assisted NH_3 gas sensing based on cuprous oxide nanoparticles anchored on reduced graphene oxide nanosheets. *J. Mater. Sci.: Mater. Electron.* 29: 3317–3325.

96 Ding, Y., Guo, X., Du, B. et al. (2021). Low-operating temperature ammonia sensor based on Cu_2O nanoparticles decorated with *p*-type MoS_2 nanosheets. *J. Mater. Chem. C* 9 (14): 4838–4846.

97 Wang, Y., Zhou, Y., Wang, Y. et al. (2021). Conductometric room temperature ammonia sensors based on titanium dioxide nanoparticles decorated thin black phosphorus nanosheets. *Sens. Actuators, B* 349: 130770.

98 Zhou, Y., Wang, Y., Wang, Y. et al. (2021). MXene Ti_3C_2Tx-derived nitrogen-functionalized heterophase TiO_2 homojunctions for room-temperature trace ammonia gas sensing. *ACS Appl. Mater. Interfaces* 13 (47): 56485–56497.

99 Duk-Dong, L. and Dae-Sik, L. (2001). Environmental gas sensors. *IEEE Sens. J.* 1 (3): 214–224.

100 Zhou, Y., Wang, Y., Wang, Y. et al. (2020). The impact of carrier gas on room-temperature trace nitrogen dioxide sensing of ZnO nanowire-integrated film under UV illumination. *Ceram. Int.* 46 (10): 16056–16061.

101 Chen, M., Wang, Z., Han, D. et al. (2011). High-sensitivity NO_2 gas sensors based on flower-like and tube-like ZnO nanomaterials. *Sens. Actuators, B* 157 (2): 565–574.

102 Sonker, R., Sabhajeet, S., Singh, S. et al. (2015). Synthesis of ZnO nanopetals and its application as NO_2 gas sensor. *Mater. Lett.* 152: 189–191.

103 Shendage, S., Patil, V., Vanalakar, S. et al. (2017). Sensitive and selective NO_2 gas sensor based on WO_3 nanoplates. *Sens. Actuators, B* 240: 426–433.

104 Zhu, Z., Lin, S., Wu, C. et al. (2018). Synthesis of TiO_2 nanowires for rapid NO_2 detection. *Sens. Actuator, A* 272: 288–294.

105 Sharma, A., Tomar, M., and Gupta, V. (2013). WO_3 nanoclusters-SnO_2 film gas sensor heterostructure with enhanced response for NO_2. *Sens. Actuators, B* 176: 675–684.

106 Shishiyanu, S., Shishiyanu, T., and Lupan, O. (2005). Sensing characteristics of tin-doped ZnO thin films as NO_2 gas sensor. *Sens. Actuators, B* 107 (1): 379–386.

107 Ruiz, A., Sakai, G., Cornet, A. et al. (2003). Cr-doped TiO_2 gas sensor for exhaust NO_2 monitoring. *Sens. Actuators, B* 93 (1–3): 509–518.

108 Sharma, A., Tomar, M., and Gupta, V. (2013). A low temperature operated NO_2 gas sensor based on TeO_2/SnO_2 *p-n* heterointerface. *Sens. Actuators, B* 176: 875–883.

109 Han, D., Zhai, L., Gu, F. et al. (2018). Highly sensitive NO_2 gas sensor of ppb-level detection based on In_2O_3 nanobricks at low temperature. *Sens. Actuators, B* 262: 655–663.

110 Ding, Y., Guo, X., Kuang, D. et al. (2021). Hollow Cu_2O nanospheres loaded with MoS_2/reduced graphene oxide nanosheets for ppb-level NO_2 detection at room temperature. *J. Hazard. Mater.* 416: 126218.

111 Wang, Y., Zhou, Y., Ren, H. et al. (2020). Room-temperature and humidity-resistant trace nitrogen dioxide sensing of few-layer black phosphorus nanosheet by incorporating zinc oxide nanowire. *Anal. Chem.* 92 (16): 11007–11017.

112 Dyer, I. (2012). Measurement of humidity. *Anaesth Intensive Care* 13 (3): 121–123.

113 Zhang, D., Sun, Y., Li, P. et al. (2016). Facile fabrication of MoS_2-modified SnO_2 hybrid nanocomposite for ultrasensitive humidity sensing. *ACS Appl. Mater. Interfaces* 8 (22): 14142–14149.

114 Tsai, F. and Wang, S. (2014). Enhanced sensing performance of relative humidity sensors using laterally grown ZnO nanosheets. *Sens. Actuators, B* 193: 280–287.

115 Qi, Q., Zhang, T., Yu, Q. et al. (2008). Properties of humidity sensing ZnO nanorods-base sensor fabricated by screen-printing. *Sens. Actuators, B* 133 (2): 638–643.

116 Wang, Z., Shi, L., Wu, F. et al. (2011). The sol-gel template synthesis of porous TiO_2 for a high performance humidity sensor. *Nanotechnology* 22 (27): 275502.

117 Zhang, Y., Fu, W., Yang, H. et al. (2008). Synthesis and characterization of TiO_2 nanotubes for humidity sensing. *Appl. Surf. Sci.* 254 (17): 5545–5547.

118 Wang, Q., Pan, Y., Huang, S. et al. (2011). Resistive and capacitive response of nitrogen-doped TiO_2 nanotubes film humidity sensor. *Nanotechnology* 22 (2): 025501.

119 Wu, R., Sun, Y., Lin, C. et al. (2006). Composite of TiO_2 nanowires and Nafion as humidity sensor material. *Sens. Actuators, B* 115 (1): 198–204.

120 Kuang, Q., Lao, C., Wang, Z. et al. (2007). High-sensitivity humidity sensor based on a single SnO_2 nanowire. *J. Am. Chem. Soc.* 129 (19): 6070–6071.

121 Zhong, Y., Li, W., Zhao, X. et al. (2019). High-response room-temperature NO_2 sensor and ultrafast humidity sensor based on SnO_2 with rich oxygen vacancy. *ACS Appl. Mater. Interfaces* 11 (14): 13441–13449.

122 Pascariu, P., Airinei, A., Olaru, N. et al. (2016). Microstructure, electrical and humidity sensor properties of electrospun NiO-SnO_2 nanofibers. *Sens. Actuators, B* 222: 1024–1031.

123 Song, X., Qi, Q., Zhang, T. et al. (2009). A humidity sensor based on KCl-doped SnO_2 nanofibers. *Sens. Actuators, B* 138 (1): 368–373.

124 Liang, Q., Xu, H., Zhao, J. et al. (2012). Micro humidity sensors based on ZnO–In_2O_3 thin films with high performances. *Sens. Actuators, B* 165 (1): 76–81.

125 Lassey, K. (2007). Livestock methane emission: from the individual grazing animal through national inventories to the global methane cycle. *Agric. For. Meteorol.* 142 (2–4): 120–132.

126 Lu, W., Ding, D., Xue, Q. et al. (2018). Great enhancement of CH_4 sensitivity of SnO_2 based nanofibers by heterogeneous sensitization and catalytic effect. *Sens. Actuators, B* 254: 393–401.

127 Bhattacharyya, P., Basu, P., Mondal, B. et al. (2008). A low power MEMS gas sensor based on nanocrystalline ZnO thin films for sensing methane. *Microelectron. Reliab.* 48 (11–12): 1772–1779.

128 Xue, D., Wang, Y., Cao, J. et al. (2019). Improving methane gas sensing performance of flower-like SnO_2 decorated by WO_3 nanoplates. *Talanta* 199: 603–611.

129 Wang, Y., Yao, M., Guan, R. et al. (2021). Enhanced methane sensing performance of NiO decorated In_2O_3 nanospheres composites at low temperature. *J. Alloys Compd.* 854: 157169.

130 Bunpang, K., Wisitsoraat, A., Tuantranont, A. et al. (2019). Highly selective and sensitive CH_4 gas sensors based on flame-spray-made Cr-doped SnO_2 particulate films. *Sens. Actuators, B* 291: 177–191.

131 Vuong, N., Hieu, N., Kim, D. et al. (2014). Ni_2O_3 decoration of In_2O_3 nanostructures for catalytically enhanced methane sensing. *Appl. Surf. Sci.* 317: 765–770.

132 Zhao, Y., Wang, S., Yuan, W. et al. (2021). Selective detection of methane by Pd-In_2O_3 sensors with a catalyst filter film. *Sens. Actuators, B* 328: 129030.

133 Tan, Y. and Lei, Y. (2020). Atomic layer deposition of Rh nanoparticles on WO_3 thin film for CH_4 gas sensing with enhanced detection characteristics. *Ceram. Int.* 46 (7): 9936–4992.

134 Wang, Y., Cui, Y., Meng, X. et al. (2021). A gas sensor based on Ag-modified ZnO flower-like microspheres: temperature-modulated dual selectivity to CO and CH_4. *Surf. Interfaces* 24: 101110.

135 Li, X., Li, Y., Sun, G. et al. (2020). Enhanced CH_4 sensitivity of porous nanosheets-assembled ZnO microflower by decoration with Zn_2SnO_4. *Sens. Actuators, B* 304: 127374.

136 Katoch, A., Kim, J., Kwon, Y. et al. (2015). Bifunctional sensing mechanism of SnO_2-ZnO composite nanofibers for drastically enhancing the sensing behavior in H_2 gas. *ACS Appl. Mater. Interfaces* 7 (21): 11351–11358.

137 Li, Z., Yan, S., Wu, Z. et al. (2018). Hydrogen gas sensor based on mesoporous In_2O_3 with fast response/recovery and ppb level detection limit. *Int. J. Hydrogen Energy* 43 (50): 22746–22755.

138 Mondal, B., Basumatari, B., Das, J. et al. (2014). ZnO-SnO_2 based composite type gas sensor for selective hydrogen sensing. *Sens. Actuators, B* 194: 389–396.

139 Lupan, O., Chai, G., and Chow, L. (2008). Novel hydrogen gas sensor based on single ZnO nanorod. *Microelectron. Eng.* 85 (11): 2220–2225.

140 Hassan, J., Mahdi, M., Chin, C. et al. (2013). A high-sensitivity room-temperature hydrogen gas sensor based on oblique and vertical ZnO nanorod arrays. *Sens. Actuators, B* 176: 360–367.

141 Chang, C., Chou, T., Chen, W. et al. (2020). Study of a WO_3 thin film based hydrogen gas sensor decorated with platinum nanoparticles. *Sens. Actuators, B* 317: 128145.

142 Ippolito, S., Kandasamy, S., Kalantar-zadeh, K. et al. (2005). Hydrogen sensing characteristics of WO_3 thin film conductometric sensors activated by Pt and Au catalysts. *Sens. Actuators, B* 108 (1–2): 154–158.

143 Wu, C., Zhu, Z., Huang, S. et al. (2019). Preparation of palladium-doped mesoporous WO_3 for hydrogen gas sensors. *J. Alloys Compd.* 776: 965–973.

144 Hu, J., Sun, Y., Xue, Y. et al. (2018). Highly sensitive and ultra-fast gas sensor based on CeO_2-loaded In_2O_3 hollow spheres for ppb-level hydrogen detection. *Sens. Actuators, B* 257: 124–135.

145 Wang, Y., Liu, B., Cai, D. et al. (2014). Room-temperature hydrogen sensor based on grain-boundary controlled Pt decorated In_2O_3 nanocubes. *Sens. Actuators, B* 201: 351–359.

146 Dhawale, D., Gujar, T., and Lokhande, C. (2017). TiO_2 Nanorods decorated with Pd nanoparticles for enhanced liquefied petroleum gas sensing performance. *Anal. Chem.* 89 (16): 8531–8537.

147 Bulakhe, R., Patil, S., Deshmukh, P. et al. (2013). Fabrication and performance of polypyrrole (Ppy)/TiO_2 heterojunction for room temperature operated LPG sensor. *Sens. Actuators, B* 181: 417–423.

148 Vaishampayan, M., Deshmukh, R., and Mulla, I. (2008). Influence of Pd doping on morphology and LPG response of SnO_2. *Sens. Actuators, B* 131 (2): 665–672.

149 Singh, S., Verma, N., Singh, A. et al. (2014). Synthesis and characterization of CuO–SnO_2 nanocomposite and its application as liquefied petroleum gas sensor. *Mater. Sci. Semicond. Process.* 18: 88–96.

150 Dhawale, D., Salunkhe, R., Fulari, V. et al. (2009). Liquefied petroleum gas (LPG) sensing performance of electron beam irradiated chemically deposited TiO_2 thin films. *Sens. Actuators, B* 141 (1): 58–64.

151 Dhawale, D., Salunkhe, R., Patil, U. et al. (2008). Room temperature liquefied petroleum gas (LPG) sensor based on *p*-polyaniline/n-TiO_2 heterojunction. *Sens. Actuators, B* 134 (2): 988–992.

152 Haridas, D., Sreenivas, K., and Gupta, V. (2008). Improved response characteristics of SnO_2 thin film loaded with nanoscale catalysts for LPG detection. *Sens. Actuators, B* 133 (1): 270–275.

153 More, A., Gunjakar, J., and Lokhande, C. (2008). Liquefied petroleum gas (LPG) sensor properties of interconnected web-like structured sprayed TiO_2 films. *Sens. Actuators, B* 129 (2): 671–677.

154 Singh, S. and Majumder, S. (2018). Synthesis, characterization and liquefied petroleum gas (LPG) sensing properties of WO_3 nano-particles. *AIP Conf. Proc.* 1953 (1): 090042.

155 Yoo, R., Cho, S., Song, M. et al. (2015). Highly sensitive gas sensor based on Al-doped ZnO nanoparticles for detection of dimethyl methylphosphonate as a chemical warfare agent simulant. *Sens. Actuators, B* 221: 217–223.

156 Lee, S., Kim, S., Lee, W. et al. (2011). Effects of textural properties on the response of a SnO_2-based gas sensor for the detection of chemical warfare agents. *Sensors* 11 (7): 6893–6904.

157 Tomchenko, A., Harmer, G., and Marquis, B. (2005). Detection of chemical warfare agents using nanostructured metal oxide sensors. *Sens. Actuators, B* 108 (1–2): 41–55.

158 Choi, N., Lee, Y., Kwak, J. et al. (2005). Chemical warfare agent sensor using MEMS structure and thick film fabrication method. *Sens. Actuators, B* 108 (1–2): 177–183.

159 Park, Y., Song, H., Lee, C. et al. (2008). Fabrication and its characteristics of metal-loaded TiO_2/SnO_2 thick-film gas sensor for detecting dichloromethane. *J. Ind. Eng. Chem.* 14 (6): 818–823.

160 Lee, S., Choi, H., Lee, S. et al. (2009). The development of SnO_2-based recoverable gas sensors for the detection of DMMP. *Sens. Actuators, B* 137 (1): 239–245.

161 Patil, L., Bari, A., Shinde, M. et al. (2012). Detection of dimethyl methyl phosphonate – a simulant of sarin: The highly toxic chemical warfare-using platinum activated nanocrystalline ZnO thick films. *Sens. Actuators, B* 161 (1): 372–380.

162 Alali, K., Liu, J., Aljebawi, K. et al. (2019). Electrospun *n-p* WO_3/CuO heterostructure nanofibers as an efficient sarin nerve agent sensing material at room temperature. *J. Alloys Compd.* 793: 31–41.

163 Gao, Y., Xiao, T., Li, Q. et al. (2022). Flexible microstructured pressure sensors: design, fabrication and applications. *Nanotechnology* 33 (32): 322002.

164 Arshak, K., Morris, D., Arshak, A. et al. (2006). Development of high sensitivity oxide based strain gauges and pressure sensors, *J. Mater. Sci.: Mater. Electron.* 17: 767–778.

165 Ruth, S., Feig, V., Tran, H. et al. (2020). Microengineering pressure sensor active layers for improved performance. *Adv. Funct. Mater.* 30 (39): 2003491.

166 Wang, X., Chai, Y., Wang, Z. et al. (2022). Metal Oxides/Carbon Felt Pressure Sensors with Ultra-Broad-Range High Sensitivity. *Adv. Mater. Interfaces* 9 (3): 2101663.

167 Xu, F., Li, X., Shi, Y. et al. (2018). Recent developments for flexible pressure sensors: A review. *Micromachines* 9 (11): 580.

168 Yang, Y., Guo, W., Qi, J. et al. (2010). Flexible piezoresistive strain sensor based on single Sb-doped ZnO nanobelts. *Appl. Phys. Lett.* 97 (22): 223107.

169 Ferreira, A., Silva, J., Rodrigues, R. et al. (2019). High performance piezoresistive response of nanostructured ZnO/Ag thin films for pressure sensing applications. *Thin Solid Films* 691: 137587.

170 Wang, X., Dong, L., Zhang, H. et al. (2015). Recent progress in electronic skin. *Adv. Sci.* 2 (10): 1500169.

171 Chen, Y., Hsieh, G., Chen, S. et al. (2015). Zinc oxide nanowire-poly(methyl methacrylate) dielectric layers for polymer capacitive pressure sensors. *ACS Appl. Mater. Interfaces* 7 (1): 45–50.

172 Kumaresan, Y., Ma, S., Ozioko, O. et al. (2022). Soft capacitive pressure sensor with enhanced sensitivity assisted by ZnO NW interlayers and airgap. *IEEE Sens. J.* 22 (5): 3974–3982.

173 Tan, Y., Yang, K., Wang, B. et al. (2021). High-performance textile piezoelectric pressure sensor with novel structural hierarchy based on ZnO nanorods array for wearable application. *Nano Res.* 14: 3969–3976.

174 Park, J., Song, M., Ghosh, R. et al. (2021). Highly sensitive and flexible pressure sensors using position- and dimension-controlled ZnO nanotube arrays grown on graphene films. *NPG Asia Mater.* 13 (1): 57.

175 Chey, C., Liu, X., Alnoor, H. et al. (2015). Fast piezoresistive sensor and UV photodetector based on Mn-doped ZnO nanorods. *Phys. Status Solidi RRL* 9 (1): 87–91.

176 Huo, Z., Wang, X., Zhang, Y. et al. (2020). High-performance Sb-doped *p*-ZnO NW films for self-powered piezoelectric strain sensors. *Nano Energy* 73: 104744.

177 Pan, C., Dong, L., Zhu, G. et al. (2013). High-resolution electroluminescent imaging of pressure distribution using a piezoelectric nanowire LED array. *Nat. Photonics* 7 (9): 752–758.

178 Wang, F., Jiang, J., Liu, Q. et al. (2020). Piezopotential gated two-dimensional InSe field-effect transistor for designing a pressure sensor based on piezotronic effect. *Nano Energy* 70: 104457.

179 Arshak, K., Morris, D., Arshak, A. et al. (2006). Evaluating the suitability of thick-film TiO_2 capacitors for use in a wireless pressure measurement system. *Mater. Sci. Eng., C* 26 (5–7): 1077–1081.

180 Korostynska, O., Arshak, A., Arshak, K. et al. (2011). Investigation into real-time pressure sensing properties of SnO_2, TiO_2, and TiO_2/ZnO thick films with interdigitated electrodes. *Mater. Sci. Eng., B* 176 (16): 1297–1300.

181 Kumari, K., ArunKumar, D., Meti, S. et al. (2020). Fabrication of Ag/PDMS-TiO_2 flexible piezoresistive pressure sensor. *Physica B* 597: 412386.

182 Wang, X., Tao, L., Yuan, M. et al. (2021). Sea urchin-like microstructure pressure sensors with an ultra-broad range and high sensitivity. *Nat. Commun.* 12 (1): 1776.

183 Wang, X., Chai, Y., Zhu, C. et al. (2022). Ultrasensitive and self-alarm pressure sensor based on laser-induced graphene and sea urchin-shaped Fe_2O_3 sandwiched structure. *Chem. Eng. J.* 448: 137664.

184 Liao, Q., Hou, W., Liao, K. et al. (2022). Solid-phase sintering and vapor-liquid-solid growth of BP@MgO quantum dot crystals with a high piezoelectric response. *J. Adv. Ceram.* 11: 1725–1734.

185 Huang, W., Cao, H., Deb, S. et al. (2011). A flexible pH sensor based on the iridium oxide sensing film. *Sens. Actuator, A* 169 (1): 1–11.

186 Qin, Y., Kwon, H., Howlader, M. et al. (2015). Microfabricated electrochemical pH and free chlorine sensors for water quality monitoring: recent advances and research challenges. *RSC Adv.* 5 (85): 69086–69109.

187 Manjakkal, L., Szwagierczak, D., and Dahiya, R. (2020). Metal oxides based electrochemical pH sensors: Current progress and future perspectives. *Prog. Mater Sci.* 109: 100635.

188 Gill, E., Arshak, K., Arshak, A. et al. (2007). Mixed metal oxide films as pH sensing materials. *Microsyst. Technol.* 14: 499–507.

189 Ghoneim, M., Nguyen, A., Dereje, N. et al. (2019). Recent progress in electrochemical ph-sensing materials and configurations for biomedical applications. *Chem. Rev.* 119 (8): 5248–5297.

190 Manjakkal, L., Zaraska, K., Cvejin, K. et al. (2016). Potentiometric RuO_2-Ta_2O_5 pH sensors fabricated using thick film and LTCC technologies. *Talanta* 147: 233–240.

191 Martínez-Máñez, R., Soto, J., García-Breijo, E. et al. (2005). A multisensor in thick-film technology for water quality control. *Sens. Actuator, A* 120 (2): 589–595.

192 Tsai, C., Chou, J., Sun, T. et al. (2005). Study on the sensing characteristics and hysteresis effect of the tin oxide pH electrode. *Sens. Actuators, B* 108 (1–2): 877–882.

193 Manjakkal, L., Cvejin, K., Kulawik, J. et al. (2014). Fabrication of thick film sensitive RuO_2-TiO_2 and Ag/AgCl/KCl reference electrodes and their application for pH measurements. *Sens. Actuators, B* 204: 57–67.

194 Özdemir, A., Caglar, B., Çubuk, O. et al. (2022). Facile synthesis of TiO2-coated cotton fabric and its versatile applications in photocatalysis, pH sensor and antibacterial activities. *Mater. Chem. Phys.* 287: 126342.

195 Santos, L., Neto, J., Crespo, A. et al. (2014). WO_3 nanoparticle-based conformable pH sensor. *ACS Appl. Mater. Interfaces* 6 (15): 12226–12234.

196 Fulati, A., Ali, S., Riaz, M. et al. (2009). Miniaturized pH Sensors Based on Zinc Oxide Nanotubes/Nanorods. *Sensors* 9 (11): 8911–8923.

197 Zaman, S., Asif, M., Zainelabdin, A. et al. (2011). CuO nanoflowers as an electrochemical pH sensor and the effect of pH on the growth. *J. Electroanal. Chem.* 662 (2): 421–425.

198 Sinha, S., Bhardwaj, R., Sahu, N. et al. (2020). Temperature and temporal drift compensation for Al_2O_3-gate ISFET-based pH sensor using machine learning techniques. *Microelectron. J.* 97: 104710.

199 Huang, Y., Tsai, F., and Wang, S. (2014). Preparation of TiO_2 nanowire arrays through hydrothermal growth method and their pH sensing characteristics. *Jpn. J. Appl. Phys.* 53 (6S): 06jg02.

200 Lee, C., Chiu, Y., Lou, L. et al. (2014). Integrated pH sensors and performance improvement mechanism of ZnO-based ion-sensitive field-effect transistors. *IEEE Sens. J.* 14 (2): 490–496.

201 Chin, Y., Chou, J., Sun, T. et al. (2001). A novel SnO_2/Al discrete gate ISFET pH sensor with CMOS standard process. *Sens. Actuators, B* 75 (1–2): 36–42.

202 Mishra, A., Jarwal, D., Mukherjee, B. et al. (2020). CuO nanowire-based extended-gate field-effect-transistor (FET) for pH sensing and enzyme-free/receptor-free glucose sensing applications. *IEEE Sens. J.* 20 (9): 5039–5047.

203 Khizir, H. and Abbas, T. (2022). Hydrothermal synthesis of TiO_2 nanorods as sensing membrane for extended-gate field-effect transistor (EGFET) pH sensing applications. *Sens. Actuator, A* 333: 113231.

204 Bae, T., Kim, H., Jung, J. et al. (2014). Fabrication of high-performance graphene field-effect transistor with solution-processed Al_2O_3 sensing membrane. *Appl. Phys. Lett.* 104 (15): 153506.

205 Arshak, K., Gill, E., Arshak, A. et al. (2007). Investigation of tin oxides as sensing layers in conductimetric interdigitated pH sensors. *Sens. Actuators, B* 127 (1): 42–53.

206 Lee, W., Park, Y., and Cho, Y. (2014). Hierarchically structured suspended TiO_2 nanofibers for use in UV and pH sensor devices. *ACS Appl. Mater. Interfaces* 6 (15): 12189–12195.

207 Simic, M., Manjakkal, L., Zaraska, K. et al. (2017). TiO_2-based thick film ph sensor. *IEEE Sens. J.* 17 (2): 248–255.

208 Manjakkal, L., Djurdjic, E., Cvejin, K. et al. (2015). Electrochemical impedance spectroscopic analysis of RuO_2 based thick film pH sensors. *Electrochim. Acta* 168: 246–255.

209 Copal, V., Tuico, A., Mendoza, J. et al. (2016). Development of resistance-based pH sensor using zinc oxide nanorods. *J. Nanosci. Nanotechnol.* 16 (6): 6102–6106.

210 Chen, Y., Mun, S., and Kim, J. (2013). A wide range conductometric pH sensor made with titanium dioxide/multiwall carbon nanotube/cellulose hybrid nanocomposite. *IEEE Sens. J.* 13 (11): 4157–4162.

211 Manjakkal, L., Szwagierczak, D., and Dahiya, R. (2020). Metal oxides based electrochemical pH sensors: Current progress and future perspectives. *Prog. Mater Sci.* 109: 100635.

6

Metal Oxide Semiconductors for Solar Cells

6.1 Introduction

Metal oxide semiconductors have been exploited for decades in advanced application fields like microelectronics, gas sensing, energy conversion, and storage. At the basis of their success is the possibility of fine-tuning their electronic, optical, and charge transport properties in a broad range of values, targeting their functional features for specific applications. In the past few years, the ability to shape metal oxide semiconductors at the nanoscale has opened new perspectives for their exploitation, significantly improving the performance of end-user devices [1, 2]. Metal oxide thin films with excellent optical, morphological, compositional, electrical properties, and acoustical properties meet excellent applications in these fields. These films have been prepared on a variety of substrates via various deposition methods such as magnetron sputtering, chemical bath deposition, spray pyrolysis, sol–gel spin coating, pulsed laser deposition, electrodeposition, thermal evaporation deposition, molecular beam epitaxy, and chemical vapor deposition.

The impacts of oxide semiconductors on the photovoltaic (PV) industry could be mainly categorized as a light-absorbing layer or a function layer. Specifically, the former (light-absorbing layer), represented by Cu_2O, converts the incident light into electricity, while the later (function layer) mainly acts like an auxiliary layer, for example, to transport the perovskite-generated electron or hole. The oxide-based light-absorbing layers are very attractive to the PV industry since they are abundant, chemically stable, mostly nontoxic, and low-cost. From an optical point of view, several oxide semiconductors with a proper bandgap are suitable for PV applications. However, the electronic properties, including short carrier lifetime and slow carrier mobility of most known oxide semiconductors, prevent their use as active solar cell materials. Fortunately, the development of third-generation solar cells, including dye-sensitized solar cells (DSSCs), quantum dots solar cells (QDSCs), polymer solar cells, and perovskite solar cells, accelerates a renaissance of oxide semiconductors playing a role in the PV industry, where the oxide mainly acts as a photoanode or carrier transport layer.

Metal Oxide Semiconductors: Synthesis, Properties, and Devices,
First Edition. Zhigang Zang, Wensi Cai, and Yong Zhou.

6.2 Solar Cell Principles

To bring the concept of all-oxide PV to a breakthrough, more oxide semiconductors must be investigated for PV applications, and new multicomponent oxides must be developed in conjunction with an optimized device design. This requires an understanding of the working principles of PV cells. Therefore, we will review in the following the fundamental principles of PV actions to demonstrate the basic opto-electronic processes occurring in PV cells. After decades of development, the family of solar cells has grown and divided into several types, including *p*–*n* homo- and heterojunction cells (i.e. Si, CIGS, CdTe, and Cu_2O), bulk heterojunction devices (i.e. plastic solar cells), nanostructured photoelectrochemical cells (i.e. DSCs, QDSCs, and PSCs), and ferroelectric domain boundary cells (i.e. $BiFeO_3$). Although the structures of these above-mentioned solar cells differ, the basic PV process is simple: the incident photons absorbed by the light absorber separate the electron and hole into free carriers. Specifically, a successful PV process should include the following aspects:

(1) A light absorber that absorbs photons to convert electrons from thermal equilibrium into an excited state (Figure 6.1a).
(2) Since the electron is excited to a metastable high-energy state, the remaining positive charges (holes) transition to a metastable low-energy state (Figure 6.1a).
(3) The photo-generated free electrons should be transported to a "high electron energy contact (negative potential)," while the remaining positive charges require a driving force to reach a "low-energy contact" (Figure 6.1b).
(4) For certain solar cell architectures, selective contacts are necessary for charge collection, which should prevent the recombination of positive charges with electrons from the high-energy contact while also ensuring that electrons from the high-energy state do not reach the low-energy contact (Figure 6.1c).

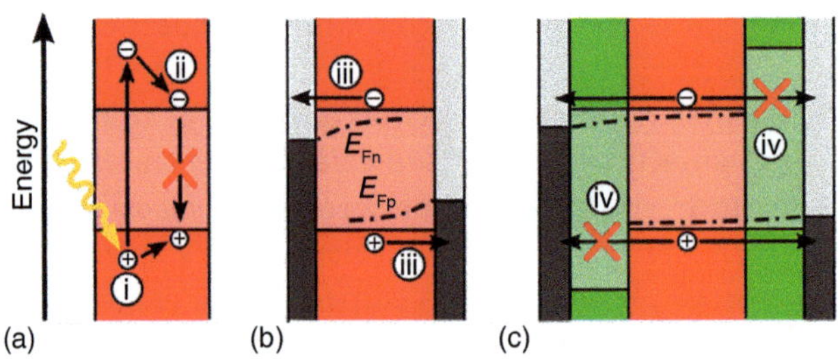

Figure 6.1 General principles of PVs. (a) Upon photon absorption, an electron is excited into an electronic state at higher energy, leaving a positive charge behind at low energy (process i). These charges are transferred into metastable high- and low-energy states (process ii). (b) A driving force is required (shown by a gradient in the quasi-Fermi levels E_{Fn} and E_{Fp}) to transport the electron to an external contact at high electron energy, while the positive charge should reach the "low-energy contact" (process iii). (c) Furthermore, the contacts should be selective for one type of charge carrier, preventing positive and negative charges from reaching the oppositely charged contact (process iv). Source: Reproduced from Rühle et al. [3]/with permission of American Chemical Society.

While the above-mentioned principles effectively generalize the basic physics of solar cells, the more important issue lies in the efficiency of the process. Therefore, the following additional criteria should be fulfilled to create an efficient single-junction solar cell:

(1) For a solar cell, its theoretical efficiency is solely determined by its bandgap. Therefore, to obtain higher efficiency, the bandgap of the absorber should be close to the optimal bandgap, i.e. 1.34 eV. This bandgap dependence of efficiency is based on detailed balance calculation, which idealizes that solar cells only absorb photons with energy above the bandgap ($h\nu > E_g$) and neglect photons with sub-bandgap energy ($h\nu < E_g$). Meanwhile, the absorbed photons are partially converted into electric power, while the rest ($h\nu - E_g$) are mostly converted into unusable heat.
(2) The minority carrier diffusion length (L_D) should be longer than the inverse of the absorption coefficient α^{-1}. This criterion, on the one hand, guarantees sufficient absorption of incident light for efficient conversion into electricity. On the other hand, a longer L_D keeps photo-generated carriers to successfully reaching their respective contacts.
(3) Band alignment is required to minimize energy loss caused by discontinuities in the valence or conduction band.

The power conversion efficiency (PCE) is the product of the open-circuit voltage (V_{OC}), short-circuit current (J_{SC}), and fill factor (*FF*). In brief, V_{OC} is influenced by recombination (including surface and bulk) and band energy alignment, whereas J_{SC} is determined by light harvesting and charge separation efficiency under large extraction fields. *FF* is determined by the device's series resistance, dark current, and charge recombination/extraction rate under low internal fields.

6.3 Metal Oxide Solar Cells

6.3.1 Cu_2O Solar Cells

Up to date, only a few light absorbers are based on oxide semiconductors, with Cu_2O being the most successful and popular one owing to the following properties: (i) availability and abundance, (ii) nontoxic nature, (iii) low cost, and (iv) an acceptable bandgap for solar energy conversion. The bandgap of Cu_2O is about 2.0 eV, which leads to a maximum theoretical conversion efficiency of ~23% as a single-junction device. In comparison, a single-junction device with a bandgap of 1.34 V has a maximum theoretical conversion efficiency of 32.9%. Nonetheless, the unique band structure of Cu_2O makes it attractive for multi-junction solar cells as a top cell and absorber for semi-transparent PV cells. Other outstanding optoelectronic properties, such as absorbance ($\alpha_{600\ nm} = 4 \times 10^3\ cm^{-1}$) [4] and hole mobility (>100 $cm^2\ V^{-1}s^{-1}$) [5], contribute to Cu_2O being a theoretically eligible candidate for PV applications. The performance of Cu_2O-based all-oxide solar cells is summarized in Table 6.1. Various synthesis methods, including thermal oxidation, anodic

Table 6.1 Survey of the Cu_2O solar cell performances, gathered from the work.

Year	*p*-Type	Method	*n*-Type	Method	J_{SC} (mA cm^{-2})	V_{OC} (V)	FF	η (%)
Homojunction								
2012	Cu_2O	ED	Cu_2O	ED	3.97	0.12	0.23	0.10
2012	Cu_2O	ED	Cu_2O	ED	4.07	0.62	0.42	1.06
2015	Cu_2O	ED	Cu_2O	ED	2.68	0.42	0.38	0.42
2016	Cu_2O	ED	Cu_2O	ED	12.40	0.29	0.15	0.89
Heterojunction								
Binary ZnO								
2006	Cu_2O	TO Cu	ZnO:Ga	VAPE	6.94	0.41	0.53	1.52
2006	Cu_2O	TO Cu	ZnO	IVS	6.78	0.59	0.50	2.01
2007	Cu_2O	ED	ZnO	ED	3.80	0.59	0.58	1.28
2011	Cu_2O	TO Cu	ZnO	PLD	10.10	0.69	0.55	3.83
2013	Cu_2O	TO Cu	ZnO	PLD	9.70	0.72	0.60	4.12
2013	Cu_2O	ED	ZnO	ED	4.50	0.54	0.60	1.43
2016	Cu_2O	TO Cu	ZnO:Cu	PLD	10.00	0.72	0.59	4.31
Binary Ga_2O_3								
2013	Cu_2O	TO Cu	Ga_2O_3	PLD	9.99	0.80	0.67	5.38
2014	Cu_2O	ED	Ga_2O_3	ALD	7.37	1.20	0.45	3.97
Nanostructured-ZnO								
2010	Cu_2O	ED	ZnO (NW)	ED	4.40	0.28	0.39	0.47
2010	Cu_2O	ED	ZnO (NR)	ED	8.20	0.29	0.36	0.88
2012	Cu_2O	HT	ZnO (NR)	HT	7.80	0.36	0.31	0.86
2015	Cu_2O	CBD	ZnO (NR)	ED	9.89	0.42	0.37	1.52
2015	Cu_2O	ED	ZnO (NF)	ED	4.38	0.43	0.42	0.77
2016	Cu_2O	ED	ZnO (NT)	ED	2.42	0.66	0.50	0.80
Nanostructured-TiO_2								
2012	Cu_2O	ED	TiO_2 (NR)	HT	8.90	0.36	0.39	1.25
2015	Cu_2O	SP	TiO_2 (NR)	SP	0.40	0.35	0.25	0.04
Ternary oxides								
2011	Cu_2O	SG	$Sn_{1-x}Co_xO$	SG	1.40	2.33	0.36	1.43
2012	Cu_2O	ED	$Zn_{1-x}Sn_xO$	ALD	7.37	0.55	0.65	2.65
2012	Cu_2O	ED	$Mg_{1-x}Zn_xO$	CVD	3.00	0.58	0.42	0.71
2015	Cu_2O	TO Cu	$Zn_{1-x}Mg_xO$	ALD	7.55	0.61	0.49	2.20
2015	Cu_2O	TO Cu	$Ga_{1-x}Al_xO$	PLD	10.95	0.84	0.66	6.10
2016	Cu_2O	TO Cu	$ZnSn_2O_3$	PLD	10.22	0.81	0.65	5.36
2016	Cu_2O	TO Cu	$Zn_{1-x}Ge_xO$	PLD	11.00	1.10	0.60	8.10

Source: Pérez-Tomás et al. [9]/with permission of Elsevier.

oxidation, electrochemical deposition, spray pyrolysis, and reactive sputtering, can be employed to produce *p*-type Cu_2O. Note that the above-mentioned synthesis methods significantly affect the grain size of Cu_2O, which, in turn, determines its mobility. For instance, high thermal oxidation, sputtering, and atomic layer deposition result in larger grain sizes and hence higher carrier mobility, while low-energy cost strategies like spray pyrolysis and electrodeposition lead to smaller grains and lower mobility. The prototype of a Cu_2O solar cell is based on a Cu/Cu_2O Schottky junction with a V_{OC} of below 0.6 V [6]. This unsatisfactory V_{OC} impedes Cu_2O solar cells from achieving an electric power conversion efficiency of over 10%, which requires a V_{OC} larger than 1 V. In addition to Schottky junction Cu_2O solar cells, homojunction Cu_2O solar cells, where the *p*–*n* junction is made of uniquely Cu_2O, also confront an unsatisfactory PCE. This dilemma is mainly caused by the low electron affinity of Cu_2O, which complicates the doping process of *n*-type Cu_2O. To date, the highest reported PCE of homojunction Cu_2O solar cells is 2.05%, as reported by Elfadill et al. [7].

The inferior performance, particularly the unsatisfactory V_{OC}, mainly originates from the unscaled work function difference between Cu_2O and metal contacts, which is attributed to a Cu-rich phase formed in the vicinity of the contact. Therefore, a heterojunction structure has been developed to achieve band alignment and improve PV performance. The typical and classical heterojunction devices mainly use ZnO as an intrinsic *n*-type semiconductor and Cu_2O as the *p*-type absorber. Based on their architecture, heterojunction Cu_2O solar cells can be classified into two types: The first type is wafer-based Cu_2O solar cells, where Cu_2O is synthesized by thermal oxidation of Cu sheets and serves as both the absorber and substrate. The second type is thin-film Cu_2O solar cells, typically deposited by electrochemical or spray pyrolysis on foreign substrates such as glass/ITO, silicon, or Ti. Although a PCE of c. 8.1% has been reported by Minami et al. [8], the imbalance between absorption coefficient and minority carrier diffusion length as mentioned earlier, continues to impact the performance of Cu_2O heterojunction solar cells. To address this issue, nanostructured interfaces have been proposed to improve charge separation (Figure 6.2).

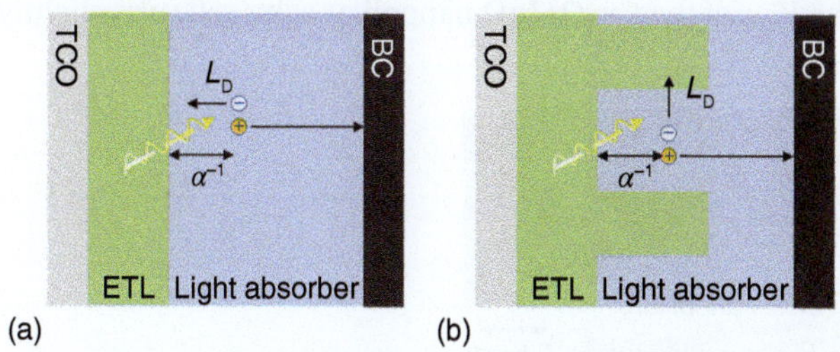

Figure 6.2 (a) Schematic drawing of an all-oxide thin-film PV cell consisting of a TCO FC, a wide bandgap electron conducting layer (ECL), a MO absorber, and a metal BC. Significant losses occur because the electron diffusion length L_D is shorter than the penetration depth α^{-1} of the light. (b) Nanostructured interfaces can be used to decouple both length scales, providing an interface for charge separation within L_D while α^{-1} is still larger than L_D.

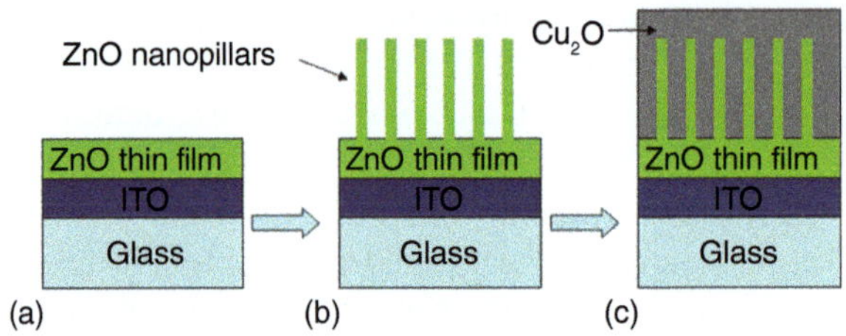

Figure 6.3 Experimental procedures for Cu_2O/ZnO nanopillar heterostructures. Source: Reproduced from Cui and Gibson [10]/with permission of American Chemical Society.

A representative work conducted by Cui and Gibson revealed the advancement of nanopillar heterojunctions consisting of *n*-type ZnO nanowire (NW) arrays embedded in *p*-type Cu_2O thin films [10]. They directly deposited ZnO nucleation films on ITO substrates (Figure 6.3a) and then constructed ZnO nanopillars through electrodeposition (Figure 6.3b). Subsequently, the Cu_2O film was also electrodeposited onto the as-prepared ZnO nanopillars. Not that a negative bias should be applied to the ZnO nanopillar substrate prior to immersing it in the copper salt bath to avoid the etching of ZnO nanopillars by the Cu_2O growth solution. This protective measure ensured minimal etching of ZnO nanopillars even after immersing them for 30 hours. The Cu_2O/ZnO nanopillar heterojunctions were completed when the Cu_2O thin film became continuous and enveloped the nanopillars (Figure 6.3c).

The morphology of the as-prepared ZnO nanopillars was investigated using top and cross-sectional views in Figure 6.4a,b, indicating an average length and diameter of around 2 μm and 100 nm for the ZnO nanopillars, respectively. Furthermore, the cross-sectional SEM image of the ZnO/Cu_2O heterojunction confirmed the full coverage of Cu_2O on the ZnO nanopillars (Figure 6.4b). The rectangular region in Figure 6.4c indicated that some nanopillars were pulled out of the Cu_2O crystals during the sample preparation for SEM measurements. The close contact and perfect coverage between ZnO nanopillars and Cu_2O crystals were facilitated by the electrodeposition techniques, which were superior to vapor-phase methods in filling the deep interpillar voids. The enhanced contact area between the *n*-type ZnO and *p*-type Cu_2O, resulting from the presence of nanopillars, led to a significant increase in the short-circuit density (J_{SC}) from 4.3 to 8.2 mA cm^{-2}, as demonstrated in Figure 6.5. However, the V_{OC} of the Cu_2O/ZnO nanopillar solar cells was slightly

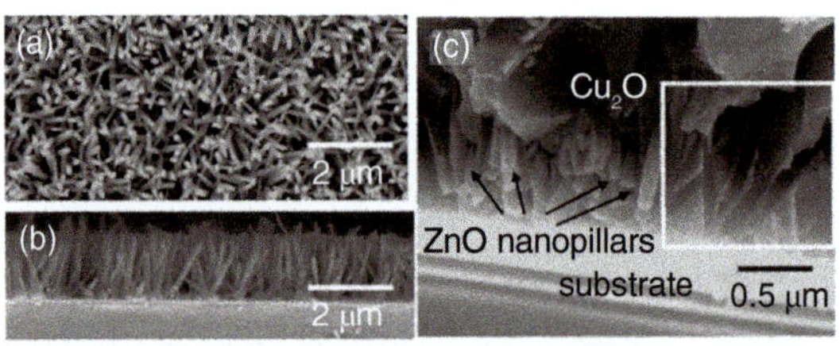

Figure 6.4 SEM images of ZnO nanopillars and Cu_2O/ZnO nanopillar heterostructures. (a) Top view of ZnO nanopillars, (b) cross-sectional view of ZnO nanopillars and (c) cross-sectional view of the sample shown in panel. Source: Reproduced with permission from Cui and Gibson [10]. © 2010, American Chemical Society.

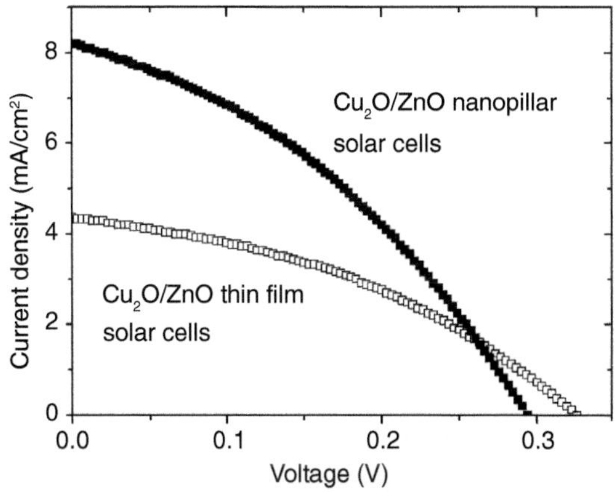

Figure 6.5 Current density vs. voltage of Cu_2O/ZnO nanopillar solar cells under AM 1.5 G. Source: Reproduced from Cui and Gibson [10]/with permission of American Chemical Society.

Figure 6.6 SEM images of several ZnO structures. (a) ZnO NWs with Cu_2O. Source: Reproduced with permission from Musselman et al. [11]. © 2010, John Wiley & Sons, Inc. (b) ZnO nanofibrous networks. Source: Reproduced with permission from Zhang et al. [14]. © 2015, Springer Nature/CC BY 4.0. (c) Top and (f–g) cross-sectional views of ZnO NRs. Source: Reproduced with permission from Chen et al. [12]. © 2015, American Chemical Society. (d) ZnO NRs with Cu_2O. Source: Reproduced with permission from Jia et al. [13]. © 2012, Springer Nature. (e) ZnO NTs. Source: Reproduced with permission from Zhang et al. [15]. © 2004, AIP Publishing LLC.

lower than the counterparts with Cu_2O/ZnO thin-film solar cells, possibly due to the dominance of different exposed crystal faces in the NWs. Apart from nanopillars, *n*-type ZnO has also been synthesized in form of NWs [11], nanorods (NRs) [12, 13], nanofibrous networks [14], and nanotubes (NTs) [15], as shown in Figure 6.6, where the highest PCE reported so far for nanostructured ZnO solar cells was 1.52% [12].

In addition to the interface recombination caused by lattice mismatch, the low built-in potential resulting from nonideal band alignment (Figure 6.7) also affects the V_{OC} of ZnO/Cu_2O heterojunction solar cells [16]. This issue can be addressed by tailoring the band-edge offset of the ZnO *n*-type window [5, 17]. Additionally, the

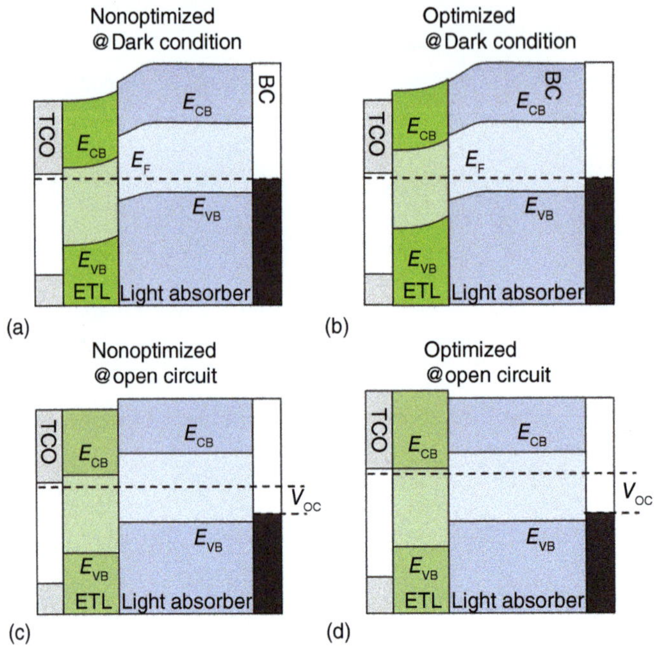

Figure 6.7 Band alignment affects V_{OC} values. (a) Nonoptimized and (b) optimized band structure in dark conditions. (c) Nonoptimized and (d) optimized band structure at open circuit conditions.

inclusion of a buffer layer, such as Ga_2O_3, offers an improved conduction-band offset in relation to the Cu_2O layer, allowing for modulation of the band alignment and reduction of Cu^{2+}-related defects in the vicinity of the interface. In 2014, Lee et al. fabricated a Cu_2O-based thin film solar module (Figure 6.8a) consisting of a 3 × 6 array of fully isolated 3 mm × 5 mm size cells, of which the atomic layer deposited on the Ga_2O_3 buffer layer was only 10 nm (Figure 6.8b) [18]. Pairing Cu_2O with the as-prepared Ga_2O_3 layer provided a near-ideal conduction band offset, effectively reducing interface recombination. Accordingly, the V_{OC} was significantly enhanced to 1.2 V with an outstanding PCE of 3.97%. Similarly, Minami et al. adopted a pulsed

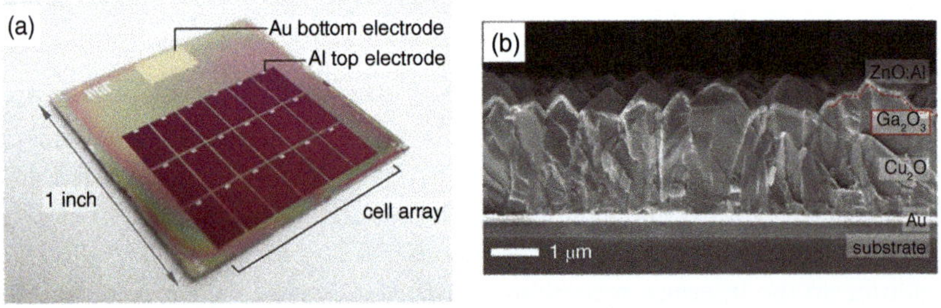

Figure 6.8 (a) A photograph of the fabricated Cu_2O solar module. (b) A cross-sectional SEM image of the device, where the red dashed line indicates the ALD Ga_2O_3 buffer layer. Source: Reproduced with permission from Lee et al. [18]. 2014, John Wiley & Sons, Inc.

laser deposition (PLD) method to fabricate Cu_2O solar cells with Ga_2O_3 buffer layers, which yielded a PCE of 5.38% and a V_{OC} of 0.80 V [19].

In addition to the buffer layer approach, the utilization of ternary oxides as the *n*-type window layer has been explored to reduce recombination current and enhance the power conversion efficiency (PCE) of Cu_2O solar cells. These ternary oxides include $ZnGa_2O_4$, $MgIn_2O_4$, $ZnSnO_3$, Zn_2SnO_4, $GaInO_3$, Zn_2SiO_4, Zn_2GeO_4, Zn_2FeO_4, $CuInO_2$, $CuGaO_2$, and $AgInO_2$. In 2014, Minami et al. reported precise control of the chemical composition of $(Al_2O_3)_{1-x}$ and $(Ga_2O_3)_x$ to form *n*-type $Al_xGa_{1-x}O$ films, resulting in PCE exceeding 6% [20]. Later in 2016, the same group tuned the Ge content in $Zn_xGe_{1-x}O$ films to maximize solar cell efficiency [21].

To summarize, Cu_2O solar cells offer potential cost-effectiveness due to the abundance and low cost of copper. Additionally, they are environmentally friendly as they are composed of nontoxic materials. However, Cu_2O solar cells currently exhibit lower efficiency compared to other types of solar cells, mainly due to their low absorption coefficient and high recombination rate.

6.3.2 CuO Solar Cells

Compared to Cu_2O, cupric oxide (CuO) is considered a more suitable material for PV applications because it is an intrinsically *p*-type semiconductor with a bandgap of c. 1.4 eV. Apart from the bandgap, other optoelectronic parameters of CuO differ from those of Cu_2O, as summarized in Table 6.2. The outstanding bandgap of CuO theoretically contributes to solar light absorption across the entire visible spectrum, where the Shockley–Queisser limit is c. 31% [32]. However, despite its promising properties, CuO faces challenges that limit its performance as a solar cell material. Some of the key challenges associated with CuO include low carrier concentration, high series resistance [33], short carrier lifetime [32], and limited carrier diffusion length due to defects [34]. These factors have thus far restricted the maximum reported PCE of CuO-based solar cells to only around 1–2% on *n*-Si substrates [35, 36].

Moreover, pure CuO-based all-oxide solar cells have exhibited inferior performance, with a PCE of even less than 1%. In 2016, Bhardwaj et al. synthesized an*n*-type $Zn_xMg_{1-x}O$ window layer using PLD, contributing a PCE of only 0.25% for the as-prepared *p*-type CuO solar cells [37]. Similarly, Morasch achieved an extremely low PCE of approximately 0.1% by sputtering a CuO thin film on a sprayed *n*-type TiO2/FTO/glass substrate [34]. These results have led researchers to explore alternative approaches, particularly those involving CuO nanostructures, to enhance the efficiency of CuO-based solar cells. One such approach involved coating CuO NWs with ZnO, resulting in a PCE of 0.1% [38]. Zhao et al. synthesized radial CuO/ZnO core/shell NW arrays, which achieved a PCE of approximately 0.17% [39]. Another strategy to improve CuO solar cell performance was heteroepitaxy on Cu_2O substrates. By integrating optically active copper oxide nanopowders onto thin films, an improved PCE of up to 2.88% was obtained compared to CuO thin films (0.03%) [40]. Wijesundera fabricated *p*-CuO/*n*-Cu_2O heterojunctions using an electrodeposition technique on a titanium substrate in an acetate bath [41]. Further refinements in the electrodeposition technique enhanced the PCE to 0.64%

Table 6.2 Parameters of Cu_2O and CuO.

Parameter	Cu_2O	Ref.	CuO	Ref.
Phase	Cubic $a = 4.27$ Å	[22]	Monoclinic $a = 4.68$ Å, $b = 3.42$ Å, $c = 5.13$ Å	[22]
Bandgap energy (eV)	2.0–2.6	[23]	1.4	
Electron affinity (eV)	3.20	[24]	4.07	[3]
Dielectric permittivity (relative)	7.11	[24]	18.10	[3]
CB effective density of states (cm^{-3})	2.02×10^{17}	[24]	3.0×10^{19}	[3]
VB effective density of states (cm^{-3})	1.1×10^{19}	[24]	5.5×10^{20}	[3]
Conduction valence bands (eV)	NA	NA	−4.27 to −2.70	[25]
Valence band edge (eV)	−4.32 to −4.0	[26]	−5.34 to −4.80	[27]
Electron mobility ($cm^2 V^{-1}s^{-1}$)	200	[24]	0.1	[3]
Hole mobility ($cm^2 V^{-1}s^{-1}$)	80	[24]	10.0	[3]
Hole effective mass	$0.69m_0$	[28]	$0.54m_0$	[27]
Electron effective mass	$0.99m_0$	[28]	$0.14m_0$	[27]
Shallow uniform donor density (cm^{-3})	0	[3]	0	[3]
Shallow uniform acceptor density (cm^{-3})	1×10^{18}	[24]	1×10^{16}	[3]
Conductivity (Ω^{-1} cm)	$10^{-6} - 10^{-8}$	[29]	1.05×10^{-2}	[27]
Hole concentration (cm^{-3})	10^{16}	[30]	6.69×10^{16}	[27]
Hall mobility ($cm^2 V^{-1}s^{-1}$)	30	[30]	0.98	[27]

Source: Adapted from Sawicka-Chudy et al. [31].

for the formed *p*-CuO/*n*-Cu_2O heterojunctions [42]. These studies highlight the various strategies employed to overcome the limitations of pure CuO thin films and improve the efficiency of CuO-based solar cells.

6.3.3 Co_3O_4 Solar Cells

The abundant and nontoxic Co_3O_4 is an optical *p*-type semiconductor with bandgaps of 1.5 and 2.2 eV [43, 44]. In the Co–O system, a mixed valence compound $[Co^{II}Co^{III}{}_2O_4]$ with a normal spinel structure makes Co_3O_4 the most stable phase. The main contribution to the conduction band is offered by the Co(II) 3d orbitals, while the valence band (VB) has a strong O 2*p* character. According to Barreca et al. [43], the presence of Co(III) centers in Co_3O_4 gives rise to a sub-band located within the energy gap. Therefore, the optical bandgap of 1.5 eV corresponds to the

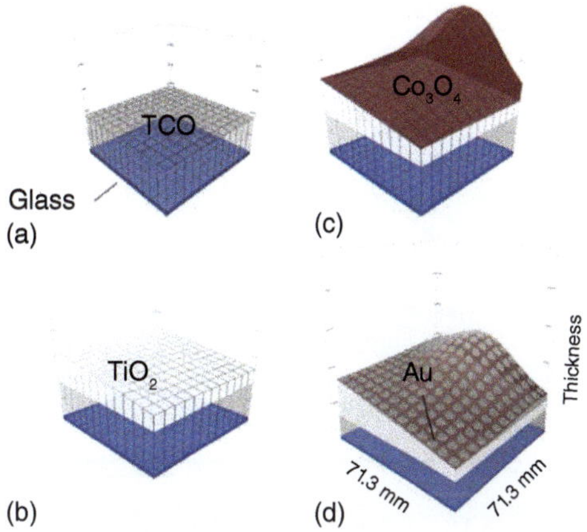

Figure 6.9 Schematic diagrams of the device structure. Source: Reprinted with permission from Kupfer et al. [45]. Copyright: permission of John Wiley & Sons.

onset of O(−II) → Co(III) excitations, while the true inter-band transition rather corresponds to c. 2 eV.

Various methods have been utilized for the preparation of Co_3O_4, including chemical vapor deposition (CVD), PLD, radio frequency sputtering, electrodeposition, magnetron sputtering, spray pyrolysis, plasma oxidation, electron-beam evaporation, and sol–gel processes. Among these synthesis techniques, CVD allows an accurate tailoring of system compositions, microstructures, and morphologies by selecting appropriate precursor compounds and deposition conditions.

The initial attempt at using Co_3O_4 as a light absorber for all-oxide PV cells was accomplished by Kupfer et al. [45], where a thickness-gradient Co_3O_4 film was prepared by PLD on a compact TiO_2 layer sprayed on FTO (Figure 6.9). This work showed that PLD temperatures significantly influenced device performance by improving crystallinity and morphology. After optimization, the as-prepared Co_3O_4 all-oxide solar cells reached a V_{OC} of 0.43 V and a J_{SC} of 0.195 mA cm^{-2}. However, despite the relatively high V_{OC} for a material with a 1.5 eV bandgap, the J_{SC} remained very low in the sub-mA range, resulting in a poor PCE of 0.01–0.001%. Element doping such as Li dopant into the spinel structure would further increase the V_{OC}, while the J_{SC} was still lower than 1 mA cm^{-2}.

6.4 Metal Oxide Functional Layers in Solar Cells

6.4.1 Metal Oxide Photoelectrodes in DSSCs

Dye-sensitized solar cells (DSSCs) draw inspiration from the mechanism of current generation in dye-sensitized semiconductors, which can be dated back to the nineteenth century with the work on silver halide emulsions using erythrosine dyes [46]. By dye sensitization, a broad-band semiconductor that only absorbs UV light can be extended to the visible and even near-infrared regions.

The sandwich structure of a typical DSSC consists of a photoelectrode (made by metal oxides) positioned between two conducting glass plates [such as

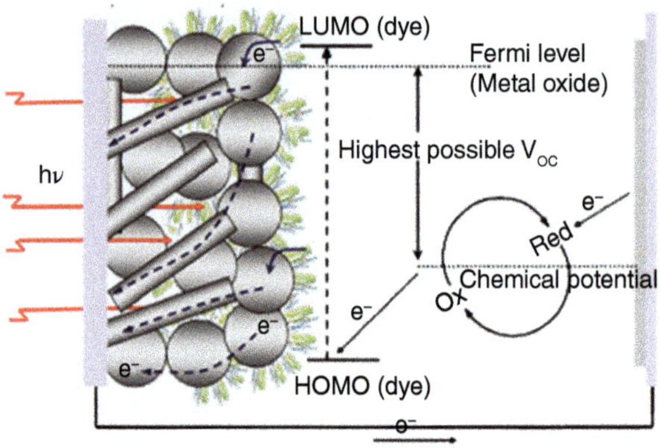

Figure 6.10 A Schematic of dye solar cell. Source: Reproduced from Jose et al. [47]/with permission of John Wiley & Sons.

Fluorine-doped Tin Oxide (FTO) or Indium tin oxide (ITO)] in the presence of an electrolyte. The bottom FTO is coated with a thin layer of metal oxide to protect the leakage of the liquid electrolyte, while the top FTO is typically coated with a thin platinum catalyst layer, as shown in Figure 6.10. Briefly, the working principle or process of DSSCs involves carrier generation, electron transport, and collection, which are facilitated by a suitable combination of three materials: metal oxides, dyes, and electrolytes. Carrier generation occurs when a dye is optically excited, causing an electron to transfer to the conduction band of a nanostructured wide-bandgap metal oxide, such as TiO_2 or ZnO, while a redox couple is present in the electrolyte. The betatopic or oxidized dye molecule is subsequently regenerated to its ground state by accepting one electron from the redox couple. Finally, the generated carriers are collected by the FTO to complete the entire light harvesting process.

In early studies, most of the work was focused on single-crystal semiconductor oxides, which provided a flat "perfect" electrode with favorable electron transport characteristics because of fewer defects. However, these single-crystal electrodes had a significant drawback of small specific surface area, leading to insufficient dye adsorption and consequently low PCE of less than 0.1% for cells based on monolayer dye sensitization [48]. Attempts were made to address the issue of insufficient light absorption by adsorbing multilayer dyes to improve the light capture rate, but this approach resulted in limitations in charge separation and transport, and the efficiency still did not exceed 1% [22].

To overcome the problems of single-crystal semiconductors, Matsumura et al. used porous polycrystalline ZnO semiconductors to adsorb dyes, which had a high specific surface area, leading to increased dye adsorption and a PCE of 1.5%. The efficiency was later improved by further increasing the surface roughness of ZnO, reaching 2.5% [49]. In 1991, based on the development of ruthenium dyes with iodine electrolytes, Gratzel's group made a significant breakthrough by combining nanocrystalline TiO_2 with ruthenium complexes, leading to an increase in efficiency of 7.1% [50]. Inspired by this research, various research groups extensively investigated different components of the cell, including sensitizers, electrolytes, counter electrodes, and others. Through rational molecular structure design, metal complexes and pure organic dyes have been applied as sensitizers, and various types of electrolytes, such as gel polymers and molten salt electrolytes, have been studied. Spectroscopic and electrochemical analysis techniques have also been applied to characterize nanocrystalline dye-sensitized solar cells. As a result, the performance of DSSCs has been continuously improved, with PCE exceeding 14% using multiple sensitizers for simultaneous sensitization, meeting the requirements for commercialization.

Although DSSCs still have a gap in PCE compared to commercial silicon solar cells (26.3%) or multi-junction stacked solar cells (>40%), they possess unique advantages as a PV technology with great application prospects. First, their low cost and simple fabrication process reduce the overall cost of solar cells to 1/5 to 1/10 of conventional technologies. Second, DSSCs exhibit a PCE exceeding 25% under diffuse light and indoor conditions, thanks to the light-absorbing properties of dyes. Furthermore, the advantages of color tunability, flexibility, and transparency give DSSCs a unique position. As a result, they find applications in portable devices, indoor lighting PVs, and PV building integration, playing a crucial role in future-oriented industrial development.

Since the exciton generated in DSSCs is separated into free carriers at the dye-photoelectrode interface, the requirements of this interface and the photoelectrodes are mainly derived from two driving forces for the charge separation. The first is that the conduction band of the metal oxide should be lower than the lowest unoccupied molecular orbital (LUMO) of the dye to energetically favor the charge separation. The second one is that the density of electronic energy states in the conduction band of a crystal should be larger than the molecular orbital of a dye to entropically favor charge separation. Based on these criteria, numerous binary metal oxides, including TiO_2, ZnO, Fe_2O_3, ZrO_2, Nb_2O_5, Al_2O_3, and CeO_2, as well as ternary compounds like $SrTiO_3$ and Zn_2SnO_4, have been identified as suitable photoelectrodes in DSSCs.

6.4.1.1 TiO_2 Electrodes

Among all photoelectrode materials, mesoporous TiO_2 is the most widely used semiconductor for DSSCs [27]. From viewpoint of the electronic configuration and recombination probability, the VB of TiO_2 is composed primarily of O 2p orbitals hybridized with Ti 3d states, whereas the conduction band (CB) is made up of pure Ti 3d orbital. Therefore, the electrons in VB and CB have different parities, reducing the possibility of electron transfer to the VB and suppressing the recombination probability. Naturally, TiO_2 has three crystalline polymorphs: rutile ($E_g = 3.05$ eV), anatase ($E_g = 3.23$ eV), and brookite ($E_g = 3.26$ eV). Among them, rutile is the most common and stable TiO_2 polymorph. Although anatase has a Fermi level 0.1 eV higher than that of rutile, several studies have confirmed that rutile scatters light more effectively and is more chemically stable [51, 52]. Accordingly, similar V_{OC} values have been documented for anatase and rutile TiO_2 DSSCs.

Since 1991, several strategies, including the regulation of TiO_2 film light scattering properties, suppression of charge recombination, improvement of interface

Table 6.3 PV performance of TiO_2-based DSSCs.

Method	Composition	η (%)	V_{OC} (V)	J_{SC} (mA cm^{-2})	FF	Ref.
Alloying	$(TiO_2)_{0.95}(ZrO_2)_{0.05}$	5.8	0.57	11.2	0.72	[53]
	$(TiO_2)_{0.9}(GeO_2)_{0.1}$	5.7	0.75	11.4	0.69	[54]
	$(TiO_2)_{0.8}(Nb_2O_5)_{0.2}$	1.9	0.65	6.8	0.42	[55]
Doping	$(TiO_2)_{0.99}(ZrO_2)_{0.01}$	8.1	0.71	16.5	0.69	[56]
	TiO_2:0.1% Nb_2O_5	1.8	0.84	3.1	0.69	[47]
Core/shell	TiO_2/ZrO_2	3.6	0.78	9.1	0.60	[57]
	TiO_2/Nb_2O_5	4.7	0.73	11.4	0.56	[58]
	TiO_2/ZnO	4.5	0.62	11.7	0.52	[59]
	TiO_2/SiO_2	4.4	0.71	10.6	0.58	[57]
	TiO_2/Al_3O_4	5.6	0.76	12.1	0.61	[57]
	TiO_2/MgO	4.5	0.72	11.7	0.54	[60]
	$TiO_2/SrTiO_3$	4.4	0.71	10.2	0.58	[61]
1D structure	TiO_2 NT	8.1	0.79	15.6	0.66	[62]
	TiO_2 NT	10.2	0.89	17.7	0.65	[63]
	TiO_2 NT	10.3	0.70	20.7	0.71	[64]
	TiO_2 NWs	2.3	0.55	8.00	0.52	[65]

energetics, and modulation of particle morphology, have been invented and continuously renovated to enhance the efficiency of TiO_2-based DSSCs [51]. A representative work reported by Grätzel and coworkers listed the following features for high performance TiO_2-based DSSCs [52]: (i) a well-sintered TiO_2 layer with a thickness of 100–200 nm on FTO to prevent the electrolyte from reaching the FTO, (ii) a thick layer of mesoporous TiO_2 with a particle size of 20–25 nm over the first monolayer, and (iii) a thin layer (thickness 2–5 mm) of TiO_2 with a particle size of c. 400 nm to improve light scattering on the top of mesoporous particle films.

On the other hand, the density of trap states influences the kinetics of electron transport, guiding the development of several routes to solve this dilemma, such as alloying, doping, core–shell structure, and one-dimensional (1D) structure, as summarized in Table 6.3.

Eguchi et al. prepared various alloying compositions in the TiO_2-Nb_2O_5 system, starting from pure TiO_2 to pure Nb_2O_5, and observed a steadily positive relation of *FF* and Nb_2O_5 content [55]. The *FF* increase generally indicated suppressed carrier recombination, with increasing Nb_2O_5 content. The V_{OC} of the composite electrode cell was also enhanced by increasing Nb_2O_5 content, while the J_{SC} was strongly related to the TiO_2 content. The improved V_{OC} value was attributed to the higher CB energy of Nb_2O_5 compared to TiO_2. However, J_{SC} was affected by the photoelectrode area, where the composite was smaller than TiO_2 NPs. Kitiyanan and Yoshikawa prepared TiO_2–ZrO_2 and TiO_2–GeO_2 composite electrodes [53]. The TiO_2–ZrO_2 composite electrodes showed an increase in all PV parameters, whereas the *FF* decreased in the TiO_2–ZrO_2 composite compared to cells fabricated using pure TiO_2.

For direct excitation in the semiconductor, the band positions determine the bandgap and, therefore, the threshold energy of the absorbed photons. Therefore, the band structure is highly related to the performance of DSSCs, especially the V_{OC} values, which are determined by the energy difference between the CB edge and the redox potential of the ion conductor. Accordingly, it is highly desirable to design semiconductors with tunable band edge positions and bandgaps, which can be conducted by element doping. Durr et al. systematically investigated the electrical properties of $Ti_{1-x}Zr_xO_2$ as a function of x and especially their influence on the PV properties of DSSCs [56]. A monotone increase in lattice constant and V_{OC} was observed with increasing Zr content. The latter was traced back to the change in the band edge positions.

Another approach to improving the electrical transport properties is to form a shell of wider bandgap material on TiO_2, inspired by the bandgap engineering of compound semiconductors. If the bandgap positions follow the typical configuration shown in Figure 6.11a, the resulting structure confines electrons in the core material. Electron confinement decreases the probability of electron transfer to the electrolyte, thereby increasing the *FF* and J_{SC}. Palomares et al. coated layers of ZrO_2, Al_2O_3, and SiO_2 (Figure 6.11b) onto 15-nm TiO_2 particles and observed a significant increase of *FF* and J_{SC} for core/shell electrodes compared with bare TiO_2 electrodes [57]. The best result was reported with an optimum shell thickness for Al_2O_3 shell, whose conduction band energy (~−3.23 eV) lied between that of ZrO_2 (~−3.13 eV) and SiO_2 (~−2.97 eV). Although *FF* increased with increasing shell thickness, J_{SC} decreased.

The random network of TiO_2 NPs with massive grain-boundary densities contributes to significant carrier recombination in DSSCs. A typical NP network in DSSCs usually has a porosity of c. 60%, resulting in an average diffusion length of 15–20 mm [66]. This limited electron transport hinders the electrode thickness for DSSCs and negatively affects the overall conversion efficiency. Therefore, modifying the random electron diffusion process into a unidirectional diffusion path can improve the diffusion coefficient (Figure 6.11c,d and, therefore, the transport properties. Since the diffusion coefficient of an electron in anatase TiO_2 NP films (c. 10^{-8} to 10^{-4} $cm^2\ s^{-1}$) is several orders of magnitude lower than that in single-crystal anatase TiO_2 (0.5 $cm^2\ s^{-1}$) [67], the fabrication of 1D TiO_2 related DSSCs, such as NR DSSCs and NW/NP composite DSSCs, has recently been reported to enhance the electron transport rate in anodes [1, 68, 69]. So et al. systematically investigated the impact of TiO_2 NTs' length and structures on the optoelectronic properties of TiO_2 photoelectrodes and the final performance of the corresponding DSSCs [62, 63]. They found that the electronic properties of TiO_2 NTs can be drastically improved by inducing a nanotwinned grain structure in the NT wall. This structure can be exclusively induced in "single-walled" NTs through high temperature treatment in pure oxygen atmospheres. NTs with a twinned grain structure within the tube

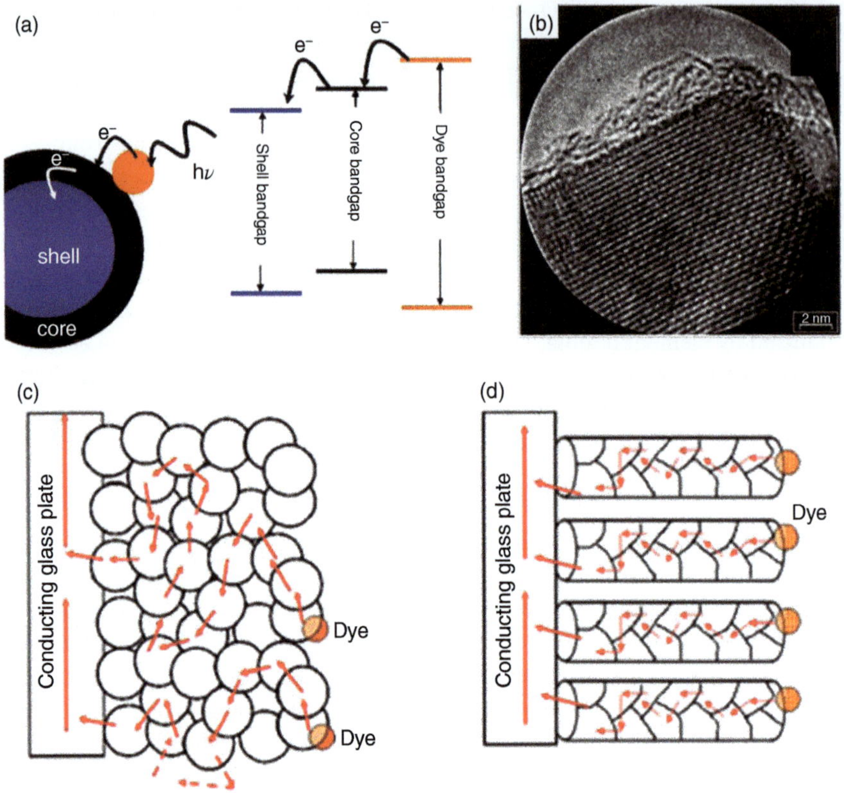

Figure 6.11 (a) Core/shell model and the corresponding energy diagram. (b) A HRTEM image of TiO_2 coated by Al_2O_3. Schematic diagrams of electron diffusion between (c) sintered spherical nanoparticles and (d) one-dimensional (1-D) nanostructures. Source: Reproduced from Palomares et al. [57]/with permission of American Chemical Society.

wall showed strongly enhanced conductivity and photogenerated charge transport compared to classic NTs.

The mixture of nanostructured TiO_2 is also an approach to accelerating electron transport. Maheswari and Venkatachalam prepared photoanodes using TiO_2 NPs, TiO_2 NWs, and various composites, where the composites were prepared by sol–gel/hydrothermal methods and mixed at different weight ratios: 100 wt% NPs without NWs, 90 wt% NPs with 10 wt% NWs, 50 wt% TNPs with 50 wt% NWs, and 10 wt% NPs with 90 wt% NWs at a total weight of 5 g. DSSCs with a TiO_2 NWs/TiO_2 NPs bilayer film achieved a higher maximum output power compared to TiO_2 NPs DSSCs. This relatively high efficiency was attributed to light trapping, a crack-free surface, and a fast electron transfer pathway provided by the TiO_2 NWs/TiO_2 NPs bilayer film [65].

6.4.1.2 ZnO Electrodes

ZnO has been widely used as a photoanode material in DSSCs. It exhibits a unique combination of potentially interesting properties such as high bulk electron

Table 6.4 Nanostructures of ZnO films, Preparation Methods, and PV parameters.

Nanostructure	Method	η (%)	V_{OC} (V)	J_{SC} (mA cm^{-2})	FF	Ref.
core–shell	ALD	1.80	0.68	2.64	0.5	[72]
NPs	Chemical bath deposition	5.40	0.57	16.09	0.59	[73]
Nanoplates	Hydrothermal method	1.90	0.55	8.4	0.41	[74]
Nanoflakes	Hydrothermal method	3.64	0.63	11.6	0.50	[75]
Hexagonal clubs	Hydrothermal method	4.28	0.58	11.50	0.65	[76]
NTs	Electrodeposited method	1.60	0.69	11.51	0.22	[77]
Nanofibers	Sputtering	1.66	0.49	8.45	0.40	[78]
Clusters	Hydrothermal method	2.22	0.58	7.62	0.50	[79]
Panel	Chemical bath deposition	5.59	0.70	11.09	0.72	[80]
Nanoflowers	Hydrothermal method	1.90	0.65	5.50	0.53	[81]
Nanofiber mats	Bulk radical polymerization	1.34	0.60	3.58	0.62	[82]
Nanosheets	Chemical bath deposition	7.07	0.58	19.53	0.63	[83]

mobility (115–155 cm^2 V^{-1}s^{-1}) [70], electron diffusivity, and probably the richest variety of nanostructures based on a very wide range of synthesis routes [71]. These outstanding properties contribute to effective electron transport and reduced recombination rates, making ZnO an attractive alternative to TiO_2 as photoanode material for DSSCs. We have summarized the performance of ZnO-based solar cells in Table 6.4. Up to now, various synthetic routes have been developed to prepare nanostructured ZnO photoanodes, including chemical bath deposition, hydrothermal, electrodeposition, ALD, bulk radical polymerization, and sputtering. These methods enable the fabrication of diverse ZnO nanostructures, such as core-shell, NPs, nanoplates, nanoflakes, hexagonal clubs, NTs, nanofibers, clusters, panels, nanoflowers, nanofiber mats, and nanosheets, as shown in Figure 6.12. Nonetheless, the reported highest efficiency of ZnO-based DSSCs was only 7.5% [84], which was still far from their TiO_2 counterparts (14.3%) [85]. The low conversion efficiencies of ZnO-based systems are primarily attributed to the dissolution of ZnO into Zn^{2+} by the adsorbed acidic dye, followed by the formation of agglomerates (insulating layer) composed of Zn^{2+} and dye molecules (such as N3, N719, and black dye), and eventually followed by the blocking of injected electrons from dye molecules to semiconductors through insulating layers [86]. The isoelectric point of ZnO is ~9 and that of TiO_2 is ~6.51, which implies that ZnO is more basic than TiO_2 and is therefore more prone to being attacked by an acidic dye [87]. The agglomerates are thus detrimental to the performance of all ZnO-based DSSCs, because they fill up nanopores of ZnO electrodes and reduce the electron injection efficiency of the dye adsorbed on ZnO. The presence of carboxyl protons in commonly used dyes can etch ZnO surfaces during dye loading. It was observed that the dissolution of ZnO by acidic carboxylic groups of dyes took place at the crystal

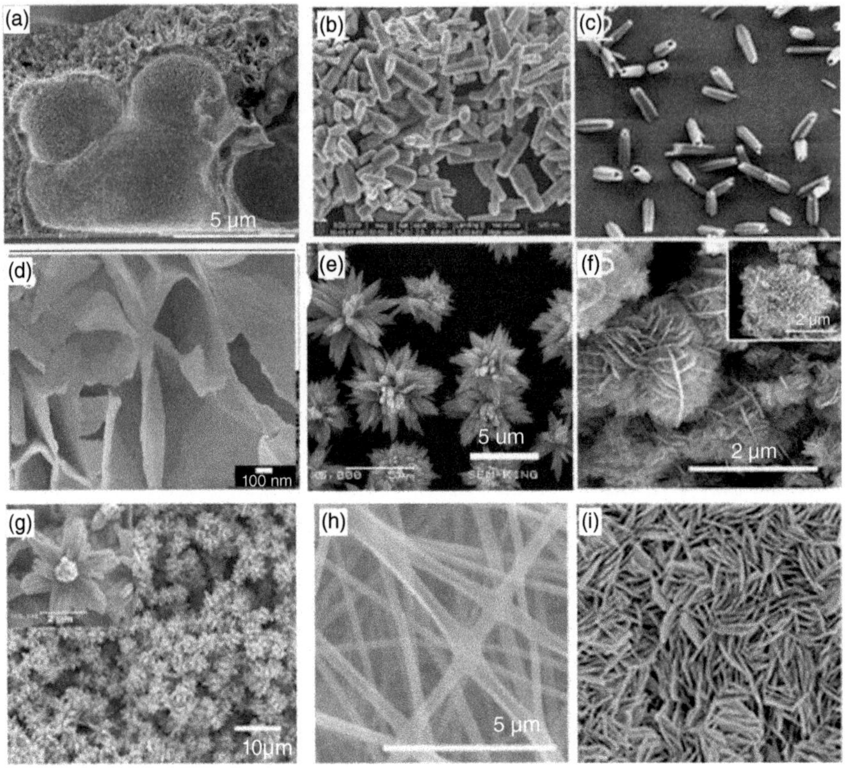

Figure 6.12 SEM images of ZnO nanostructures. (a) Bilayer urchin-like ZnO-NWs/TiO_2 core–shell. Source: Reproduced with permission from Karam et al. [72]. © 2018, American Chemical Society. (b) Hexagonal clubs. Source: Reproduced with permission from Lee et al. [76]. 2011, Royal Society of Chemistry. (c) NTs. Source: Reproduced with permission from Abd-Ellah et al. [77]. 2013, American Chemical Society. (d) Nanoflakes. Source: Reproduced with permission from Mou et al. [75]. 2011, ELSEVIER. (e) Hedgehog-like needle-clusters. Source: Reproduced with permission from Qu et al. [79]. 2014, Royal Society of Chemistry. (f) Panels. Source: Reproduced with permission from Shi et al. [80]. 2013, John Wiley & Sons, Inc. (g) Nanoflowers. Source: Reproduced with permission from Jiang et al. [81]. 2007, AIP Publishing. (h) Nanofiber mats. Source: Reproduced with permission from Kim et al. [82]. (i) Nanosheets. Source: Reproduced with permission from Shin et al. [83].

surface, leading to a formation of Zn^{2+} dye complexes and thereby preventing an efficient electron injection.

To address this issue, a core–shell structure was developed by applying a buffer layer onto the ZnO surface to prevent the formation of Zn^{2+}/dye agglomerates or complexes, thereby improving the stability of ZnO in acidic dye solutions. SiO_2 has been demonstrated to be an exceptionally effective shell material on ZnO, which prevents the formation of Zn^{2+}/dye aggregates by virtue of a strong interaction between Si^{4+} and O^{2-} ions [88]. Another representative material for surface modification of ZnO is TiO_2, which effectively prevents the dissolution of ZnO molecules on the surface, inhibiting the formation of Zn^{2+}/dye agglomerates, as shown in Figure 6.12a [72]. In addition to SiO_2 and TiO_2, other buffer layers, including Al_2O_3 [25], have been developed for deposition onto ZnO nanocrystals and NWs.

Other factors, such as NPs size, ZnO facets, selective redox couples, electron conduction, and stability of ZnO electrodes in contact with electrolytes, also determine the performance of ZnO-based DSSCs. A crucial requirement for ZnO is a high surface area-to-volume ratio, enabling sufficient dye loading to effectively capture light and achieve high J_{SC} values [89]. Sakai et al. conducted a systematic investigation into the size effect of ZnO on DSSCs performance, and the results showed that ZnO particles with a size of 40 nm led to a higher efficiency in DSSCs compared to those with a particle size of 0.2–2 μm [90], because of the larger specific surface area of smaller NPs. In aspect of facets, numerous works have been focused on ZnO $\{10\bar{1}0\}$, while other facets such as $\{10\bar{1}\bar{1}\}$ are rarely investigated. Chang et al. selectively prepared single-crystalline ZnO and ZnO-NRs by a hot-injection method with either mainly exposed $\{10\bar{1}\bar{1}\}$ or $\{10\bar{1}0\}$ facets, respectively. The results showed that DSSCs assembled with C218 dye-coated ZnO with $\{10\bar{1}\bar{1}\}$ facet photoanodes demonstrated the highest PCE of 4.36%, which was three times of the efficiency obtained from the ZnO-NRs with $\{10\bar{1}0\}$ counterparts [91]. Grain boundaries and defects in ZnO mesoporous films limit the direct transport of electrons in NPs in comparison with NRs. In a previous study, Virkki et al. concluded that NRs exhibited faster electron transport compared to NPs [92]. NTs and NWs with a 1D nanostructure could accelerate electron transfer and reduce recombination [93]. The growth of various nanostructures has improved the aforementioned factors to some extent. However, further investigations are necessary to explore different synthesis techniques and their resulting products in order to enhance the overall performance of DSSCs.

6.4.1.3 Nb_2O_5 Electrodes

In principle, Nb_2O_5 has a wide bandgap of c. 3.5 eV with a CB edge more negative than that of TiO_2, making it a candidate for use as a mesoporous layer in DSSCs [94]. As previously mentioned, V_{OC} of a DSSC is dependent on the Fermi level of the oxide, where a more negative CB edge promises a higher V_{OC} value. Furthermore, Nb_2O_5 offers advantages such as superior chemical stability confronted by ZnO and UV stability in TiO_2 material. The potential of Nb_2O_5 as a mesoporous layer was first demonstrated in 1998 by Sayama et al. who conducted a series of experiments to explore the potential of common metal oxides, including TiO_2, Nb_2O_5, ZnO, SnO_2, In_2O_3, WO_3, Ta_2O_5, and ZrO_2, as porous semiconductor films in DSSCs [94]. It was found that cells based on Nb_2O_5 had the next highest incident monochromatic photon-to-current efficiency (18%) compared to TiO_2 cells and showed the highest V_{OC} among them. To date, the highest PCE achieved by Nb_2O_5 films has surpassed 6% using hydrothermally grown NRs [95]. The photoelectrochemical parameters of some DSSCs based on mesoporous Nb_2O_5 films are summarized in Table 6.5.

Although Nb_2O_5 has numerous outstanding properties for applications in DSSCs, DSSCs based on various Nb_2O_5 nanostructures have shown relatively poor performance in comparison with TiO_2. The first challenge responsible for the poor performance is the larger unit cell size of Nb_2O_5, which can reduce its surface area and

Table 6.5 Nanostructures of Nb_2O_5 films, preparation methods, and PV parameters.

Nanostructure	Method	η (%)	V_{OC} (V)	J_{SC} (mA cm^{-2})	FF	Ref.
NPs	Doctor-blading	1.20	0.60	3.10	0.63	[94]
NPs	Screen printing	4.68	0.66	11.57	0.61	[95]
NRs	Hydrothermal method	6.03	0.75	12.20	0.66	[95]
$Nb_3O_7(OH)$ NRs	Hydrothermal method	6.77	0.74	15.00	0.61	[95]
NWs	Electrospinning	3.05	0.77	6.68	0.59	[96]
NTs	Doctor-blading	3.15	0.74	6.23	0.68	[97]
Nanochannels	Anodization	4.48	0.64	17.6	0.40	[98]
Nanoforest	Pulsed laser	2.41	0.71	6.65	0.49	[99]

thereby the number of adsorbed dyes. The smaller amount of dye coverage limits the light harvesting and generation of photocurrent, leading to lower conversion efficiencies.

Therefore, several strategies for preparation of Nb_2O_5 films have been invented to increase the thickness and surface area and improve the amount of dye adsorbed. Sayama et al. decreased the sintering temperature during the preparation of Nb_2O_5 powders, resulting in a high roughness of the film for absorbing a higher mount of dye and therefore a higher photocurrent [94]. Furthermore, treating the sintered Nb_2O_5 NPs with niobium alkoxides, followed by its calcination, also increased the roughness of the film, and resulted in 1.2 times higher dye loading along with a better conductivity between particles, thus improving J_{SC}, FF, and overall PCEs [94]. Ghosh et al. fabricated vertically aligned bundles of Nb_2O_5 nanocrystals by PLD, and applied the as-prepared Nb_2O_5 nanoforests in DSSCs [99]. The best PLD Nb_2O_5 films achieved a N3 dye molecule density of a 0.3 to 0.4×10^{20} molecules cm^{-3} density, which was 1 to 5 times higher than that of previously reported Nb_2O_5 photoanodes, indicating that dye loading plays a dominant role in light absorption and photocurrent generation of DSSCs.

To free tune the size of Nb_2O_5 spheres, Jin et al. prepared mesoporous Nb_2O_5 microspheres assembled by NPs through a simple solvothermal process and a subsequent annealing treatment, without using any template or catalyst [100]. By controlling the concentration of reactant $NbCl_5$, the diameter of Nb_2O_5 spheres could be tuned from 200 to 900 nm. The microspheres exhibited a higher surface area upon increasing their diameter, which was beneficial for more effective light harvesting. Moreover, its superior light scattering properties improved the light-harvesting efficiency and resulted in devices with efficiencies up to 2.97%. However, the increase in microsphere size led to a slower electron transport rate and a faster recombination rate, resulting in lower V_{OC} values, indicating that trap sites and recombination centers in the microspheres increased due to the presence of more interfaces. Therefore, achieving an appropriate balance between photon absorption, electron transport, and charge transfer properties is crucial for obtaining the highest overall performance [101].

6.4.2 Metal Oxide Carrier Transport Layers in Perovskite Solar Cells

Generally, perovskite solar cells (PSCs) employ an electron transport layer (ETL), typically metal oxides, to conduct the photo-generated electrons to anodes and block the holes. In this regard, the modulation and selection of an appropriate ETL can significantly enhance electron transport and suppress carrier recombination (Figure 6.13a), thereby contributing to enhanced PV performance and environmental stability [102–104]. In particular, for normal *n-i-p* PSCs, ETLs play a crucial role as a substrate for the crystallization of the upper perovskite layer, thereby regulating the morphology of the perovskite film [105]. Therefore, the desired criteria for ETLs, as illustrated in Figure 6.14b, include: (i) favorable electronic properties (energy structure, trap state, and mobility), (ii) suitable optical properties (transmittance, bandgap, and refractive index), (iii) control over morphology with respect to the adjacent perovskite layer, (iv) chemical stability against moisture and polar aprotic organic solvents used for perovskite precursors, and (v) easy and cost-effective fabrication with non-toxic materials. After about ten years of development, dozens of *n*-type metal oxides have shown their potential for integration into highly efficient PSCs, as summarized in Figure 6.13c [26, 107]. This fast development of ETLs in hybrid organic–inorganic PSCs provides facility to adopt the mature modulation strategies referring to the ETLs and the ETL/perovskite interface for inorganic PSCs.

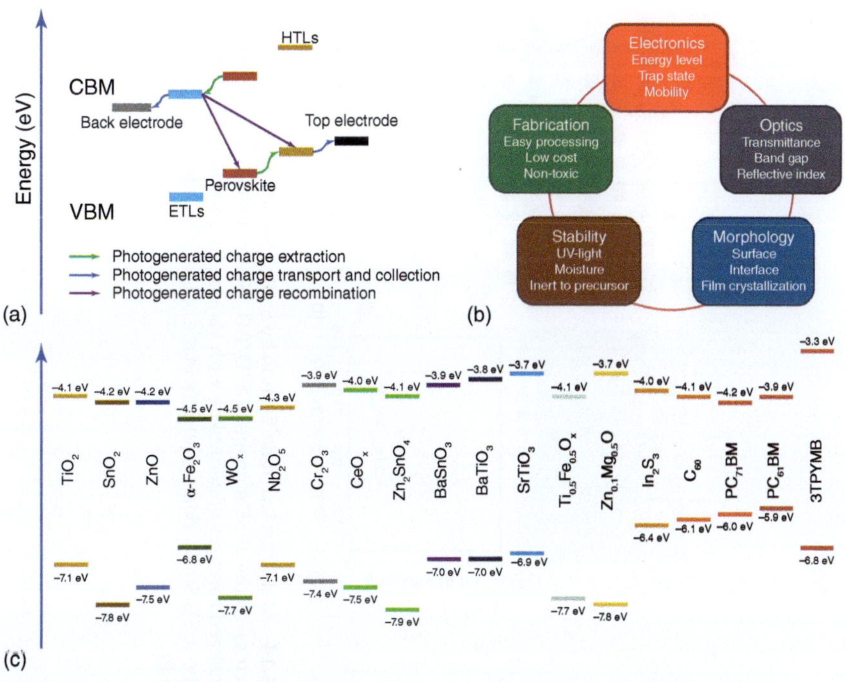

Figure 6.13 (a) Schematic diagrams of the charge transfer process in PSCs. (b) The criterion for an ideal ETL in PSCs. (c) Energy band structures of common ETL materials.

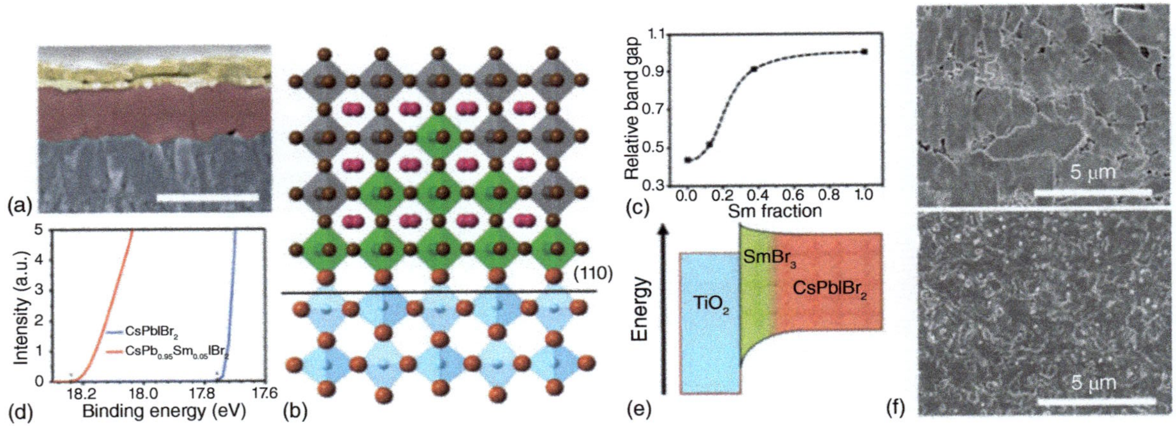

Figure 6.14 (a–f) Interface modification by $SmBr_3$. Source: Reproduced with permission from Subhani et al. [106]. © 2019, John Wiley & Sons, Inc. (a) The cross-sectional SEM image of FTO/TiO_2/$SmBr_3$/$CsPbIBr_2$/PTAA. (b) A schematic diagram depicting the SIM on the (110) crystal plane of rutile-TiO_2. (c) Normalized bandgap of perovskite with various Sm concentrations from density functional theory (DFT) calculations. (d) UPS spectra of $CsPbIBr_2$ with and without 5% $SmBr_3$ doping. (e) Schematic of the SIM-induced gradient band structure. (f) Top-view SEM images of $CsPbIBr_2$ on TiO_2 (top) and TiO_2/$SmBr_3$.

Inorganic *n*-type materials, especially binary and ternary metal oxides such as TiO_2, SnO_2, ZnO, and ZnMgO, are excellent candidates for ETLs in PSCs due to their facile fabrication and remarkable properties [24, 108, 109]. However, challenges such as severe carrier recombination at the interface, improper band alignment with the adjacent perovskite layer, and high sintering temperatures perplex the community. The following section discusses the applications and modifications of binary and ternary metal oxide ETLs.

6.4.2.1 TiO_2 ETLs

TiO_2, originally used in DSSCs, is the most successful and widely used electron transport material in both hybrid and inorganic PSCs. The conduction band minimum (CBM) of TiO_2 is −4.1 eV, lower than the LUMO of perovskite, helping electron transport from perovskite to the back electrode. Additionally, its high transmittance in the visible region guarantees sufficient light absorption by the perovskite layer [110]. The outstanding inorganic PSCs with TiO_2 are summarized in Table 6.6, classified by the X content of $CsPbX_3$. Despite its numerous advantages, the high sintering temperature (>450 °C), mismatched energy levels, and UV-light instability can adversely affect the performance of inorganic PSCs. Vast elaborate strategies were designed to tackle these flaws. Choy et al. utilized a simple low-temperature route to deposit a thick TiO_2 backbone film decorated with fullerene on the perovskite film to configure an inverted PSC. The reduced defect densities and more favorable energy-level alignment contribute to the suppressed mono- molecular Shockley–Read–Hall recombination and ion diffusion. Remarkably, the thickness of the TiO_2 does not significantly affect the performance of PSCs, thereby contributing to their robust fabrication and high reproducibility [23]. Liu and coworkers utilized a low-cost lanthanide halide, i.e. $SmBr_3$, as a modification layer inserting into the interface between $TiO_2/CsPbI_2Br$, as shown in Figure 6.14a–f [106]. From the viewpoint of element doping, the $SmBr_3$ enhances the interface by binding with the oxygen of TiO_2 and intercalates into the $CsPbIBr_2$ lattice with a gradient distribution simultaneously. The substitution of Pb by Sm contributes to band structure variation, resulting in a gradient energy band with a hole blocking function. Meanwhile, the $SmBr_3$ interface layer positively influences the growth of perovskite, resulting in a more compact and pinhole-free $CsPbIBr_2$ film on the $TiO_2/SmBr_3$ substrate, as demonstrated by the top-view SEM image (Figure 6.14f). Similarly, Zhang and coworkers found the CsBr layered between TiO_2 and $CsPbIBr_2$

Table 6.6 Efficiency record chart of inorganic PSCs with TiO_2 ETLs.

Sample	PCE (%)	V_{OC} (V)	J_{SC} (mA cm^{-2})	FF	Year ref.
$CsPbI_3$	19.03	1.14	20.23	0.83	2019 [30]
$CsPbI_2Br$	16.79	1.32	15.32	0.83	2020 [111]
$CsPbIBr_2$	10.88	1.17	12.75	0.73	2019 [106]
$CsPbBr_3$	10.85	1.63	7.73	0.86	2020 [112]

profited from the band alignment, thus suppressing carrier recombination and improving carrier transport [113]. Feng and coworkers elucidated the mechanism of the CsBr interfacial modifier through first-principles calculations [28]. The charge difference $\Delta\rho$ indicator suggested that the accumulation of charge mainly occurred on Ti and O atoms, while the depletion was underneath Cs and Br atoms. Another indicator, the charge displacement curve (ΔQ) was employed to describe the charge transfer across the interface. The positive ΔQ in CsBr/TiO_2 interface implied that the charge was transferred from $CsPbBr_3$ to TiO_2 ETL. Meanwhile, the negative ΔQ in the perovskite side of $PbBr_2$/TiO_2 interfaces explained the weaker optical absorption and V_{OC} of $PbBr_2$/TiO_2 films and solar devices. Furthermore, the more decreased interfacial bandgap of the CsBr/TiO_2 interface spurring the tunneling effect, boosting the performance of corresponding $CsPbBr_3$ PSCs.

Stability is another challenge for ETLs. To enhance the stability of ETL and corresponding inorganic PSCs, Chen and coworkers deposited a carbide-titanium oxide (c-TiO_2) on $CsPbI_2Br$ perovskite films to form *p-i-n* inorganic PSC [29]. The *in situ* c-AFM results suggested that c-TiO_2 films had a higher thermal stability than PCBM films. The stable morphology and electrical properties of c-TiO_2 contributed to the good long-term stability of the resulting PSCs.

6.4.2.2 SnO_2 ETLs

In contrast to TiO_2, which requires a high sintering temperature to remove organic materials and enhance crystallinity, SnO_2 can be fabricated at a low temperature of <200 °C, making it a promising alternative for TiO_2 as an ETLs [114]. Apart from the low-temperature process, SnO_2 has other splendid properties to be ETLs in inorganic PSCs: (i) the deep CBM and well-matched energy structure; (ii) high bulk electron mobility (240 $cm^2\ V^{-1}s^{-1}$) and high conductivity to enhance electron dynamic; (iii) wide bandgap (3.6–4.0 eV) and high transmittance to maximize the incident light to perovskite layers; (iv) excellent chemical stability [115]. Inorganic PSCs with the best performance based on SnO_2 are summarized in Table 6.7.

Given that the properties of SnO_2 are significantly affected by the processing methods and conditions, modifications and/or post-treatments of SnO_2 are necessary. However, achieving high-performance inorganic PSCs is impeded by the loss in V_{OC}, which is attributed to the energy band matching and interfacial defects. Therefore, interface engineering is developed to enhance band alignment between SnO_2 and perovskite layers, simultaneously passivating the defects.

Table 6.7 Efficiency record chart of inorganic PSCs with SnO_2 ETLs.

Sample	PCE (%)	V_{OC} (V)	J_{SC} (mA cm^{-2})	FF	Year ref.
$CsPbI_3$	18.64	1.23	18.30	0.83	2019 [116]
$CsPbI_2Br$	16.20	1.30	15.30	0.82	2019 [117]
$CsPbIBr_2$	11.04	1.36	11.70	0.69	2019 [118]
$CsPbBr_3$	10.60	1.61	7.80	0.84	2019 [119]

You and coworkers inserted an inorganic shunt-blocking layer – lithium fluoride (LiF) between SnO_2 and $CsPbI_{3-x}Br_x$ perovskites, elevating the CBM of SnO_2 and suppressing the surface defects, as shown in Figure 6.15a [116]. Improved band alignment and the passivation effect of LiF resulted in an enhancement of V_{OC} from 1.08 to 1.18 V (Figure 6.15b–c). Yip and coworkers also found significant V_{OC} losses in $CsPbI_2Br$ devices. To alleviate this dilemma, they proposed a SnO_2/ZnO bilayered ETL to facilitate a better cascade energy level alignment with a superior electron extraction capacity, as shown in Figure 6.15d [120]. Consequently, the suppressed interfacial recombination and greater electron extraction efficiency resulted in a 0.17 V increase in V_{OC}. Similarly, Zhang and coworkers employed ALD to deposit a TiO_2 layer on SnO_2 ETLs, aiming to improve the morphology of $CsPbIBr_2$ films and form a cascade energy level alignment, where the V_{OC} was improved from 1.11 to 1.27 V [121].

Apart from the mismatched energy level, other challenges persist in solution-processed SnO_2, including low crystallization and inferior electron mobility at a low annealing temperature. Tang and coworkers controlled the bandgap and energy level of SnO_2 QDs through changing the aging time of SnO_2 QDs by the synthesis process, as shown in Figure 6.15e–f [119]. By optimizing the aging time to 36 hours, the band alignment was improved, resulting in an increase in V_{OC} to 1.57 V. Very recently, Tang and coworkers implanted a $TiCl_4$ hydrolyzed TiO_xCl_{4-2x} accessory layer on SnO_2 to simultaneously passivate the oxygen-vacancy-related defects and set an intermediate energy level at ETL/perovskite interfaces [122]. Meanwhile, the third function of the TiO_xCl_{4-2x} accessory layer was to improve the $CsPbBr_3$ film morphology. Furthermore, Yip and coworkers found that an amino-functionalized polymer (PN4N) on SnO_2 ETL served three functions in the $CsPbI_2Br$ PSCs: promoting better surface wetting properties to enhance $CsPbI_2Br$ film morphology (Figure 6.15g–h) and stability, achieving more suitable energy level alignment, and passivating electronic surface trap states to reduce interfacial recombination [117].

6.4.2.3 ZnO ETLs

Similar to SnO_2, ZnO exhibits a high potential as an ETL in PSCs due to its high bulk electron mobility, favorable energy level structure, wide bandgap with high transmittance, and ease of fabrication [123, 124]. Meanwhile, the mature applications in both *p-i-n* and *n-i-p* structures are another attractive merit of ZnO [125, 126]. Besides, the diverse structures of ZnO, such as NPs and NRs, can be fabricated at a low temperature, contributing to the integration in flexible PSCs [127]. The basic nature of ZnO with surface hydroxyl groups and/or residual acetate ligands can lead to deprotonation of the methylammonium cation, causing the decomposition of hybrid perovskite films [128]. Although challenges exist in ZnO-based hybrid PSCs, ZnO ETLs show extraordinary performance in inorganic PSCs, as summarized in efficiency record chart (Table 6.8).

Tang et al. doped a classical alkali metal carbonate, Cs_2CO_3 to tune the energy levels and electrical properties of sol–gel-derived ZnO ETLs in $CsPbI_2Br$ PSCs, as shown in Figure 6.16a–d [130]. After partially substituting Zn^{2+} by Cs^+, the

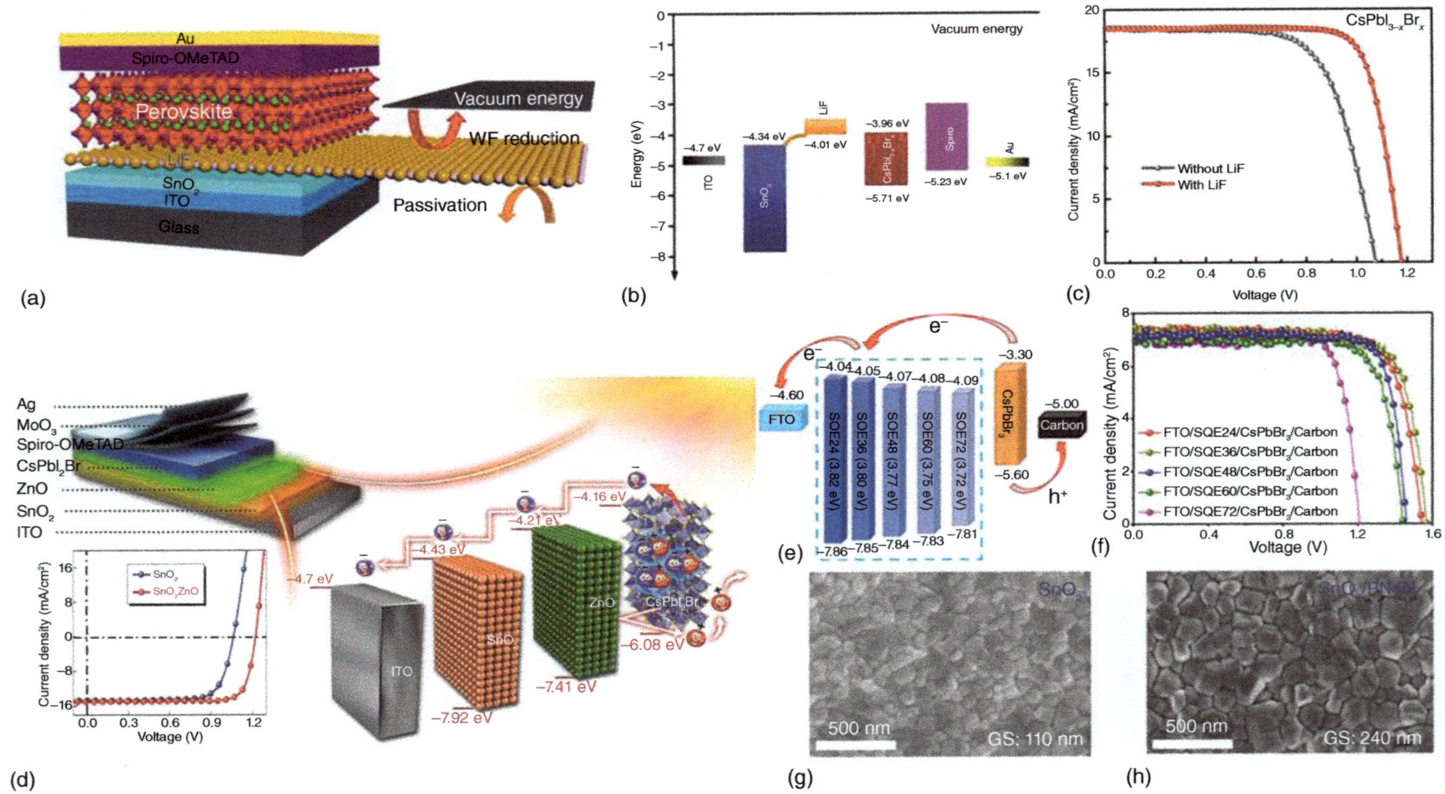

Figure 6.15 (a) The scheme of device architecture. (b) Energy band alignment of inorganic PSCs, and (c) the corresponding *I*–*V* curves. Source: Reproduced with permission from Ye et al. [116]. Copyright 2019 MDPI/Public Domain CC BY 4.0. (d) Device architecture and the corresponding energy diagrams of all-inorganic $CsPbI_2Br$ PSCs and the corresponding *I*–*V* curves. Source: Reproduced from Yan et al. [120]/with permission of Elsevier. (e) An energy diagram of all-inorganic $CsPbBr_3$ PSCs and (f) corresponding *J*–*V* curves with various ETLs. Source: Reproduced with permission from Zhao et al. [119]. Copyright 2021 MDPI/Public Domain CC BY 4.0. Top-view SEM images of $CsPbI_2Br$ films on (g)SnO_2 and (h) SnO_2/PN4N. Source: Reproduced with permission from Tian et al. [117], © 2019, John Wiley & Sons, Inc.

Table 6.8 Efficiency record chart of inorganic PSCs with ZnO ETLs.

Sample	PCE (%)	V_{OC} (V)	J_{SC} (mA cm^{-2})	FF	Year ref.
$CsPbI_3$	21.06	1.31	21.79	0.74	2019 [129]
$CsPbI_2Br$	16.42	1.28	16.34	0.79	2020 [130]
$CsPbIBr_2$	10.16	1.27	11.52	0.69	2019 [102]
$CsPbBr_3$	7.78	1.44	7.01	0.77	2018 [131]

combination of O^{2-} dangling bands and Cs^+ formed a permanent dipole moment pointing to Zn^{2+} at the ZnO surface, reducing the work function from 4.0 to 3.7 eV. Accordingly, a matched energy level alignment between ZnO: Cs_2CO_3 and $CsPbI_2Br$ was achieved, promoting the V_{OC} from 1.13 to 1.28 V. Zang and coworkers recently induced NH_4Cl to ZnO precursors to lower the work function of ZnO ETLs [102]. As a result, the band alignment between ETLs and $CsPbIBr_2$ perovskites was formed to improve the electron mobility and reduce the interface recombination (Figure 6.16e). Consequently, V_{OC} and PCE were elevated significantly compared with the control devices (Figure 6.16f).

Interface modification is another strategy to deal with interfacial recombination. Hao and coworkers developed an ambidextrous strategy through doping engineering of ZnO ETLs and inserting a TiO_2 buffer layer between ZnO/$CsPbI_3$ [129]. On the one hand, the interfacial layer relieved the energy band bending and structure disorder of $CsPbI_3$. On the other hand, the conductivity of the ZnO/TiO_2 bilayer was enhanced with increasing doping concentrations of ZnO, due to the strong interaction between the TiO_2 and ZnO layers. Yip and coworkers also integrated a ZnO/SnO_2 bilayer to enhance electron extraction efficiency at the cathode interface [132]. In *p-i-n*-structured $CsPbI_2Br$ PSCs, ZnO/C_{60} bilayer was demonstrated as an effective strategy to form band alignment and enhance electron extraction efficiency, as shown in Figure 6.17a,b [133].

Similar to the bilayer concept, novel core–shell ZnO@SnO_2 NPs were developed by Yang and coworkers to improve the PV performance of $CsPbI_2Br$ PSCs, as shown in Figure 6.17c,d [134]. The enhanced electron mobility and $CsPbI_2Br$ film morphology contributed to the high-efficiency perovskite solar cells. In addition, doping engineering in ZnO ETLs regulated the $CsPbI_2Br$ film morphology. Song and coworkers demonstrated that the thickness-insensitive Al-doped ZnO not only provided an appropriate energy level structure to improve V_{OC} but also enhanced the quality of perovskite films, as shown in Figure 6.17 [135].

6.4.2.4 MOs HTLs

Easy-fabricated *p*-type metal oxide semiconductors with proper energy structure and stability are promising HTL candidates to be integrated into PSCs. These *p*-type semiconductors including NiO_x, Co_3O_4, and MoO_x, all of which have been successfully applied to PSCs. Mai et al. developed solution processed NiO_x HTLs in $CsPbI_2Br$ and $CsPbIBr_2$ inverted systems (Figure 6.18a), confirming

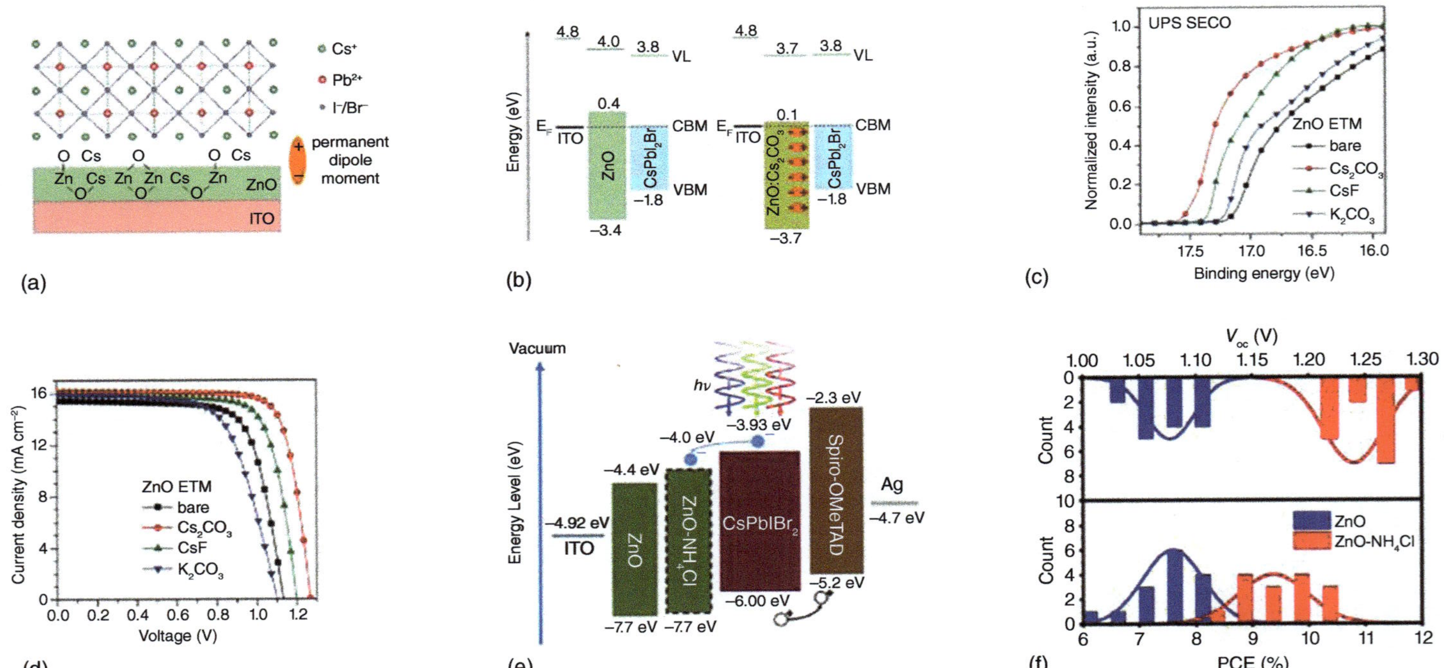

Figure 6.16 (a) Dipole formation and (b) the energy level alignments at the ETL/perovskite interface, (c) UPS spectra in the secondary electron cutoff region and (d) the *J*–*V* curves of PSCs based on different ETLs. Source: Reproduced with permission from Shen et al. [130]. Copyright 2019 American Chemical Society. (e) Energy level diagrams of the device and (f) V_{OC} and PCE statistical comparison for 15 ZnO and ZnO-NH_4Cl PSCs. Source: Reproduced from Wang et al. [102]/with permission of John Wiley & Sons.

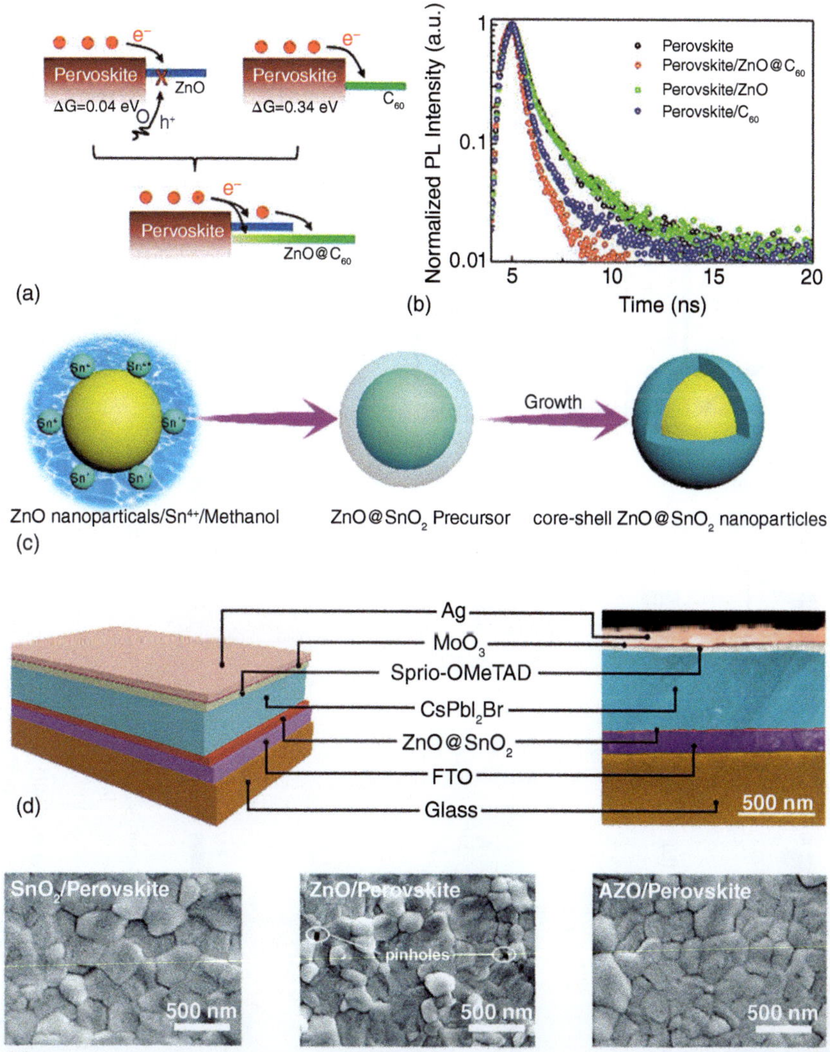

Figure 6.17 (a) Schematic view of different ETL with bandgap alignment and (b) TRPL spectra of $CsPbI_2Br$ thin films based on different ETLs. Source: Reproduced from Liu et al. [133]/with permission of American Chemical Society. (c) Synthetic diagram of core–shell $ZnO@SnO_2$ NPs and (d) schematic of the device structure of the inorganic perovskite solar cell. Source: Reproduced from Li et al. [134]. 2019, © American Chemical Society. (e) Top-view SEM images of perovskites deposited with SnO_2, ZnO and Al: ZnO films. Source: Reproduced with permission from Yang et al. [135]. © 2020, Royal Society of Chemistry.

the effectiveness of NiO_x HTLs in inorganic PSCs [136]. However, this solution process needed an ultra-high sintering temperature (400 °C), potentially corroding the conductivity of transparency conductor oxide electrodes. To overcome this dilemma, Tang and coworkers fabricated NiO_x HTLs for $CsPbI_2Br$ PSCs by DC reactive magnetron sputtering, as shown in Figure 6.18b. The O_2 gas flux and film

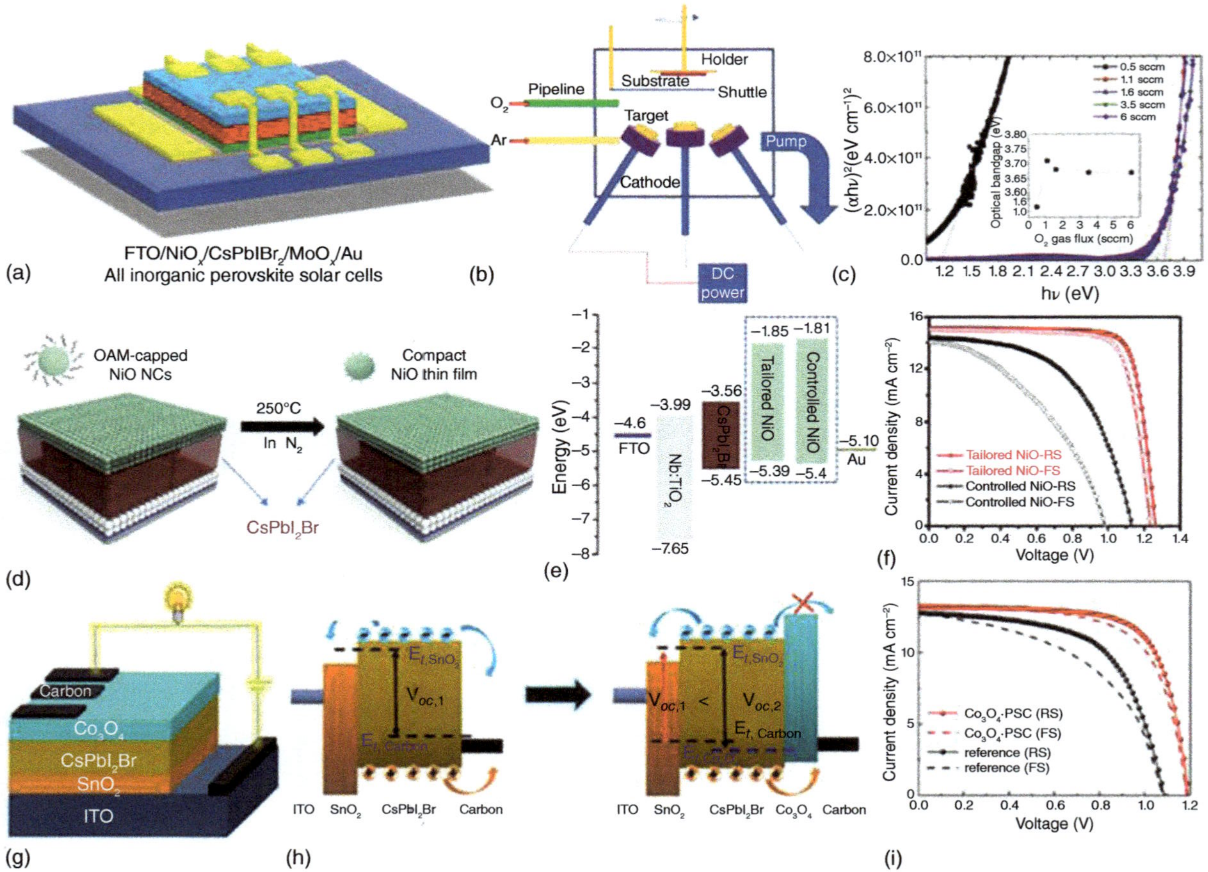

Figure 6.18 (a) Schematic illustrations of *p-i-n* CsPbIBr$_2$ PSCs with a NiO$_x$ HTL. Source: Reproduced from Liu et al. [136]/with permission of Elsevier. (b) Schematic illustrations of NiO$_x$ thin films deposited by DC reactive sputtering with a three-target sputtering system and (c) bandgap as a function of O$_2$ flux. Source: Reproduced from Pan et al. [137]/with permission of Elsevier. (d) Schematic illustrations of the fabrication process of tailored NiO HTLs. (e) Energy structures of PSCs and (f) *J–V* curves. Source: Reproduced from Chen et al. [138]/with permission of Elsevier. (g) Schematic illustrations of Co$_3$O$_4$ PSCs. (h) Energy level diagrams for reference and Co$_3$O$_4$ PSCs and (i) *J–V* curves. Source: Reproduced from Zhou et al. Zhou et al. [139]/with permission of John Wiley & Sons.

thickness were precisely controlled to optimize the target voltage, transmittance, deposition rate, and resistivity [137]. Since NiO_x needs a high-energy deposition process, it is unsuitable to deposit it on organic perovskite films with fragile formation. However, this thermal obstacle could be relieved in $CsPbX_3$ PSCs, which are stable under 250 °C. Accordingly, several *n-i-p* inorganic PSCs based on NiO_x HTLs are developed with good PV performance and stability. Chen et al. deposited oleylamine (OAM)-tailored NiO QDs on $CsPbI_2Br$ films to fabricate all-inorganic PSCs, as shown in Figure 6.18e [138]. The tailored NiO QDs films exhibited compact, pinhole-free morphologies, and excellent charge-extraction capabilities, which were contributed by band alignment with adjacent perovskites. These merits enhanced the PCEs and stability. Apart from NiO_x, other *p*-type metal oxides like MoO_3 were also integrated into the $CsPbX_3$ PSCs as an accessory of HTL. Hao and coworkers inserted a thin MoO_3 layer between Spiro-OMeTAD and Ag electrodes to accelerate hole extraction and transport to electrodes [140]. Liu and coworkers exploited a new HTL i.e. Co_3O_4, and applied it to $CsPbI_2Br$ PSCs, as shown in Figure 6.18g–i [139]. Compared with the Fermi level of the carbon electrodes, the deeper VBM of Co_3O_4 restrained the interface recombination and thus enhanced hole extraction efficiency, resulting in an improved performance.

Metal oxide semiconductors have demonstrated their potential in solar cells, especially third-generation solar cells such as DSSCs and PSCs. However, further studies should pay more attention to the contact stability and interface carrier dynamics to improve the performance of the related solar cells.

References

1 Mor, G., Varghese, O., Paulose, M. et al. (2006). A review on highly ordered, vertically oriented TiO_2 nanotube arrays: fabrication, material properties, and solar energy applications. *Sol. Energy Mater. Sol. Cells* 90 (14): 2011–2075.

2 Pan, Z., Dai, Z., and Wang, Z. (2001). Nanobelts of semiconducting oxides. *Science* 291 (5510): 1947–1949.

3 Rühle, S., Anderson, A., Barad, H. et al. (2012). All-oxide photovoltaics. *J. Phys. Chem. Lett.* 3 (24): 3755–3764.

4 Paracchino, A., Brauer, J., Moser, J. et al. (2012). Synthesis and characterization of high-photoactivity electrodeposited Cu_2O solar absorber by photoelectrochemistry and ultrafast spectroscopy. *J. Phys. Chem. C* 116 (13): 7341–7350.

5 Nishi, Y., Miyata, T., and Minami, T. (2013). The impact of heterojunction formation temperature on obtainable conversion efficiency in n-ZnO/p-Cu_2O solar cells. *Thin Solid Films* 528: 72–76.

6 Olsen, L., Addis, F., and Miller, W. (1982). Experimental and theoretical studies of Cu_2O solar cells. *Sol. Cells* 7 (3): 247–279.

7 Elfadill, N., Hashim, M., Chahrour, K. et al. (2016). Preparation of p-type Na-doped Cu_2O by electrodeposition for a p-n homojunction thin film solar cell. *Semicond. Sci. Technol.* 31 (6): 065001.

8 Minami, T., Nishi, Y., and Miyata, T. (2016). Efficiency enhancement using a $Zn_{1-x}Ge_x$–O thin film as an n-type window layer in Cu_2O-based heterojunction solar cells. *Appl. Phys Express* 9 (5): 052301.

9 Pérez-Tomás, A., Mingorance, A., Tanenbaum, D. et al. (2018). Metal oxides in photovoltaics: all-oxide, ferroic, and perovskite solar cells. In: *The Future of Semiconductor Oxides in Next-Generation Solar Cells*, 267–356. Elsevier.

10 Cui, J. and Gibson, U. (2010). A simple two-step electrodeposition of Cu_2O/ZnO nanopillar solar cells. *J. Phys. Chem. C* 114 (14): 6408–6412.

11 Musselman, K., Wisnet, A., Iza, D. et al. (2010). Strong efficiency improvements in ultra-low-cost inorganic nanowire solar cells. *Adv. Mater.* 22 (35): 254–258.

12 Chen, X., Lin, P., Yan, X. et al. (2015). Three-dimensional ordered ZnO/Cu_2O nanoheterojunctions for efficient metal–oxide solar cells. *ACS Appl. Mater. Interfaces* 7 (5): 3216–3223.

13 Jia, W., Dong, H., Zhao, J. et al. (2012). P-Cu_2O/n-ZnO heterojunction fabricated by hydrothermal method. *Appl. Phys. A.* 109 (3): 751–756.

14 Zhang, L., Sun, H., Xie, L. et al. (2015). Inorganic solar cells based on electrospun ZnO nanofibrous networks and electrodeposited Cu_2O. *Nanoscale Res. Lett.* 10 (1): 1–13.

15 Zhang, B.P., Binh, N.T., Wakatsuki, K. et al. (2014). Formation of highly aligned ZnO tubes on sapphire (0001) substrates. *Appl. Phys. Lett.* 84 (20): 4098–4100.

16 Scheer, R. (2009). Activation energy of heterojunction diode currents in the limit of interface recombination. *J. Appl. Phys.* 105 (10): 104505.

17 Minami, T., Miyata, T., and Nishi, Y. (2016). Relationship between the electrical properties of the n-oxide and p-Cu_2O layers and the photovoltaic properties of Cu_2O-based heterojunction solar cells. *Sol. Energy Mater. Sol. Cells* 147: 85–93.

18 Lee, Y., Chua, D., Brandt, R. et al. (2014). Atomic layer deposited gallium oxide buffer layer enables 1.2 V open-circuit voltage in cuprous oxide solar cells. *Adv. Mater.* 26 (27): 4704–4710.

19 Minami, T., Nishi, Y., and Miyata, T. (2013). Effect of the thin Ga_2O_3 layer in n +-ZnO/n-Ga_2O_3/p-Cu_2O heterojunction solar cells. *Thin Solid Films* 549: 65–69.

20 Minami, T., Nishi, Y., and Miyata, T. (2015). Heterojunction solar cell with 6% efficiency based on an n-type aluminum–gallium–oxide thin film and p-type sodium-doped Cu_2O sheet. *Appl. Phys Express* 8 (2): 022301.

21 Minami, T., Nishi, Y., and Miyata, T. (2016). Cu_2O-based solar cells using oxide semiconductors. *J. Semicond.* 37 (1): 014002.

22 Willig, F., Eichberger, R., Sundaresan, N. et al. (1990). Experimental time scale of Gerischer's distribution curves for electron-transfer reactions at semiconductor electrodes. *J. Am. Chem. Soc.* 112 (7): 2702–2707.

23 Zhao, Y., Zhang, H., Ren, X. et al. (2018). Thick TiO_2-based top electron transport layer on perovskite for highly efficient and stable solar cells. *ACS Energy Lett.* 3 (12): 2891–2898.

24 Guo, Z., Teo, S., Xu, Z. et al. (2019). Achievable high: V_{oc} of carbon based all-inorganic $CsPbIBr_2$ perovskite solar cells through interface engineering. *J. Mater. Chem. A* 7 (3): 1227–1232.

25 Law, M., Greene, L., Radenovic, A. et al. (2006). ZnO–Al_2O_3 and ZnO–TiO_2 core–shell nanowire dye-sensitized solar cells. *J. Phys. Chem. B* 110 (45): 22652–22663.

26 Liu, H., Chen, Z., Wang, H. et al. (2019). A facile room temperature solution synthesis of SnO_2 quantum dots for perovskite solar cells. *J. Mater. Chem. A* 7 (17): 10636–10643.

27 Shakeel, A.M., Pandey, A., and Abd, R.N. (2017). Advancements in the development of TiO_2 photoanodes and its fabrication methods for dye sensitized solar cell (DSSC) applications. A review. *Renewable Sustainable Energy Rev.* 77: 89–108.

28 Qian, C., Deng, Z., Yang, K. et al. (2018). Interface engineering of $CsPbBr_3/TiO_2$ heterostructure with enhanced optoelectronic properties for all-inorganic perovskite solar cells. *Appl. Phys. Lett.* 112 (9): 093901.

29 Zhang, S., Chen, W., Wu, S. et al. (2019). Hybrid inorganic electron-transporting layer coupled with a halogen-resistant electrode in $CsPbI_2Br$-based perovskite solar cells to achieve robust long-term stability. *ACS Appl. Mater. Interfaces* 11 (46): 43303–43311.

30 Liu, X., Li, B., Zhang, N. et al. (2018). Multifunctional RbCl dopants for efficient inverted planar perovskite solar cell with ultra-high fill factor, negligible hysteresis and improved stability. *Nano Energy* 53: 567–578.

31 Sawicka-Chudy, P., Sibiński, M., Rybak-Wilusz, E. et al. (2020). Review of the development of copper oxides with titanium dioxide thin-film solar cells. *AIP Adv.* 10 (1): 010701.

32 Dimopoulos, T., Peić, A., Müllner, P. et al. (2013). Photovoltaic properties of thin film heterojunctions with cupric oxide absorber. *J. Renewable Sustainable Energy* 5 (1): 011205.

33 Masudy-Panah, S., Radhakrishnan, K., Tan, H. et al. (2015). Titanium doped cupric oxide for photovoltaic application. *Sol. Energy Mater. Sol. Cells* 140: 266–274.

34 Morasch, J., Wardenga, H., Jaegermann, W. et al. (2016). Influence of grain boundaries and interfaces on the electronic structure of polycrystalline CuO thin films. *Phys. Status Solidi* 213 (6): 1615–1624.

35 Masudy-Panah, S., Dalapati, G., Radhakrishnan, K. et al. (2015). P -CuO/n -Si heterojunction solar cells with high open circuit voltage and photocurrent through interfacial engineering. *Prog. Photovoltaics Res. Appl.* 23 (5): 637–645.

36 Masudy-Panah, S., Radhakrishnan, K., Ru, T. et al. (2016). In situ codoping of a CuO absorber layer with aluminum and titanium: the impact of codoping and interface engineering on the performance of a CuO-based heterojunction solar cell. *J. Phys. D: Appl. Phys.* 49 (37): 375601.

37 Bhardwaj, R., Barman, R., and Kaur, D. (2016). Improved photovoltaic effect in CuO/$Zn_{1-x}Mg_xO$ heterojunction solar cell by pulsed laser deposition. *Mater. Lett.* 185: 230–234.

38 Wang, P., Zhao, X., and Li, B. (2011). ZnO-coated CuO nanowire arrays: fabrications, optoelectronic properties, and photovoltaic applications. *Opt. Express* 19 (12): 11271.

39 Zhao, X., Wang, P., Gao, Y. et al. (2014). CuO/ZnO core/shell nanowire arrays and their photovoltaics application. *Mater. Lett.* 132: 409–412.

40 Bhaumik, A., Haque, A., Karnati, P. et al. (2014). Copper oxide based nanostructures for improved solar cell efficiency. *Thin Solid Films* 572: 126–133.

41 Wijesundera, R. (2010). Fabrication of the CuO/Cu_2O heterojunction using an electrodeposition technique for solar cell applications. *Semicond. Sci. Technol.* 25 (4): 045015.

42 Jayathilaka, C., Kapaklis, V., Siripala, W. et al. (2015). Improved efficiency of electrodeposited p-CuO/n-Cu_2O heterojunction solar cell. *Appl. Phys Express* 8 (6): 065503.

43 Barreca, D., Massignan, C., Daolio, S. et al. (2001). Composition and microstructure of cobalt oxide thin films obtained from a novel cobalt(II) precursor by chemical vapor deposition. *Chem. Mater.* 13 (2): 588–593.

44 Patil, P., Kadam, L., and Lokhande, C. (1996). Preparation and characterization of spray pyrolysed cobalt oxide thin films. *Thin Solid Films* 272 (1): 29–32.

45 Kupfer, B., Majhi, K., Keller, D. et al. (2015). Thin film Co_3O_4/TiO_2 heterojunction solar cells. *Adv. Energy Mater.* 5 (1): 1401007.

46 Moser, J. (1887). Notiz über Verstärkung photoelektrischer Ströme durch optische Sensibilisirung. *Monatsh. Chem. verw. Teile anderer Wiss.* 8 (1): 373–373.

47 Jose, R., Thavasi, V., and Ramakrishna, S. (2009). Metal oxides for dye-sensitized solar cells. *J. Am. Ceram. Soc.* 92 (2): 289–301.

48 Alonso, V., Beley, M., Chartier, P. et al. (1981). Dye sensitization of ceramic semiconducting electrodes for photoelectrochemical conversion. *Rev. Phys. Appliquée* 16 (1): 5–10.

49 Matsumura, M., Matsudaira, S., Tsubomura, H. et al. (1980). Dye sensitization and surface structures of semiconductor electrodes. *Ind. Eng. Chem. Prod. Res. Dev.* 19 (3): 415–421.

50 O'Regan, B. and Grätzel, M. (1991). A low-cost, high-efficiency solar cell based on dye-sensitized colloidal TiO_2 films. *Nature* 353 (6346): 737–740.

51 Barbé, C., Arendse, F., Comte, P. et al. (2005). Nanocrystalline titanium oxide electrodes for photovoltaic applications. *J. Am. Ceram. Soc.* 80 (12): 3157–3171.

52 Wang, Q., Ito, S., Grätzel, M. et al. (2006). Characteristics of high efficiency dye-sensitized solar cells. *J. Phys. Chem. B* 110 (50): 25210–25221.

53 Kitiyanan, A. and Yoshikawa, S. (2005). The use of ZrO_2 mixed TiO_2 nanostructures as efficient dye-sensitized solar cells' electrodes. *Mater. Lett.* 59 (29–30): 4038–4040.

54 Kitiyanan, A., Kato, T., Suzuki, Y. et al. (2006). The use of binary TiO_2–GeO_2 oxide electrodes to enhanced efficiency of dye-sensitized solar cells. *J. Photochem. Photobiol., A* 179 (1–2): 130–134.

55 Eguchi, K., Koga, H., Sekizawa, K. et al. (2000). Nb_2O_5-based composite electrodes for dye-sensitized solar cells. *J. Ceram. Soc. Japan* 108 (12): 1067–1071.

56 Dürr, M., Rosselli, S., Yasuda, A. et al. (2006). Band-gap engineering of metal oxides for dye-sensitized solar cells. *J. Phys. Chem. B* 110 (43): 21899–21902.

57 Palomares, E., Clifford, J., Haque, S. et al. (2003). Control of charge recombination dynamics in dye sensitized solar cells by the use of conformally deposited metal oxide blocking layers. *J. Am. Chem. Soc.* 125 (2): 475–482.

58 Zaban, A., Chen, S., Chappel, S. et al. (2000). Bilayer nanoporous electrodes for dye sensitized solar cells. *Chem. Commun.* 22: 2231–2232.

59 Roh, S., Mane, R., Min, S. et al. (2006). Achievement of 4.51% conversion efficiency using ZnO recombination barrier layer in TiO_2 based dye-sensitized solar cells. *Appl. Phys. Lett.* 89 (25): 253512.

60 Jung, H., Lee, J., Nastasi, M. et al. (2005). Preparation of nanoporous MgO-coated TiO_2 nanoparticles and their application to the electrode of dye-sensitized solar cells. *Langmuir* 21 (23): 10332–10335.

61 Diamant, Y., Chen, S., Melamed, O. et al. (2003). Core–shell nanoporous electrode for dye sensitized solar cells: the effect of the $SrTiO_3$ shell on the electronic properties of the TiO_2 core. *J. Phys. Chem. B* 107 (9): 1977–1981.

62 So, S., Kriesch, A., Peschel, U. et al. (2015). Conical-shaped titania nanotubes for optimized light management in DSSCs reach back-side illumination efficiencies >8%. *J. Mater. Chem. A* 3 (24): 12603–12608.

63 So, S., Hwang, I., Yoo, J. et al. (2018). Inducing a nanotwinned grain structure within the TiO_2 nanotubes provides enhanced electron transport and DSSC efficiencies >10%. *Adv. Energy Mater.* 8 (33): 1800981.

64 Yen, Y., Ko, W., Chen, J. et al. (2013). Enhancing the performance of dye-sensitized solar cells based on TiO_2 nanotube/nanoparticle composite photoanodes. *Electrochim. Acta* 105: 142–148.

65 Maheswari, D. and Venkatachalam, P. (2015). Enhancing the performance of dye-sensitized solar cells based on organic dye sensitized TiO_2 nanoparticles/nanowires composite photoanodes with ionic liquid electrolyte. *Measurement* 60: 146–154.

66 Benkstein, K., Kopidakis, N., van de Lagemaat, J. et al. (2003). Influence of the network geometry on electron transport in nanoparticle networks. *MRS Proc.* 789, N13.5.

67 Forro, L., Chauvet, O., Emin, D. et al. (1994). High mobility n-type charge carriers in large single crystals of anatase (TiO_2). *J. Appl. Phys.* 75 (1): 633–635.

68 Adachi, M., Jiu, J., and Isoda, S. (2007). Synthesis of morphology-controlled titania nanocrystals and application for dye-sensitized solar cells. *Curr. Nanosci.* 3 (4): 285–295.

69 Thavasi, V., Singh, G., and Ramakrishna, S. (2008). Electrospun nanofibers in energy and environmental applications. *Energy Environ. Sci.* 1 (2): 205.

70 Kaidashev, E., Lorenz, M., von Wenckstern, H. et al. (2003). High electron mobility of epitaxial ZnO thin films on c-plane sapphire grown by multistep pulsed-laser deposition. *Appl. Phys. Lett.* 82 (22): 3901–3903.

71 Vittal, R. and Ho, K. (2017). zinc oxide based dye-sensitized solar cells: a review. *Renewable Sustainable Energy Rev.* 70: 920–935.

72 Karam, C., Habchi, R., Tingry, S. et al. (2018). Design of multilayers of urchin-like ZnO nanowires coated with TiO_2 nanostructures for dye-sensitized solar cells. *ACS Appl. Nano Mater.* 1 (7): 3705–3714.

73 Lee, C., Chou, C., Chen, C. et al. (2014). Zinc oxide-based dye-sensitized solar cells with a ruthenium dye containing an alkyl bithiophene group. *J. Power Sources* 246: 1–9.

74 Akhtar, M., Khan, M., Jeon, M. et al. (2008). Controlled synthesis of various ZnO nanostructured materials by capping agents-assisted hydrothermal method for dye-sensitized solar cells. *Electrochim. Acta* 53 (27): 7869–7874.

75 Mou, J., Zhang, W., Fan, J. et al. (2011). Facile synthesis of ZnO nanobullets/nanoflakes and their applications to dye-sensitized solar cells. *J. Alloys Compd.* 509 (3): 961–965.

76 Lee, C., Lin, J., Wang, Y. et al. (2011). Synthesis of hexagonal ZnO clubs with opposite faces of unequal dimensions for the photoanode of dye-sensitized solar cells. *Phys. Chem. Chem. Phys.* 13 (47): 20999–21008.

77 Abd-Ellah, M., Moghimi, N., Zhang, L. et al. (2013). Effect of electrolyte conductivity on controlled electrochemical synthesis of zinc oxide nanotubes and nanorods. *J. Phys. Chem. C* 117 (13): 6794–6799.

78 Lupan, O., Guérin, V., Ghimpu, L. et al. (2012). Nanofibrous-like ZnO layers deposited by magnetron sputtering and their integration in dye-sensitized solar cells. *Chem. Phys. Lett.* 550: 125–129.

79 Qu, J., Yang, Y., Wu, Q. et al. (2014). Hedgehog-like hierarchical ZnO needle-clusters with superior electron transfer kinetics for dye-sensitized solar cells. *RSC Adv.* 4 (22): 11430–11437.

80 Shi, Y., Zhu, C., Wang, L. et al. (2013). Asymmetric ZnO panel-like hierarchical architectures with highly interconnected pathways for free-electron transport and photovoltaic improvements. *Chem. A Eur. J.* 19 (1): 282–287.

81 Jiang, C., Sun, X., Lo, G. et al. (2007). Improved dye-sensitized solar cells with a ZnO-nanoflower photoanode. *Appl. Phys. Lett.* 90 (26): 263501.

82 Kim, I., Hong, J., Lee, B. et al. (2007). Dye-sensitized solar cells using network structure of electrospun ZnO nanofiber mats. *Appl. Phys. Lett.* 91 (16): 163109.

83 Lin, C., Lai, Y., Chen, H. et al. (2011). Highly efficient dye-sensitized solar cell with a ZnO nanosheet-based photoanode. *Energy Environ. Sci.* 4 (9): 3448.

84 Memarian, N., Concina, I., Braga, A. et al. (2011). Hierarchically assembled ZnO nanocrystallites for high-efficiency dye-sensitized solar cells. *Angew. Chem.* 123 (51): 12529–12533.

85 Kakiage, K., Aoyama, Y., Yano, T. et al. (2015). Highly-efficient dye-sensitized solar cells with collaborative sensitization by silyl-anchor and carboxy-anchor dyes. *Chem. Commun.* 51 (88): 15894–15897.

86 Horiuchi, H., Katoh, R., Hara, K. et al. (2003). Electron injection efficiency from excited N_3 into nanocrystalline ZnO films: effect of (N^{3-}–Zn^{2+}) aggregate formation. *J. Phys. Chem. B* 107 (11): 2570–2574.

87 Anta, J., Guillén, E., and Tena-Zaera, R. (2012). ZnO-based dye-sensitized solar cells. *J. Phys. Chem. C* 116 (21): 11413–11425.

88 Shin, Y., Lee, J., Park, J. et al. (2007). Enhanced photovoltaic properties of SiO_2-treated ZnO nanocrystalline electrode for dye-sensitized solar cell. *Chem. Lett.* 36 (12): 1506–1507.

89 Giannouli, M., Govatsi, K., Syrrokostas, G. et al. (2018). Factors affecting the power conversion efficiency in ZnO DSSCs: nanowire vs. nanoparticles. *Materials* 11 (3): 411.

90 Sakai, N., Usui, R., and Murakami, T. (2013). Optimum particle size of ZnO for dye-sensitized solar cells. *Chem. Lett.* 42 (8): 810–812.

91 Chang, J., Ahmed, R., Wang, H. et al. (2013). ZnO nanocones with high-index $\{10\bar{1}1\}$ facets for enhanced energy conversion efficiency of dye-sensitized solar cells. *J. Phys. Chem. C* 117 (27): 13836–13844.

92 Virkki, K., Tervola, E., Ince, M. et al. (2018). Comparison of electron injection and recombination on TiO_2 nanoparticles and ZnO nanorods photosensitized by phthalocyanine. *R. Soc. Open Sci.* 5 (7): 180323.

93 Ragam, M., Sankar, N., and Ramachandran, K. (2013). Charge transport in 1-D nanostructured CdS dye sensitized solar cell. *J. Nanosci.* 2013: 1–7.

94 Sayama, K., Sugihara, H., and Arakawa, H. (1998). Photoelectrochemical properties of a porous Nb_2O_5 electrode sensitized by a ruthenium dye. *Chem. Mater.* 10 (12): 3825–3832.

95 Zhang, H., Wang, Y., Yang, D. et al. (2012). Directly hydrothermal growth of single crystal $Nb_3O_7(OH)$ nanorod film for high performance dye-sensitized solar cells. *Adv. Mater.* 24 (12): 1598–1603.

96 Le Viet, A., Jose, R., Reddy, M. et al. (2010). Nb_2O_5 photoelectrodes for dye-sensitized solar cells: choice of the polymorph. *J. Phys. Chem. C* 114 (49): 21795–21800.

97 Liu, X., Yuan, R., Liu, Y. et al. (2016). Niobium pentoxide nanotube powder for efficient dye-sensitized solar cells. *New J. Chem.* 40 (7): 6276–6280.

98 Abdul, R.R., Zoolfakar, A., Subbiah, J. et al. (2014). Highly ordered anodized Nb_2O_5 nanochannels for dye-sensitized solar cells. *Electrochem. Commun.* 40: 20–23.

99 Ghosh, R., Brennaman, M., Uher, T. et al. (2011). Nanoforest Nb_2O_5 photoanodes for dye-sensitized solar cells by pulsed laser deposition. *ACS Appl. Mater. Interfaces* 3 (10): 3929–3935.

100 Jin, X., Liu, C., Xu, J. et al. (2014). Size-controlled synthesis of mesoporous Nb_2O_5 microspheres for dye sensitized solar cells. *RSC Adv.* 4 (67): 35546–35553.

101 Nunes, B., Faustino, L., Muller, A. et al. (2019). Nb_2O_5 dye-sensitized solar cells. In: *Nanomaterials for Solar Cell Applications*, 287–322. Elsevier.

102 Wang, H., Cao, S., Yang, B. et al. (2020). NH_4Cl-modified ZnO for high-performance $CsPbIBr_2$ perovskite solar cells via low-temperature process. *Sol. RRL* 4 (1): 1900363.

103 Yang, D., Yang, R., Wang, K. et al. (2018). High efficiency planar-type perovskite solar cells with negligible hysteresis using EDTA-complexed SnO_2. *Nat. Commun.* 9 (1): 3239.

104 Wei, J., Guo, F., Wang, X. et al. (2018). SnO_2-in-polymer matrix for high-efficiency perovskite solar cells with improved reproducibility and stability. *Adv. Mater.* 30 (52): 1805153.

105 Zhou, Y., Li, X., and Lin, H. (2020). To be higher and stronger—metal oxide electron transport materials for perovskite solar cells. *Small* 16 (15): 1902579.

106 Subhani, W., Wang, K., Du, M. et al. (2019). Interface-modification-induced gradient energy band for highly efficient $CsPbIBr_2$ perovskite solar cells. *Adv. Energy Mater.* 9 (21): 1803785.

107 Wojciechowski, K., Leijtens, T., Siprova, S. et al. (2015). C_{60} as an efficient n-type compact layer in perovskite solar cells. *J. Phys. Chem. Lett.* 6 (12): 2399–2405.

108 Deng, F., Li, X., Lv, X. et al. (2020). Low-temperature processing all-inorganic carbon-based perovskite solar cells up to 11.78% efficiency via alkali hydroxides interfacial engineering. *ACS Appl. Energy Mater.* 3 (1): 401–410.

109 Mali, S., Patil, J., and Hong, C. (2020). Simultaneous improved performance and thermal stability of planar metal ion incorporated $CsPbI_2Br$ all-inorganic perovskite solar cells based on MgZnO nanocrystalline electron transporting layer. *Adv. Energy Mater.* 10 (3): 1902708.

110 Han, G., Song, Y., Jin, Y. et al. (2015). Reduced graphene oxide/mesoporous TiO_2 nanocomposite based perovskite solar cells. *ACS Appl. Mater. Interfaces* 7 (42): 23521–23526.

111 Han, Y., Zhao, H., Duan, C. et al. (2020). Controlled n-doping in air-stable $CsPbI_2Br$ perovskite solar cells with a record efficiency of 16.79%. *Adv. Funct. Mater.* 30 (12): 1909972.

112 Duan, J., Wang, Y., Yang, X. et al. (2020). Alkyl-chain-regulated charge transfer in fluorescent inorganic $CsPbBr_3$ perovskite solar cells. *Angew. Chem. Int. Ed.* 59 (11): 4391–4395.

113 Zhu, W., Zhang, Z., Chai, W. et al. (2019). Band alignment engineering towards high efficiency carbon-based inorganic planar $CsPbIBr_2$ perovskite solar cells. *ChemSusChem* 12 (10): 2318–2325.

114 Cao, S., Wang, H., Li, H. et al. (2020). Critical role of interface contact modulation in realizing low-temperature fabrication of efficient and stable $CsPbIBr_2$ perovskite solar cells. *Chem. Eng. J.* 394: 124903.

115 Jiang, Q., Zhang, X., and You, J. (2018). SnO_2: A wonderful electron transport layer for perovskite solar cells. *Small* 14 (31): 1801154.

116 Ye, Q., Zhao, Y., Mu, S. et al. (2019). Cesium Lead inorganic solar cell with efficiency beyond 18% via reduced charge recombination. *Adv. Mater.* 31 (49): 1905143.

117 Tian, J., Xue, Q., Tang, X. et al. (2019). Dual interfacial design for efficient $CsPbI_2Br$ perovskite solar cells with improved photostability. *Adv. Mater.* 31 (23): 1901152.

118 Wang, H., Li, H., Cao, S. et al. (2020). Interface modulator of ultrathin magnesium oxide for low-temperature-processed inorganic $CsPbIBr_2$ perovskite solar cells with efficiency over 11%. *Sol. RRL* 279 (24): 2000226.

119 Zhao, Y., Duan, J., Yuan, H. et al. (2019). Using SnO_2 QDs and $CsMBr_3$ (M = Sn, Bi, Cu) QDs as charge-transporting materials for 10.6%-efficiency all-inorganic $CsPbBr_3$ perovskite solar cells with an ultrahigh open-circuit voltage of 1.610 V. *Sol. RRL* 3 (3): 1800284.

120 Yan, L., Xue, Q., Liu, M. et al. (2018). Interface engineering for all-inorganic $CsPbI_2Br$ perovskite solar cells with efficiency over 14%. *Adv. Mater.* 30 (33): 1802509.

121 Zhu, W., Chai, W., Zhang, Z. et al. (2019). Interfacial TiO_2 atomic layer deposition triggers simultaneous crystallization control and band alignment for efficient $CsPbIBr_2$ perovskite solar cell. *Org. Electron.* 74: 103–109.

122 Zhou, Q., Duan, J., Wang, Y. et al. (2020). Tri-functionalized TiO_xCl_{4-2x} accessory layer to boost efficiency of hole-free, all-inorganic perovskite solar cells. *J. Energy Chem.* 50: 1–8.

123 Liu, D. and Kelly, T. (2014). Perovskite solar cells with a planar heterojunction structure prepared using room-temperature solution processing techniques. *Nat. Photonics* 8 (2): 133–138.

124 Tiwana, P., Docampo, P., Johnston, M. et al. (2011). Electron mobility and injection dynamics in mesoporous ZnO, SnO_2, and TiO_2 films used in dye-sensitized solar cells. *ACS Nano* 5 (6): 5158–5166.

125 You, J., Meng, L., Song, T. et al. (2016). Improved air stability of perovskite solar cells via solution-processed metal oxide transport layers. *Nat. Nanotechnol.* 11 (1): 75–81.

126 Tseng, Z., Chiang, C., Chang, S. et al. (2016). Surface engineering of ZnO electron transporting layer via Al doping for high efficiency planar perovskite solar cells. *Nano Energy* 28: 311–318.

127 Liang, L., Huang, Z., Cai, L. et al. (2014). Magnetron sputtered zinc oxide nanorods as thickness-insensitive cathode interlayer for perovskite planar-heterojunction solar cells. *ACS Appl. Mater. Interfaces* 6 (23): 20585–20589.

128 Yang, J., Siempelkamp, B., Mosconi, E. et al. (2015). Origin of the thermal instability in CH3NH3PbI3 thin films deposited on ZnO. *Chem. Mater.* 27 (12): 4229–4236.

129 Yue, M., Su, J., Zhao, P. et al. (2019). Optimizing the performance of $CsPbI_3$-based perovskite solar cells via doping a ZnO electron transport layer coupled with interface engineering. *Nano-Micro Lett.* 11 (1): 91.

130 Shen, E., Chen, J., Tian, Y. et al. (2020). Interfacial energy level tuning for efficient and thermostable $CsPbI_2Br$ perovskite solar cells. *Adv. Sci.* 7 (1): 1901952.

131 Chen, W., Zhang, J., Xu, G. et al. (2018). A semitransparent inorganic perovskite film for overcoming ultraviolet light instability of organic solar cells and achieving 14.03% efficiency. *Adv. Mater.* 30 (21): 1800855.

132 Zhang, B., Zhou, Y., Xue, Q. et al. (2019). The energy-alignment engineering in polytriphenylamines-based hole transport polymers realizes low energy loss and high efficiency for all-inorganic perovskite solar cells. *Sol. RRL* 3 (9): 1900265.

133 Liu, C., Li, W., Zhang, C. et al. (2018). All-inorganic $CsPbI_2Br$ perovskite solar cells with high efficiency exceeding 13%. *J. Am. Chem. Soc.* 140 (11): 3825–3828.

134 Li, Z., Wang, R., Xue, J. et al. (2019). Core-Shell ZnO@SnO_2 nanoparticles for efficient inorganic perovskite solar cells. *J. Am. Chem. Soc.* 141 (44): 17610–17616.

135 Yang, X., Yang, H., Hu, X. et al. (2020). Low-temperature interfacial engineering for flexible $CsPbI_2Br$ perovskite solar cells with high performance beyond 15%. *J. Mater. Chem. A* 8 (10): 5308–5314.

136 Liu, C., Li, W., Chen, J. et al. (2017). Ultra-thin MoO_x as cathode buffer layer for the improvement of all-inorganic $CsPbIBr_2$ perovskite solar cells. *Nano Energy* 41: 75–83.

137 Pan, L., Liu, C., Zhu, H. et al. (2020). Fine modification of reactively sputtered NiO_X hole transport layer for application in all-inorganic $CsPbI_2Br$ perovskite solar cells. *Sol. Energy* 196: 521–529.

138 Chen, W., Zhang, S., Liu, Z. et al. (2019). A tailored nickel oxide hole-transporting layer to improve the long-term thermal stability of inorganic perovskite solar cells. *Sol. RRL* 3 (11): 1900346.

139 Zhou, Y., Zhang, X., Lu, X. et al. (2019). Promoting the hole extraction with Co_3O_4 nanomaterials for efficient carbon-based $CsPbI_2Br$ perovskite solar cells. *Sol. RRL* 3 (4): 1800315.

140 Zhou, L., Guo, X., Lin, Z. et al. (2019). Interface engineering of low temperature processed all-inorganic $CsPbI_2Br$ perovskite solar cells toward PCE exceeding 14%. *Nano Energy* 60: 583–590.

7

Metal Oxide Semiconductors for Ultraviolet Photodetectors

7.1 Introduction

Ultraviolet (UV) light refers to electromagnetic radiation with a typical wavelength ranging from 10 to 400 nm, corresponding to the large energy distribution of 3.1–124 eV [1]. According to wavelength, UV light is classified into four subregions, including UVA (320–400 nm), UVB (280–320 nm), UVC (100–280 nm), and extreme ultraviolet (EUV) (10–120 nm) [2, 3]. It has been well known that most solar UV radiation consisting of UVB, UVC, and EUV is absorbed by the ozone layer near the ground, resulting in their absence in the solar spectrum [4]. As a result, compared with visible and near-infrared light, UV light may not be affected by solar radiation, exhibiting the distinctive characteristics of UV in various fields, such as UV communication, environment monitoring, space exploration, industrial detection, medicine, and healthcare [5–7].

In the last century, despite the advance of scientific and technological fields based on UV radiation, the exploration and development of UV photodetector relatively lag behind because most of the reported absorbing materials with narrow bandgaps cannot respond to the high-energy UV light as well as original wide-bandgap semiconductors, i.e. GaN and SiC, are associated with high manufacturing costs and low efficiency [8, 9]. Metal oxide semiconductors with wide bandgaps, which feature feasible processability, low manufacturing cost, and excellent performance, have been regarded as excellent candidates for UV absorbers and applied in UV photodetectors. Apart from the exploration of absorbers in metal oxide semiconductors, the design and optimization of device structures are critical to the enhancement of detection performance. Thus, it is of great importance to review the reported UV photodetectors based on metal oxide semiconductors as well as summarize their technical strategies and advantages in order to promote the development of high-performance UV photodetectors.

In this chapter, a comprehensive review of UV photodetectors based on metal oxide semiconductors and their detection properties is provided. First, the device structures, including photoconductors, Schottky photodiodes, avalanche photodiodes, metal-semiconductor-metal (MSM) photodetectors, *p*-(*i*)-*n* photodiodes, and phototransistors, as well as corresponding features are introduced. Then, key parameters to evaluate the performance of UV photodetectors are presented.

Metal Oxide Semiconductors: Synthesis, Properties, and Devices,
First Edition. Zhigang Zang, Wensi Cai, and Yong Zhou.

Furthermore, UV detection properties of a variety of metal oxide semiconductors, consisting of ZnO, TiO_2, SnO_2, Ga_2O_3, NiO, and ZnMgO, are discussed and summarized. In addition, the influence of their morphologies on UV detection performance is also included in this chapter. Finally, conclusions and outlooks on metal oxide semiconductors for UV photodetectors are presented and highlighted, respectively.

7.2 Device Structures of UV Photodetectors

Generally, UV photodetectors are divided into two main types, namely vacuum UV photodetectors and solid-state UV photodetectors. It has been reported that the cores of vacuum UV photodetectors are photomultiplier tubes (PMTs) and integrated systems based on PMTs [10]. Although vacuum UV photodetectors exhibit advantages of large-area detection and fast response, the drawbacks of PMTs, such as high power consumption, heavy weight, low efficiency, and complex manufacturing process, may impede their further commercialization in the next generation [11]. In contrast, solid-state UV photodetectors based on semiconductors are known to offer several advantages and are identified as potential candidates for efficient UV photodetectors. According to the working mechanisms in UV detection, solid-state UV photodetectors can be classified into photoconductors, Schottky photodiodes, avalanche photodiodes, MSM photodetectors, *p*-(*i*)-*n* photodiodes, and phototransistors, as shown in Figure 7.1. Their detailed structures as well as advantages and drawbacks are described in this subchapter.

7.2.1 Photoconductors

The basic structure of a photoconductor involves a semiconductor as light-absorbing layer and two conductors as electrodes, resulting in two Ohmic contacts at the different sides of the semiconductor, as shown in Figure 7.2a [12]. The electrodes are

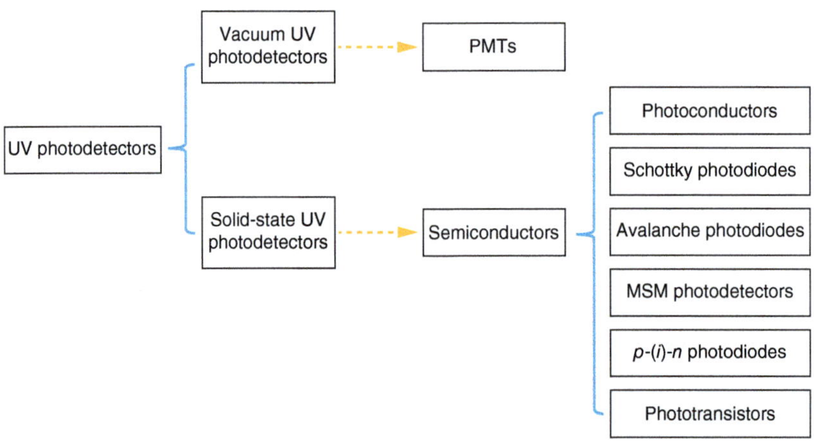

Figure 7.1 Categories of UV photodetectors.

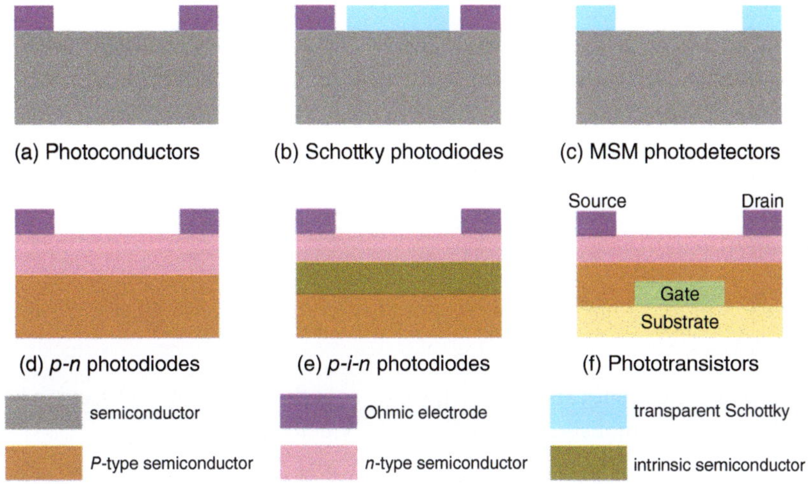

Figure 7.2 Schematic structures of typical UV photodetectors. (a) Photoconductors. (b) Schottky photodiodes. (c) MSM photodetectors. (d) *p–n* photodiodes. (e) *p-i-n* photodiodes. (f) Phototransistors.

deposited directly on the semiconductors, allowing for the fabrication of devices with a photoconductor structure. The working principle of the photodetectors is regarded as the photoconductor effect, in which the free carriers excited by incident photons with higher energy than the bandgaps of absorbers enable them to drift within the semiconductors when the external voltages are applied to the two electrodes [13]. The external voltages applied to the photoconductors facilitate the separation of the photon-generated excitons into electrons and holes. Besides, the external voltages benefit the drift of electrons and holes to the opposite electrodes, leading to the generation of light current (I_{light}). However, in the darkness, the external voltages may allow part of electrons in semiconductors to flow from one electrode to another, exhibiting a small dark current (I_{dark}). Thus, the photocurrent (I_{ph}) of photodetectors, which is evaluated as $I_{ph} = I_{light} - I_{dark}$, is primarily contributed by the photogenerated majority carriers conducted between two Ohmic contacts.

Apart from majority carriers, the minority carriers are found to diffuse to one of the electrodes. To maintain the charge balance, additional majority carriers are generated until the recombination of the minority carriers occurs, leading to an increased photocurrent, which is confined as photoconductor gain (G). During the gain process of majority carriers described above, the amount of increased majority carriers is mainly determined by the ratio of a minority carrier lifetime to a majority carrier drifting time in the Ohmic contacts. Thus, a longer minority carrier lifetime benefits in the achievement of a higher gain, but the response time of photoconductors, which depends on the minority carriers' lifetime, becomes longer. Generally, despite the advantages of photoconductors, such as a facile manufacturing process, a simple device structure, and a high gain, the large dark current and slow response speed induced by the existence of external voltages

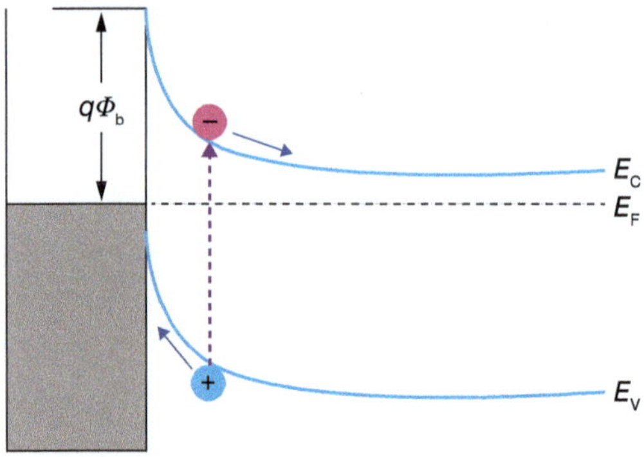

Figure 7.3 A schematic diagram of an operating Schottky junction in Schottky photodiodes.

applied between electrodes and low energy barrier in the Ohmic contact seem to be formidable for applications such as high-speed UV detection.

7.2.2 Schottky Photodiodes

The dominant structure of Schottky photodiodes combines an Ohmic contact with a transparent Schottky contact, as shown in Figure 7.2b. In contrast to Ohmic contacts, Schottky contacts in UV photodetectors consist of semiconductors and metal with different work functions from those of corresponding semiconductors, resulting in internal electric fields in Schottky contacts, as shown in Figure 7.3 [14]. When the photons with larger energy ($h\nu$) than the bandgaps of semiconductors stimulate the semiconductors, the excited electron–hole pairs may spontaneously separate to free electrons and holes resulting from the driven effect of internal fields. Thus, the generation of light current arises in the Schottky photodiodes without the external field. Besides, the internal fields in Schottky contacts act as an energy barrier to hinder the diffusion of reversed carriers and decrease the storage of minority carriers, resulting in the suppression of dark current [15, 16]. Consequently, the photocurrent of Schottky photodiodes is found to be larger than that of photoconductors, and in the meantime, the internal fields enhance the response speed of detection.

Apart from low dark current and rapid response, the possible self-powered operation, high responsivity, and the absence of high-temperature manufacturing requirements are the attractive advantages of Schottky photodiodes. However, in terms of Schottky photodiodes, it is necessary for the incident UV light to penetrate the transparent Schottky "window" and then be absorbed by the interlayers, so the transparent metal electrodes with small thickness need to be deposited on the semiconductors, which may complicate the fabrication process and weaken the characteristics of metal-semiconductor Schottky junctions. The use of semitransparent metal electrodes, on the other hand, may cause a reduced UV intensity and the decreased sensitivity of UV detection [17].

7.2.3 MSM Photodetectors

Similar to Schottky photodiodes, the main structures of MSM photodetectors are Schottky junctions, but an MSM photodiode contains two Schottky contacts to form

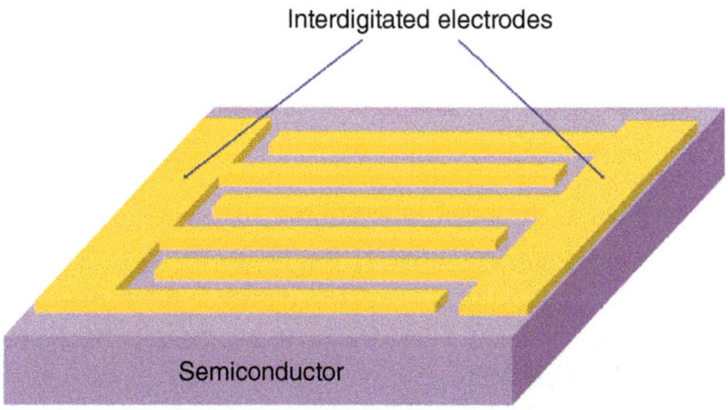

Figure 7.4 A schematic diagram showing an MSM detector with interdigitated fingers.

back-to-back connected Schottky junctions [18, 19]. A typical structure of MSM photodetectors is shown in Figure 7.2c, where the double metal electrodes are deposited on the top of semiconductors to achieve MSM configuration of interdigitated fingers (Figure 7.4). Note that MSM photodetector can only operate with the assistance of external bias voltages. Under UV illuminations, the external voltages enable the separation and driving of the excited photocarriers to electrodes, which operate similarly to photoconductors. Owing to the rectifying nature originated from Schottky junctions, the dark current of MSM photodetectors is smaller than that of photoconductors. In addition, because of the low capacitance per unit area and limited drifting time, MSM photodetectors exhibit rapid response speed.

Furthermore, the photocarriers trapped at the surface cause an asymmetric charge field, which may change the potential distribution near the Schottky junctions. And the minority carriers are found to be trapped near one electrode, which can eventually create an increased majority carrier injection from another electrode via a tunneling effect, giving rise to the presence of the photocurrent gain in MSM photodetectors [20]. The low-cost and easily controllable manufacturing processes ensure the large-scale applications of MSM photodetectors in UV detections. The interdigital electrodes are identified as the main reason for low external quantum efficiency (EQE), which may be solved by depositing semitransparent metal electrodes on the top of semiconductors.

7.2.4 *p*-(*i*)-*n* Photodiodes

For semiconductors, doping is regarded as an effective strategy to modulate their carrier types (*p*: positive; *n*: negative) and concentrations [21]. A typical structure of *p*-(*i*)-*n* photodiodes is schematically shown in Figure 7.2d, in which the interface and contact of a *p*-type and *n*-type semiconductor is defined as a "*p*–*n* junction." Owing to the difference in work functions, the carriers in a *p*–*n* junction have to diffuse to meet the requirement of Fermi-level alignment. As a result, the carriers in a space charge region (depleted region) are found to reduce dramatically, leading to the creation of a built-in electric field. Similar to Schottky junctions, the *p*–*n* junction exhibits a rectifying behavior, modulating the dark current as an asymmetric current–voltage (*I*–*V*) character [22]. Under UV illuminations, photons with larger energy than that of semiconductors' bandgaps are absorbed, creating electron–hole pairs. They may diffuse to the space charge region and then be separated into free electrons and holes by the built-in electric field. With the assistance of external bias voltages, the separated electrons and holes are propelled toward opposite electrodes, resulting in an increased light current. In addition, the energy barriers near the *p*–*n* junction enable to suppress the dark current in the devices. Thus, *p*–*n* photodiodes without bias voltages exhibit high specific detectivity (D^*), maximum linearity, and sensitivity due to the reduced dark current [23–26]. When applying a reversed bias voltage to the *p*–*n* photodiodes, the decrease in width in the depleted region benefits in reduced transit time and diode capacitance, resulting in fast response speeds.

To further improve the sensitivity and response time of UV detection, an intrinsic (*i*) layer is inserted between a *p*–*n* junction in order to increase the width of the depleted region, as shown in Figure 7.2e [27, 28]. Besides, the incorporated *i* layer enables the enhancement of UV absorptance. By optimizing the thickness of the *i* layer, the sensitivity, response time, and EQE of photodiodes can be increased when applying a reverse bias voltage. However, the current gain of *p*-(*i*)-*n* photodiodes is smaller than that of photoconductors and MSM photodetectors. Besides, the response time and detective spectra of *p*-(*i*)-*n* photodiodes depend primarily on the doping types and semiconductor components, which are interrelated.

7.2.5 Avalanche Photodiodes

When sufficiently large reverse bias voltages are applied in photodiodes, the increased electric field may facilitate the interaction of excited electrons and holes to atoms in semiconductors. As a result, the valance band of atoms may be broken with more free carriers being generated [29]. The continuous interaction of atoms with carriers and the generation of multiple carriers, resembling an avalanche effect, result in a significant increase in the light current of photodiodes, providing a large current gain. As a specific type of photodiode, avalanche diodes possess similar advantages as *p*-(*i*)-*n* photodiodes, including high response speed and sensitivity. But the large reverse bias voltages may cause the presence of considerable noise, which can potentially decrease the signal-to-noise ratio (SNR) of UV detection in avalanche photodiodes.

7.2.6 Phototransistors

Similar to conventional electronic transistors, phototransistors are typical three-terminal devices consisting of sources, drains, and gates, as shown in Figure 7.2f. In real UV phototransistors, heterojunctions of semiconductors are employed, in which one layer acts as the UV absorber and another one serves as a charge transport channel. Under UV illuminations, photogenerated electron–hole pairs from the absorption layer are separated by a source-gate bias voltage [29]. Therefore, one type of carrier is driven into the charge transport channel while the other type is trapped in the origin region because of the alignment of energy levels in the heterojunction. Subsequently, the carriers in the transport channel are drifted by a source-drain field, resulting in the presence of photocurrent in phototransistors. Compared with photoconductors and photodiodes mentioned above, phototransistors exhibit higher responsivity and sensitivity due to their high current gain but slower response speed and complex manufacturing. Therefore, UV phototransistors are usually utilized in specific fields with requirements of high responsivity and sensitivity.

7.3 Important Parameters of UV Photodetectors

To evaluate the performance of UV photodetectors with different structures and materials, a variety of parameters are chosen and proposed. Herein, some important parameters of UV photodetectors are summarized as follows [30–32]:

Responsivity (*R*): the ratio of photocurrent (I_{ph}) or photovoltage (V_{ph}) to incident light power (P_{in}) within the effective area of detectors. The unit of responsivity is $A\,W^{-1}$ or $V\,W^{-1}$. It represents the photoelectric conversion capability of the devices and can be described as:

$$R - \frac{I_{ph}\ or\ V_{ph}}{P_{in}}$$

External quantum efficiency (EQE): the ratio of the number of output electron–hole pairs as incident photons in the devices, respectively. EQE is calculated as:

$$\mathrm{EQE} = \frac{I_{ph}/e}{P_{in}/h\nu} = R\frac{hc}{e\lambda},$$

where e and h are the elementary charge and the Planck's constant, respectively. ν is the frequency of incident light, c is the speed of light, and λ represents the wavelength of incident light.

Gain (*G*): the ratio between the number of electrons collected by the electrodes and the incident photons. The gain (G) can be used to evaluate the increase of photocurrent in a device.

Response time/response speed: the time for the measured light current to increase from 10% to 90% and reduce from 90% to 10%, respectively, which is presented as rise time (τ_r) and decay time (τ_d). A short response time or fast response speed is generally required in the applications of UV detection.

Signal-to-noise ratio (SNR): the ratio of signal current to noise current.

Specific detectivity (*D):** an important parameter in photodetectors that is related to the responsivity and the noise current (i_n) of devices, and can be calculated as follows:

$$D^* = \frac{(A\Delta f)^{1/2}R}{i_n},$$

where the A and Δf represent the effective area and the electrical bandwidth, respectively. i_n is the noise current. The unit of specific detectivity is $\mathrm{cm\,Hz^{1/2}\,W^{-1}}$ (Jones).

Noise equivalent power (NEP): the minimum optical power required to achieve a SNR of 1 and a bandwidth of 1 Hz, which is calculated as:

$$\mathrm{NEP} = \frac{(A\Delta f)^{1/2}}{D^*} = \frac{i_\mathrm{n}}{R}$$

A smaller NEP corresponds to a larger responsivity value for UV photodetectors.

Linear dynamic range (LDR): the range where the photo-generated current is linearly related to the incident light intensity. LDR can be described as follows:

$$\mathrm{LDR} = 20\,\mathrm{lg}\,\frac{I_\mathrm{ph}}{I_\mathrm{d}}$$

Beyond LDR, the intensity of the light signal cannot be detected precisely. Thus, a large LDR is required for UV photodetectors to detect both weak and strong light signals.

7.4 Materials and Performance of UV Photodetectors

Metal oxides are the focus of current nanotechnology research because of their special composition, chemical, and physical properties, and they are the most common minerals on earth. The main advantages of metal oxide UV detectors are their broad response range, strong UV absorption, low cost, and feasible manufacturing processes. Photodetectors based on metal oxide nanostructures and heterogeneous interfaces have high responsivity and photoconductivity gains, where the high gain is usually attributed to oxygen adsorption and light absorption. The performance of UV photodetectors with various structures based on different metal oxides, including ZnO, TiO_2, SnO_2, Ga_2O_3, NiO, and ZnMgO, is presented and summarized in this section to illustrate the potential of metal oxides in UV detection.

7.4.1 ZnO-Based UV Photodetectors

Zinc oxide is a typical II–VI semiconductor material with a hexagonal fibrillated structure. Due to its ease of fabrication, low cost, and unique physical and chemical properties, ZnO has been used as a building block for many electronic and optoelectronic devices, including solar cells and field-effect transistors [33–35]. The application of ZnO in UV detection has gained widespread interest due to its wide bandgap of 3.37 eV at room temperature and large exciton binding energy [36]. In addition to bulk ZnO, rationally designed ZnO nanostructures with complex shapes have shown promise for UV detectors. In addition, doping Mg in ZnO allows for the tuning of the bandgap, making it possible to prepare UV detectors with different cut-off wavelengths. Thus, ZnO-based photoconductors, Schottky photodiodes, MSM photodiodes, and *p–n* junction photodetectors have been successfully developed.

The energy conversion efficiency of a photoconductor from UV to an electrical signal is closely related to the Ohmic contact, since a certain amount of photocurrent can be dissipated as Joule heat at the contact resistance, weakening the intensity of the light response signal [37]. As a result, the development of low-specific resistance Ohmic contacts on ZnO-based optoelectronic devices has been extensively studied. Various electrode materials for ZnO photoconductors have been developed, including Ti/Au, Al/Au, Ni/Au, and Al [38]. Many research groups have attempted to improve ZnO photoconductors by optimizing growth techniques or applying post-growth treatments. For example, Takeuchi et al. prepared ZnO-based photoconductors on a grain layer through gas evaporation in Ar/O_2. They found that by increasing the ZnO grain size from 7.6 to 40 nm and controlling the air pressure during the growth process, the photosensitivity of the photoconductors improved [39].

Owing to the existence of a two-step UV response, response speed is of great concern in ZnO-based UV photoconductors [40]. The two-step response consists of fast and slow responses. The former is due to the generation or recombination of electron–hole pairs, while the latter is due to the photochemical absorption and desorption of carriers by oxygen. Clearly, the slow response is influenced by the stoichiometric composition and crystal quality. Initially, due to limitations in growth technology, the response speed at early stages was often several s or even tens of minutes. However, this limitation has been overcome by UV photoconductors with MOCVD epitaxially grown ZnO films. In these cases, the rise time and fall time were measured to be 1 and 1.5 μs, respectively. Later, a fast-response ZnO UV photoconductor with interdigital electrodes was fabricated on *c*-plane-preferred ZnO thin films by radio frequency sputtering. According to the transient response characteristics, an optical response speed with a rise time of 100 ns and a fall time of 1.5 μs was obtained [41]. Further improving the quality of ZnO film by post-growth annealing shortened the rise time to 45 ns.

In 2001, Liang et al. fabricated an MSM photodiode using silver as a Schottky contact, obtaining a low-frequency photoresponsivity of 1.5 $A\,W^{-1}$ at a bias voltage of 5 V [42]. The dark current of this device under 5 V bias is about 1 nA. The photoresponse of the detector showed a fast response component with a rise time of 12 ns and a fall time of 50 ns. Jiang et al. studied an Au-ZnO-Au MSM photodiode using RF sputtering, which exhibited a dark current of 1 nA and a voltage of 3 V [43]. The photoresponsivity of the detector was 0.337 $A\,W^{-1}$ at 3 V, with a high UV/visible suppression ratio of more than four orders of magnitude. The 10–90% response times were 20 ns on ascent and 250 ns on descent. Furthermore, the transient response was found to be independent of the distance between electrodes, suggesting that the carrier transit time could not be the main contributor to the response time. Ultimately, the response speed is attributed to hole capture. In 2008, Jin et al. prepared UV photodetectors with sol–gel-deposited ZnO quantum dots (QDs) as the foundation among MSM structures, achieving a responsivity of up to 61 $A\,W^{-1}$, a dark current of about 0.2 nA, and a light-dark suppression ratio of more than 6 orders of magnitude [44]. However, such devices required a high operating voltage of 120 V, with a rise time of more than 25 s and a fall time of more than 120 s.

Nasiri et al. used a flame spray pyrolysis method to directly condense ZnO aerosol spray on cross-finger electrodes, resulting in ZnO QD UV detectors with an MSM structure, where the QDs showed a porous membrane structure [86]. At a bias of 5 V, the device exhibited an ultra-high light-dark suppression ratio (3.4×10^5) and an extremely strong capability for detecting weak light (minimum $0.1\,\mathrm{W\,cm^{-2}}$), with a responsivity of $20\,\mu\mathrm{W\,cm^{-2}}$. However, the rise time (250 s) and the fall time (150 s) were relatively slow. Using a similar method, researchers also studied the application of UV detectors with ZnO QDs porous films at a very low voltage, showing a responsiveness of $14\,\mathrm{mA\,W^{-1}}$ at a bias of 0.2 mV. Researchers believe that porous ZnO QDs films can absorb a large amount of oxygen and produce a large depletion layer. Under illuminations, a substantial amount of oxygen can be desorbed, resulting in a reduction of the depletion layer and potential barrier. This change in the depletion layer caused by the adsorption/desorption process is the main reason for the strong detection ability of the device.

In recent years, UV photodetectors based on graphene and ZnO Schottky barriers have attracted extensive attention due to their intrinsic properties. The different work functions of graphene and ZnO form rectifying potential barriers that can facilitate efficient charge transfer and Schottky barrier formation, enabling effective carrier separation. Nie and his coauthors fabricated UV photodetectors with PMMA-supported monolayer graphene (MLG) films/ZnO NR arrays [45]. The respective current–voltage (*I–V*) characteristics of the device confirmed the formation of rectifying behavior, with a turn-on voltage of 1.0 V (Figure 7.5a). When the devices were exposed to UV irradiation at reverse bias voltage (365 nm, $100\,\mu\mathrm{W\,cm^{-2}}$), the photocurrent at reverse bias increased by 3 orders of magnitude compared to the dark current, surpassing that of conventional ZnO nanomaterial-based photodetectors. As shown in Figure 7.5b, the rise and decay times of devices were estimated to be 0.7 and 3.6 ms, respectively. Zhang et al.

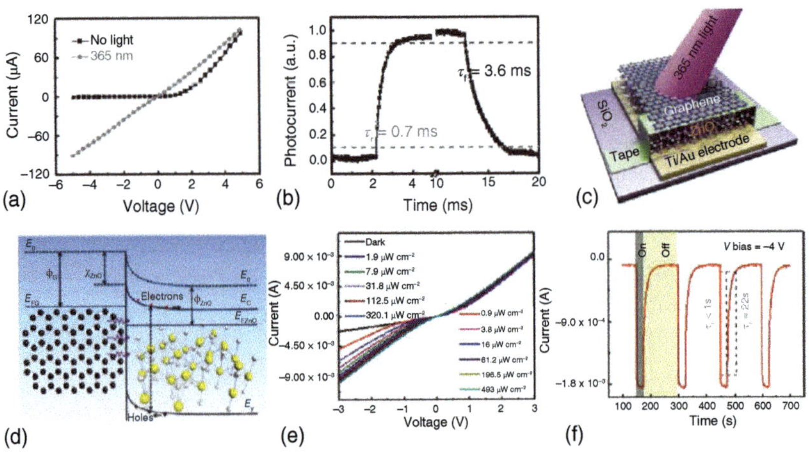

Figure 7.5 (a) *I–V* responses recorded with and without 365 nm UV illumination of photodetector. (b) Estimation of response/recovery time of the photodetector. Source: Reprinted with permission from Ref. [45]. Copyright 2013 John Wiley and Sons. (c) The schematic diagram of graphene-ZnO Schottky junction-based photodetectors. (d) A respective energy band diagram of the photodetectors under UV light. (e) *I–V* responses at different UV illumination (365 nm) intensities. (f) Estimation of response/recovery time of the photodetector. Source: Reprinted with permission from Ref. [46]. Copyright 2016 De Cruyter.

reported graphene-ZnO Schottky junction photodetectors (PDs) (Figure 7.5c) for high-performance photoconductivity, with Figure 7.5d depicting a schematic diagram of the respective bands of graphene-ZnO Schottky junctions under UV exposure [46]. Figure 7.5e shows the I–V characteristics in the absence and presence of UV illumination at different intensities. The response and recovery times of manufactured photodetectors were less than 1 and 22 s, respectively, as shown in Figure 7.5f. The authors pointed out that the relatively slow response to UV light was related to the presence of defect states, such as O or Zn vacancies in ZnO, or the overabundance of Zn and O atoms in ZnO. In addition, the authors reported a high response of 3×10^4 A W^{-1} and a specific detection rate of 4.33×10^{14} Jones under reverse bias conditions.

Pal et al. combined gold nanoparticles with ZnO thin films to prepare highly sensitive ultrafast UV detectors, wherein the photoactive layer exhibited superior UV light absorption [34]. Under appropriate UV light irradiation, ZnO and Au nanoparticles generated carriers, thereby increasing the rate of photocurrent generation. The performance of the prepared devices also depended on the pyroelectric effect generated in ZnO, which was caused by its non-centrosymmetric crystal structure. This pyroelectric effect, in addition to the increase of overall photodetection capability, also greatly reduced the response time of the device, making it suitable for ultrafast operation in ZnO-based UV photodetectors. Tian et al., prepared a high-quality ZnO thin film on a quartz substrate using a radio frequency magnetron sputtering technique and constructed a UV detector with an MSM structure on the ZnO film [47]. To further increase the UV responsivity of devices, Pt nanoparticles (Pt NPs) were deposited on the surface of the ZnO film. As shown in Figure 7.6, the absorption of ZnO thin films was partially enhanced in the UV band. It was revealed that Pt NPs played a pivotal role in augmenting the performance of the photodetectors by facilitating surface plasmon resonance.

The UV light detection mechanisms were also studied and compared for ZnO-based photodetectors and ZnO/PEDOT:PSS hybrid photodiodes [48]. As shown in Figure 7.7, ZnO thin films were deposited on glass substrates using a spray pyrolysis method. The deposition temperature ranged from 350 to 425 °C, and the physical properties of the ZnO thin films were investigated. It was observed that

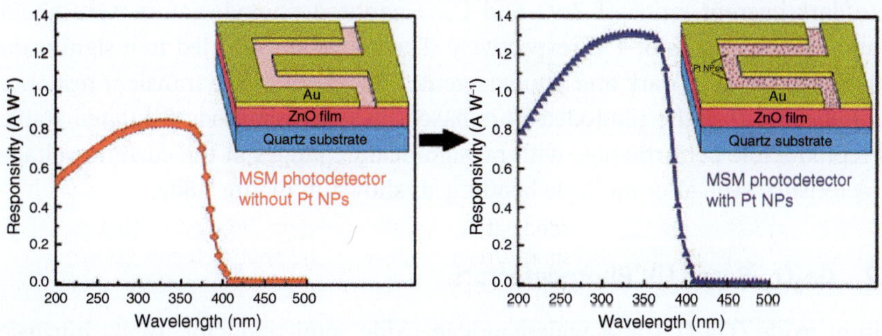

Figure 7.6 Variation of photodetector responsivity of UV photodetectors without and with Pt NPs. Source: Reprinted with permission from Ref. [47]. Copyright 2014 American Chemical Society.

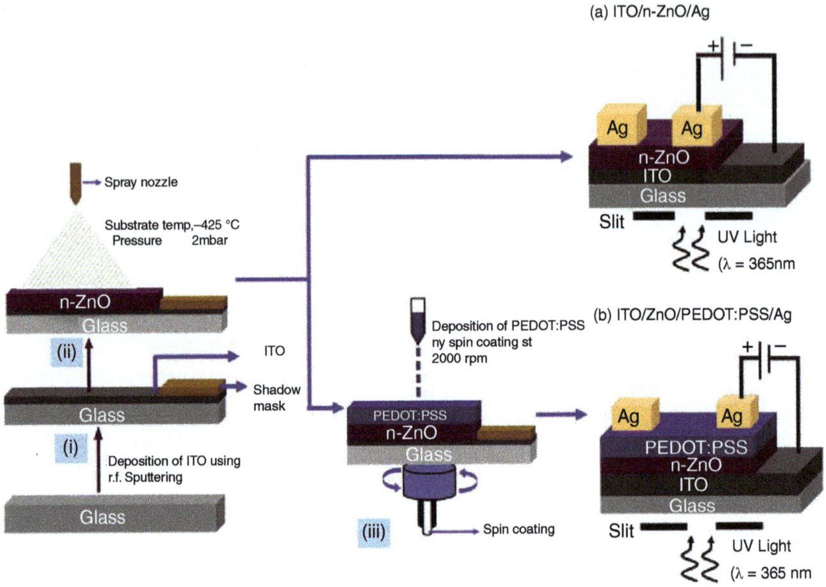

Figure 7.7 Schematic device structures. (a) ITO/ZnO/Ag photodetectors and (b) ITO/ZnO/PEDOT:PSS/Ag photodiodes. Source: Reprinted with permission from Ref. [48]. Copyright 2020 Springer Nature.

the ZnO-based photodiodes exhibited a high photoresponsivity value of 0.25 A W^{-1} with a fast photoresponse switching speed.

The inherent defects in wide-bandgap semiconductors may decrease the response time of UV photodetectors. An effective method to compensate for these inherent defects is doping. Besides, doping metal ions provides a possible way to change the bandgap and properties of the ZnO lattice [49]. Khosla et al. fabricated MSM UV detectors with gallium-doped zinc oxide (GZO) nanorods on flexible substrates, as shown in Figure 7.8a,b [50]. The electrical measurement performance of Ga-doped ZnO (GZO) nanorods was significantly improved compared to that of pure ZnO nanorods. Figure 7.8c shows the performance of UV detectors, from which a greatly improved overall performance was obtained in GZO nanorods photodetectors. The photo/dark current ratios of ZnO and GZO nanorod photodetectors were about 8.5 and 570.2 at a bias of 1 V, respectively. The addition of Ga led to a significant improvement in both dark and photocurrent. The results of the transient response tests indicated that the photodetectors based on GZO nanorods exhibited stable and reproducible performance, with no significant changes in the current–voltage characteristics even after multiple bending, as shown in Figure 7.8d,e.

7.4.2 Ga_2O_3-Based UV Photodetectors

Gallium oxide (Ga_2O_3) is a wide-bandgap oxide semiconductor. In its intrinsic state, the conductivity of gallium oxide is very low and similar to that of insulators. However, gallium oxide is usually *n*-type due to the existence of internal oxygen

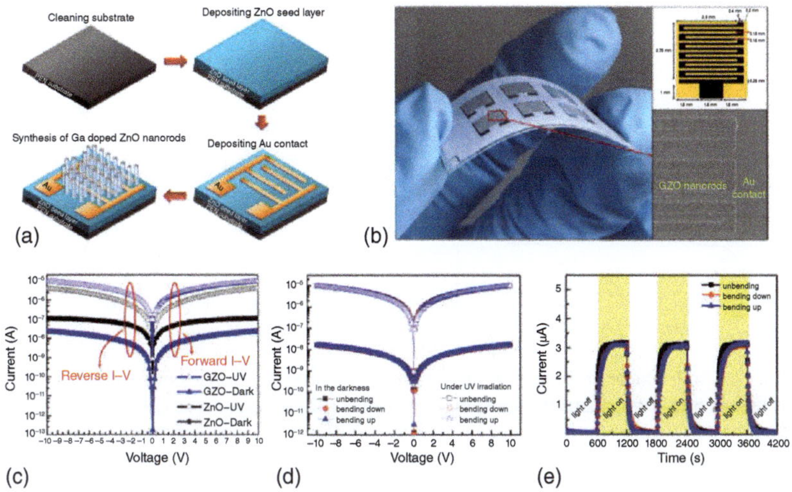

Figure 7.8 (a) Schematic illustrations of the fabrication process for ZnO and Ga-doped ZnO (GZO) nanorods metal-semiconductor-metal photodetectors. (b) Schematic process of photodetectors fabrication. (c) Current–voltage properties of UV photodetectors constructed from pure ZnO and gallium-doped ZnO nanorods. (d) Current–voltage properties of GZO nanorods PD under different bending conditions. (e) Transient switching photoresponse curves. Source: Reprinted with permission from Ref. [50]. Copyright 2020 American Chemical Society.

vacancy. Gallium oxide exhibits homopolymorphism and contains five isomers, namely α-Ga_2O_3, β-Ga_2O_3, γ-Ga_2O_3, θ-Ga_2O_3, and ε-Ga_2O_3. The most common ones are hexagonal α-Ga_2O_3 and monoclinic β-Ga_2O_3. Among them, β-Ga_2O_3 is the most stable one with a high melting point, a high voltage withstand value, and a wide optical bandgap, while gallium oxide of other structures can be converted into β-Ga_2O_3 under specific conditions [51, 52]. Most semiconductor devices based on Ga_2O_3 use stable β-type structure. As a wide-bandgap semiconductor, Ga_2O_3 exhibits high transmittance in UV band and over 90% in the visible light range, making it a suitable material for preparing UV-transparent conductive films.

Among all types of UV detectors, photoconductive detectors have been widely studied because of their simple preparation. However, the photocurrent attenuates due to carrier recombination caused by surface and internal defects in the semiconductor, leading to a reduction in signal detection efficiency. Researchers have explored various types of Ga_2O_3 to improve detection performance through different approaches. In 2017, Chen et al. achieved the epitaxial synthesis of an α-Ga_2O_3 single crystal layer on a ZnO substrate using laser molecular beam epitaxy (MBE), resulting in α-Ga_2O_3/ZnO heterojunction detectors with high avalanche gain [53]. These detectors operated without the need for external bias voltage and exhibited outstanding performance, including pA-level dark current, a UV/Vis detection ratio in the range of 10^3, and a high sensitivity of 9.66×10^{12} Jones. Under bias conditions, the detector demonstrated a dual-frequency response peaking at 255 and 365 nm, with respective values of 0.50 and 0.071 $A\,W^{-1}$ (at a test voltage

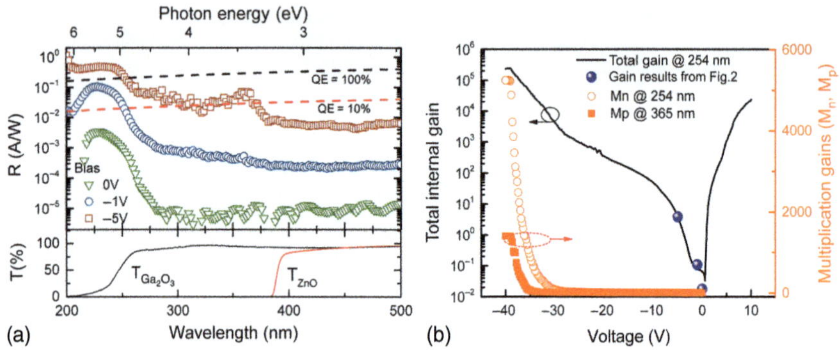

Figure 7.9 (a) Spectral photoresponse of a 400 μm diameter Schottky diode under different biases and the transmittance spectra of the Ga_2O_3 epilayer and ZnO substrate. (b) Photoconductive gain for the device under 254 nm illumination as a function of applied bias and the avalanche multiplication gain for 254 and 365 nm light irradiation, respectively; Source: Reprinted with permission from Ref. [53]. Copyright 2017 American Chemical Society.

of −5 V), as shown in Figure 7.9a. At a bias voltage of −40 V, the device exhibited avalanche multiplication, achieving a photoresponsivity as high as 1×10^4 A W^{-1} for 254 nm deep UV light and a total gain exceeding 10^5 (Figure 7.9b).

In 2018, Zhang et al. reported photoconductive solar-blind UV detectors based on high-quality β-Ga_2O_3 thin films obtained via metal-organic chemical vapor deposition (MOCVD) using N_2O as the reactive gas [54]. Compared to β-Ga_2O_3 thin film detectors prepared by predecessors, the reported device exhibited a higher photoresponse of 26.1 A W^{-1}, a larger on/off ratio of 10^4, and faster response speed (rise time of 0.48 s and attenuation time of 0.18 s). It is believed that the high photoresponsivity and fast response of β-Ga_2O_3 solar-blind UV photodetectors are attributed to the reduction of photogenerated carrier scattering/trapping centers in β-Ga_2O_3 films grown in N_2O atmosphere.

In 2019, Li et al. deposited Ga_2O_3 thin films with a highly preferred orientation on hexagonal SiC (6H-SiC) substrates by magnetron sputtering [55]. In previous studies, most Ga_2O_3 thin films were poorly crystallized polycrystals or expensive single crystals. This work successfully achieved a compromise between good optoelectronic properties and fabrication cost between single crystals and polycrystals using simple fabrication conditions. In detail, this included the choice of optimal sputtering parameters (RF power: 120 W) to prepare a 400 nm thick film and then an annealing treatment at 800 °C to obtain Ga_2O_3 films with a highly preferred orientation in the (−201) crystal plane and good crystallization (FWHM = 0.108°). The photoconductive UV detectors with Al electrodes had high responsivity, detection rate, and quantum efficiency of 2.6 A W^{-1}, 1.6×10^{12} Jones, and 1265% under 10 V bias, respectively, and the response time was 0.26 s.

In 2020, Zhou et al. reported a sun-blind UV detector based on polycrystalline Ga_2O_3 films prepared by annealing amorphous Ga_2O_3 to polycrystalline ones, which avoided the use of high-temperature conditions in the direct growth of polycrystalline Ga_2O_3 films [56]. First, amorphous Ga_2O_3 was grown by PLD at room temperature in an oxygen atmosphere and annealed to obtain polycrystalline

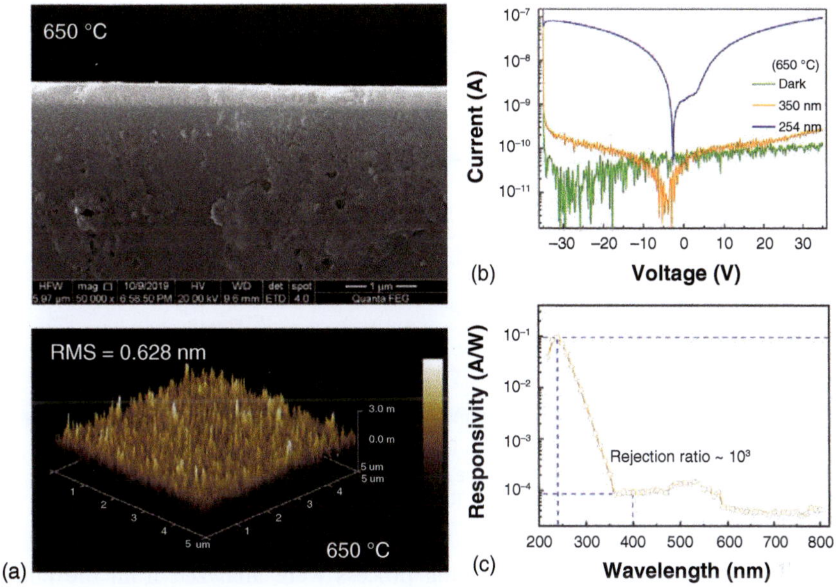

Figure 7.10 (a) Scanning electron microscope and atomic force microscope images of bulk b-Ga_2O_3. (b) Current–voltage curves at different wavelengths, and (c) the responsivity curve. Source: Reprinted with permission from Ref. [56]. Copyright 2020 Elsevier.

Ga_2O_3 thin films with a relatively stable structure. Then, a photoconductive high-temperature UV detector based on Au interdigital electrodes was fabricated. The effect of different annealing temperatures on the performance of the device was studied, and it was found that the polycrystalline Ga_2O_3 thin film detectors annealed at 650 °C exhibited the best solar-blind UV photoelectric detection performance with good thermal stability. More prominently, the device also showed a fast response speed (0.16 s), low dark current (0.1 nA), and a high light-to-dark current ratio (97). The morphology and device performance are shown in Figure 7.10a. Clearly, some properties are even better than previously reported high-temperature UV detectors with high-quality crystalline Ga_2O_3 films or Ga_2O_3 bulk materials (Figure 7.10b,c).

Using homogeneous or heterogeneous substrates to obtain high-quality single-crystal bulk Ga_2O_3 based on floating zone method, edge-defined film-fed growth (EFG), and pulling method and applying it to UV detection is also one of the focused research areas. In 2017, Mu et al. used an optimized EFG method to grow bulk β-Ga_2O_3 single crystals with good crystal quality and effectively solved the cracking and polycrystalline problems in the growth of monocrystalline bulk materials by using a high-quality, 1-in. seed layer, as shown in Figure 7.11a,b [57]. The prepared β-Ga_2O_3 single crystals showed a very wide transmission range from 0.26 to 7.33 μm, corresponding to an estimated bandgap of 4.73 eV. The X-ray rocking curve in Figure 7.11c proves the high crystal quality of β-Ga_2O_3 single crystals. In addition, the authors also proposed a simple mechanical peeling method to directly obtain large-size and high-quality β-Ga_2O_3 wafers, providing the advantages of being ultra-smooth, low-cost, one-step, and avoiding cutting, grinding, and polishing of the wafer surface. The mechanism of the stripping

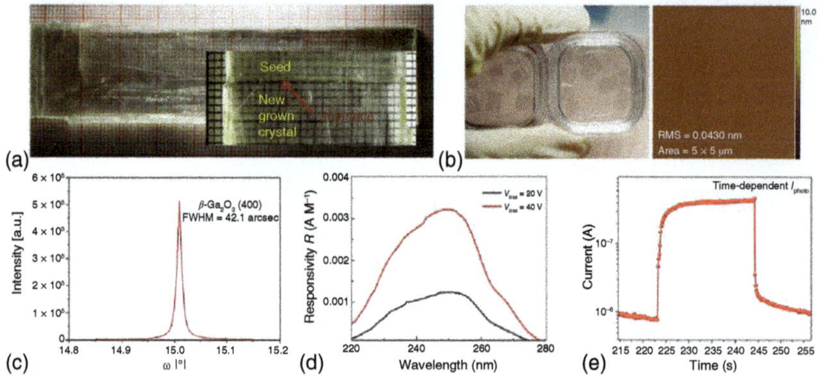

Figure 7.11 (a) The morphology of bulk β-Ga_2O_3. (b) The morphology and AFM image of β-Ga_2O_3 wafers. (c) The rocking curve of β-Ga_2O_3 wafers. (d) The response test of β-Ga_2O_3 wafer-based devices. (e) The response time test of the device. Source: Reprinted with permission from Ref. [57]. Copyright 2017 Royal Society of Chemistry.

method and the effect of the annealing process were analyzed from the point of view of crystallography. UV detectors with Au/Ti interdigital electrodes were prepared on this wafer, which possessed a responsivity of 0.003 A W^{-1} and rise/fall time of 4.4/0.14 s, respectively, under a bias voltage of 40 V (Figure 7.11d,e).

UV detectors made of thin films and bulk materials have attracted absolute attention and development due to their rich and diverse preparation methods. However, the responsivity obtained by these two types of materials is often low, and the manufacturing cost is generally high. In comparison, micro- and nano-materials have obtained excellent photosensitive properties in device applications due to their high specific surface area, Debye length similar to their size, and the high stability brought by their high crystalline quality. In 2019, Wang et al. synthesized β-Ga_2O_3 nanowires *in situ* on flexible glass fiber fabric substrates using chemical vapor deposition (CVD), which had diameters varying from 40 to 280 nm and lengths ranging from one to several tens of microns at high temperatures, as shown in Figure 7.12a [58]. Photoconductive detectors based on such nanowires showed excellent performance (Figure 7.12b). Under 254 nm UV light irradiations, the detectors showed a photo-dark current ratio of 260, a photoresponsivity of 0.71 A W^{-1}, and response decay time of 0.19 s, as shown in Figure 7.12c,d. In addition, the device exhibited good repeatability and could withstand different degrees of bending without compromising its performance. The device is simple to prepare and can be used as a flexible UV detector with good working stability at high temperatures, suggesting promising application prospects.

Lu et al. prepared deep UV photodetectors with an MSM structure using Ga_2O_3 nanowire network grown on a sapphire substrate with an Au catalyst treated by a CVD method, as shown in Figure 7.13a [59]. Figure 7.13b illustrates that the light current of the UV photodetectors under 254-nm irradiation is significantly higher than that under 365-nm irradiation, and the calculated DUV/visible inhibition ratio (R_{265}/R_{430}) is about 124.88, indicating a good deep UV detectivity of the detectors,

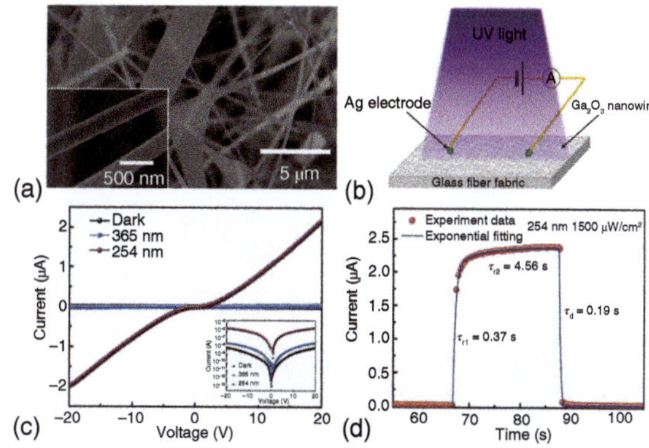

Figure 7.12 (a) The morphology of synthesized β-Ga_2O_3 nanowires. (b) Photoconductive ultraviolet detectors based on nanowires. (c) I–V curves of the device and the result after logarithmic coordinate conversion. (d) The response speed curve of the device. Source: Reprinted with permission from Ref. [58]. Copyright 2019 Elsevier.

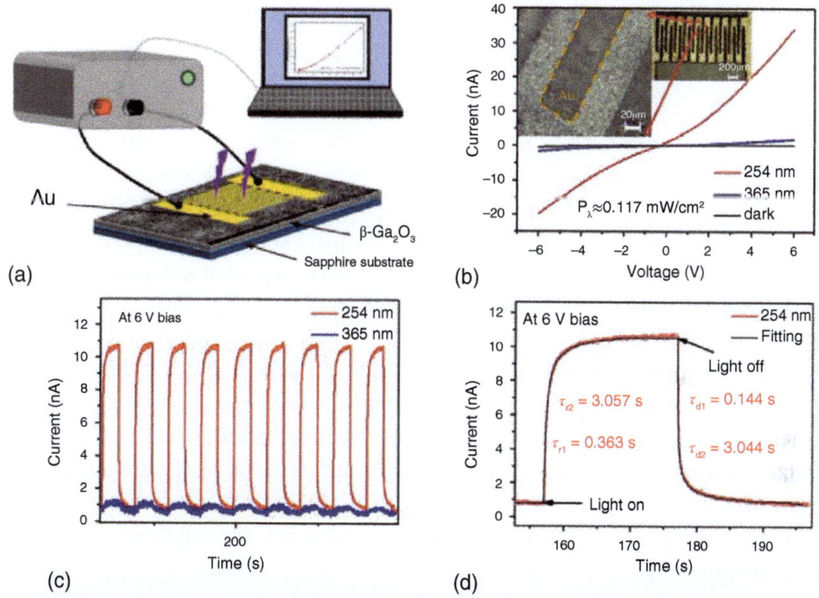

Figure 7.13 (a) Schematic diagrams of Ga_2O_3 nanowire network-based MSM photodetectors and the test system. (b) Dark current of the detector in the range of −6 to 6 V and photocurrent under illuminations of 254 and 365 nm (inserts are optical images of interdigital electrode preparation template and part of interdigital electrodes); (c), (d) Transient response of detectors to illumination at 6 V. Source: Reprinted with permission from Ref. [59]. Copyright 1996 IOP Publishing.

mainly due to the large bandgap of ~4.9 eV for Ga_2O_3 nanowire network. In the deep UV range, the photodetectors exhibited good photoelectric performance, including a responsivity of 128 mA W^{-1} and a specific detectivity of 3.676×10^{10} Jones at a driven voltage of 6 V. In addition, the response times of devices to 254 nm UV light at 6 V were 3.42 and 3.188 s, respectively, as shown in Figure 7.13c,d.

Kumar et al. fabricated quasi-two-dimensional β-Ga_2O_3 nanoflakes (NFs)-based deep UV photodetectors [60]. The thin Ga_2O_3 NFs were prepared by a low-cost and

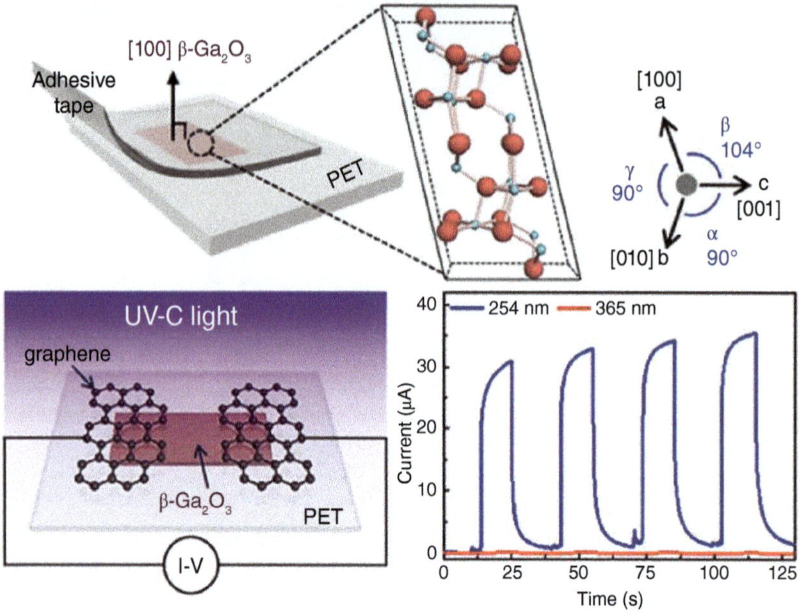

Figure 7.14 Schematics showing the crystal structure of β-Ga_2O_3 and the fabricated β-Ga_2O_3-based solar-blind UV detectors. *I–V* characteristics of the fabricated PDs in the dark and upon illuminations at different conditions. Source: Reprinted with permission from Ref. [61]. Copyright 2018 American Chemical Society.

scalable electrospinning method. The fabrication process included a deposition of electrodes on a cleaned silicon (001) substrate, followed by an electrospinning of polymerized $GaCl_3$ nanofibers, and then further annealing of fibers to develop and optimize the size of NFs. Owing to the very large surface-to-volume ratio and photon trapping, detectors based on the fabricated 2D Ga_2O_3 NFs exhibited high responsivity and photo-dark current ratio, as well as impressive rise and fall times of 87.6 and 87.1 ms, respectively.

Kim and coworkers separated a Ga_2O_3 microflake from an unintentionally doped Ga_2O_3 single-crystalline substrate via mechanical exfoliation and fabricated MSM-structured solar-blind UV detectors using graphene as electrodes [61]. As shown in Figure 7.14, the detectors demonstrated excellent performance, with a photo-responsivity of up to 29.8 A W^{-1}, a light-dark ratio of about 1×10^4, and a detectivity of about 1×10^{12} Jones. The absence of shading by the integration of graphene with β-Ga_2O_3 allows maximum exposure to incident photons, suggesting great potential in deep UV optoelectronic applications.

7.4.3 TiO_2-Based UV Photodetectors

TiO_2 is a wide-bandgap semiconductor material with low cost, non-toxicity, as well as high physical and chemical stability. The common crystal structures are anatase and rutile. Due to its excellent properties, such as good electrical conductivity and fast catalytic speed, TiO_2 has been widely studied and applied in gas-sensitive sensors, photocatalytic decomposition of water, solar cells, purification of environmental pollutants, and other fields. In recent years, TiO_2 has also been gradually introduced into the fields of UV photoelectric detection, leading to the development of a variety of UV detectors based on TiO_2.

In 2007, anatase TiO_2 MSM photodetectors were prepared by a sol–gel method on silicon substrates [62]. At a 5 V bias, the dark current and responsivity at 260 nm were about 1.9 nA and 199 $A\,W^{-1}$, respectively. However, the response time of the device was short, with a rise time of 6 s and a decay time of 15 s, which might be attributed to the increased trap density in the deposited film and the limited space between metal contacts and photoactive materials. In 2011, Jie and coworkers grew TiO_2 nanofilms by RF magnetron sputtering and produced MSM UV detectors based on such TiO_2 films [63]. The dark current of the device at a 10 V bias voltage was less than 1 nA. However, the light responsivity was low, and the spectral response peak reached 3.63 $A\,W^{-1}$ only at a 310 nm wavelength.

Moudgil et al. incorporated TiO_2 with In_2O_3 thin films on SiO_2/Si substrates to form heterojunction UV photodetectors [64]. The length, width, and gap between two fingers of interdigitated electrodes were 800, 10, and 20 μm, respectively. The energy levels of TiO_2 matched exactly with those of In_2O_3, facilitating electron transfer from In_2O_3 to TiO_2, which improved the electron–hole dissociation efficiency. Zhang et al. fabricated UV photodetectors based on TiO_2/NiO *p–n* heterojunctions, which showed an overall improvement in performance compared to devices without heterojunction (dark current of 0.033 nA at 280 nm, Figure 7.15a,b) [65]. Figure 7.15c,d presented that the achieved responsivity was 181.9 $A\,W^{-1}$, and the response time was reduced to 598 ms, respectively. Sun et al. fabricated a self-powered Fe-doped TiO_2/*n*-Si heterojunction UV-Vis detector using a simple solution process [66]. Under zero-bias illumination of 0.5 $mW\,cm^{-2}$, the responsivity of the device was 46 $mA\,W^{-1}$ (350 nm) and 60 $mA\,W^{-1}$ (600 nm), respectively. The rise and fall times of the device were less than 10 and 15 ms, respectively. Patel et al. reported Cu/TiO_2/FTO Schottky photodetectors fabricated by direct current (DC) sputtering technology [67]. At a −1 V bias, the peak responsivity and detection rate were 0.897 $A\,W^{-1}$ and 4.5×10^{12} Jones, respectively. In addition, the response time of the device was very fast, with a rise/fall time of 1.12/1.46 ms.

Serkjan et al. successfully fabricated a novel self-powered *p–n* heterojunction UV detector using a combination of hydrothermal and electrochemical methods [68]. The results showed that the prepared TiO_2 NRs/polyterthiophene (PTTh) UV photodetectors had self-powered performance, high sensitivity, good stability, and fast response. In addition, due to the thermoelectric effects between Au NPs and TiO_2 NRs, the detection performance of UV detectors can be further improved by using Au NPs as an intermediate layer. The response rates of TiO_2 NRs/PTTh and TiO_2 NRs/Au/PTTh UV photodetectors were 0.0774 and 1.894 $mA\,W^{-1}$, respectively. These results also confirmed the great potential of TiO_2 NRs/Au/PTTh UV detectors with a heterojunction structure as self-powered UV detectors.

Fang et al. introduced organic layers and silver nanowires in TiO_2-based UV detectors for ultrafast speed, dark current suppression, and self-powered photocurrent

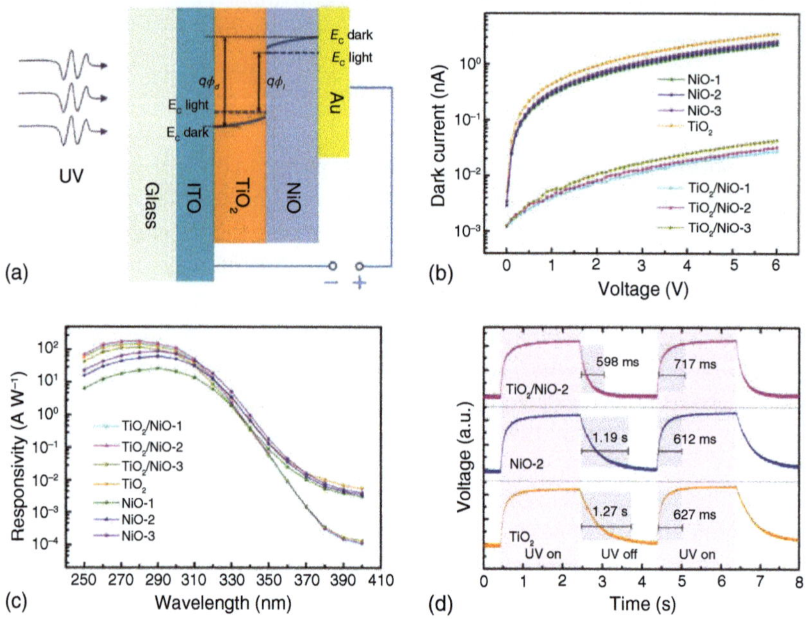

Figure 7.15 (a) A structure diagram of UV photodetectors based on TiO_2/NiO *p–n* heterojunctions. (b) *I–V* characteristics of UV photodetectors in the dark. (c) Response and (d) time-response characteristics of all UV photodetectors. NiO with thicknesses of 54, 90, and 126 nm were named as NiO-1, NiO-2, and NiO-3, respectively. Source: Reprinted with permission from Ref. [65]. Copyright 2017 IOP Publishing.

(Figure 7.16a) [69]. By introducing a $PC_{71}BM$ layer to fill the defects on the TiO_2 surface, the dark current was efficiently suppressed from 0.1 mA to 20 nA at −1 V. The introduction of PEDOT:PSS layer provided a middle energy stage between Ag NWs and $PC_{71}BM$, facilitating the transfer of holes to Ag NWs electrodes. As a result, the separation of electrons/holes became more efficient, and the photocurrent was improved, as shown in Figure 7.16b. The photocurrent of the device increased to 1.6 μA, and the device achieved the highest *R* of over 30 mA W^{-1} under 310 nm. The rise and decay times were 0.06 s and <1 μs, respectively, without any bias voltage, as shown in Figure 7.16c,d.

Li et al. reported self-powered UV PDs with mixed-dimensional heterostructure arrays of NiO nanosheets and TiO_2 nanorods [70]. The light-trapping nanostructure heterojunctions facilitate the generation of more photogenerated carriers. The built-in potential field of the NiO/TiO_2 heterojunction and the transfer path of separated photogenerated holes were adjusted by controlling the thickness of each layer and the NiO nanosheet films. The devices were measured under on/off switching of a 365 nm UV LED, showing fast UV light response characteristics. Specifically, when the light was turned off, the current dropped rapidly and approached 300 pA at zero bias. Li et al. used $CuInS_2$ (CIS) NFs films and vertically aligned TiO_2 NRs to form a nanostructured heterojunction using a simple and environmentally friendly solvothermal method (Figure 7.17a) [71]. As shown in Figure 7.17b,c, the photodetectors exhibited fast and stable photoresponse characteristics in a

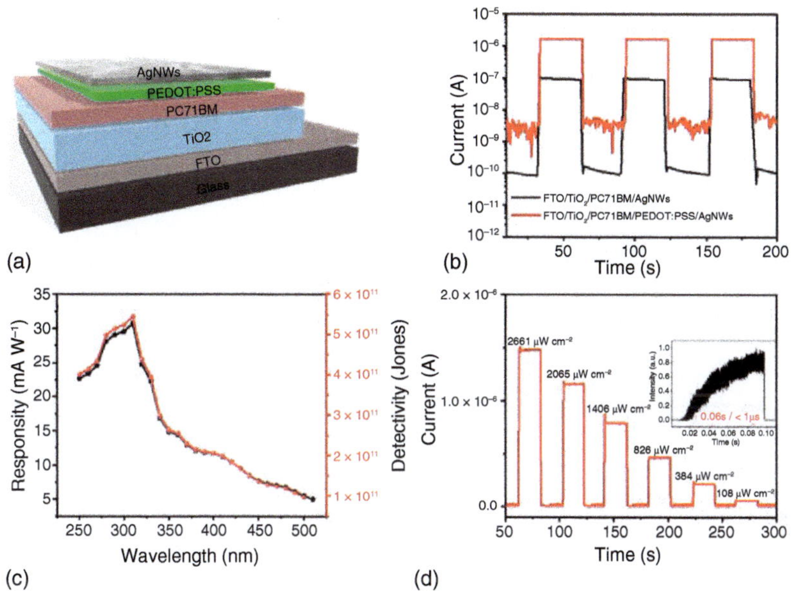

Figure 7.16 (a) The device structure of self-powered PDs. (b) $I-t$ curves of $FTO/TiO_2/PC_{71}BM$/Ag NWs and $FTO/TiO_2/PC_{71}BM$/PEDOT:PSS/Ag NWs at 0 V and under 350 nm. (c) Responsivity and detectivity of $FTO/TiO_2/PC_{71}BM$/PEDOT:PSS/Ag NWs at 0 V, 350 nm. (d) $I-t$ curves (the inset is the estimated rise and decay time from a single-pulse response curve) of $FTO/TiO_2/PC_{71}BM$/PEDOT:PSS/Ag NWs at 0 V with different power density values. Source: Reprinted with permission from Ref. [69]. Copyright 2021 American Chemical Society.

wide wavelength range from UV to visible light. The PDs with CIS deposited demonstrated stable performance with R, D^*, LDR, and sensitivity of 44.8 mA W^{-1}, 1.3×10^{12} Jones, 88.5 dB, and 2.7×10^4, respectively, as shown in Figure 7.17d.

7.4.4 WO_3-Based UV Photodetectors

Among the materials studied, tungsten oxide (WO_3) is particularly interesting as a sensor material. WO_3 is an n-type semiconductor with an energy bandgap of around ~3.2 eV and has been extensively studied for applications in gas sensors and electrochromic displays. WO_3 also exhibits strong electrochromism because of its ability to change color and modulate transmission up to 83% of the wavelength in the visible range of 400–800 nm. These electrochromic properties of WO_3 facilitate applications in screen displays and solar cells [72, 73]. Furthermore, WO_3 can be synthesized in various morphologies, such as nanobelts, nanowires, nanocubes, and monolayer films, all of which have been tested for UV photodetection.

Hak and coworkers heat-treated large-area amorphous WO_3 films to synthesize high-density, high-quality WO_3 nanorods [74]. They applied self-crosslinked WO_3 nanostructures to a practical sensor device fabrication by simply placing the electrodes without complex processes like transfer and e-beam lithography, as shown in Figure 7.18a,b. These nanorods were then employed in UV detectors, exhibiting a response time of 0.316 s and high sensitivity to UV-C (261 nm) wavelength

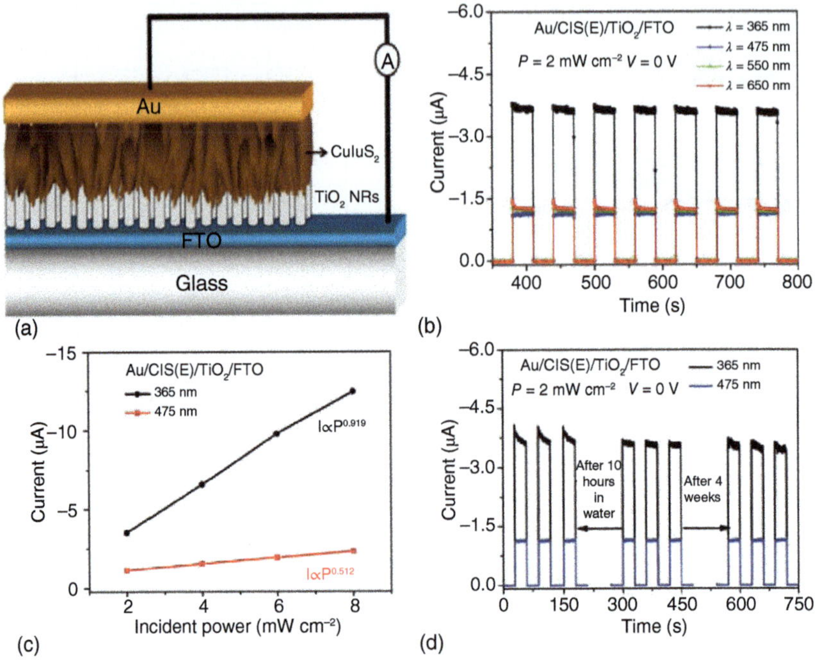

Figure 7.17 (a) A schematic diagram of Au/CIS/TiO_2/FTO heterojunction PDs. (b) *I*–*t* curves under UV, blue, green, and red light irradiations. (c) Photocurrent under UV (365 nm) and visible (475 nm) light illumination with different incident powers. (d) Stability of Au/CIS(E)/TiO_2/FTO PDs. Source: Reprinted with permission from Ref. [71]. Copyright 2021 Royal Society of Chemistry.

(Figure 7.18c). Also, by applying plasmon-based light absorption through Ag nanoparticle coating, they achieved over a 350-fold improvement in the on/off process during UV-C irradiation, as shown in Figure 7.18d.

Niederberge and Cheng presented a method to arrange ultrathin tungsten oxide nanowires in an oriented fashion over a large area using a simple evaporation-induced self-assembly process and explored their applications in UV photodetections [75]. The UV response of the device was measured under UV light of 254 and 365 nm. Upon 254 nm illumination, the resistance was significantly reduced from 122.5 to 1.2 MΩ in ~1 minute, giving an on/off resistance ratio of 10^2. The fall time and rise time of resistance were 1.3 and 166.4 minutes, respectively. A better sensing performance was achieved under 254 nm UV light, perhaps due to the enhanced light absorption of the nanowires at this particular wavelength, which was also in agreement with the observations made in single tungsten oxide nanowire-based UV photodetectors.

7.4.5 SnO_2-Based UV Photodetectors

Tin dioxide (SnO_2) possesses a comparable wide bandgap (3.6 eV) and exhibits sensitivity to a wide range of UV light. Its excellent transparency, high electrical

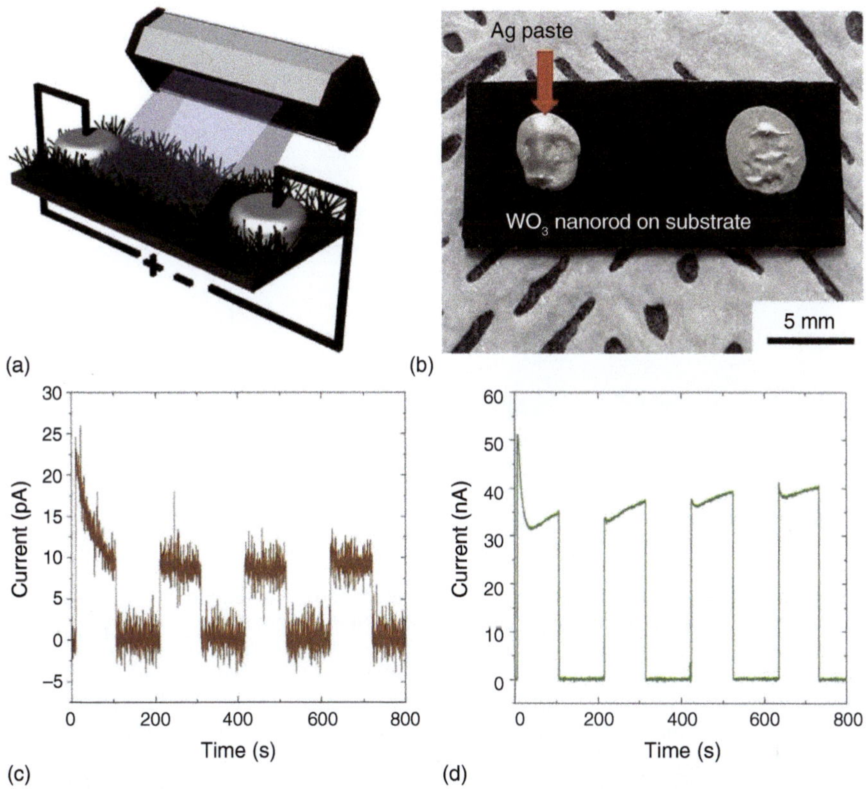

Figure 7.18 (a) The schematic illustration of performance measurement for self-crosslinked WO_3 nanorods-based UV detectors. (b) The real photograph of an UV detector. The photocurrent response of self-crosslinked WO_3 (c) without and (d) with Ag nanoparticles under UV-C ray (216 nm) irradiations. Source: Reprinted with permission from Ref. [74]. Copyright 2020 Royal Society of Chemistry.

conductivity, and chemical stability make it a promising candidate for photoanode materials in self-powered UV detectors. To prevent carrier recombination, one effective approach is to form heterojunctions with other materials.

Gao et al. used an optical tweezers system consisting of spatial light modulators, microscopes, and optical components to assemble SnO_2 nanowires in liquid environment [76]. By optically trapping the nanowires, they could be precisely assembled between two patterned electrodes, creating a photodetector. Under 275 nm deep-UV light exposure, the as-assembled photodetectors showed a high I_{ph}/I_{dark} ratio of 2.99×10^5, a large responsivity of 4.3×10^4 A W^{-1}, an excellent EQE of 1.94×10^7%, and a high detectivity of 2.32×10^{13} Jones. Passivation with a polymer further improved the photoresponse speed of the devices, reducing the rise and decay times to approximately 60 and 100 ms, respectively.

Wan and coworkers prepared a photodetector with a hollow multi-shell structure (HoMSs) using SnO_2 as the active material, as shown in Figure 7.19 [77]. Compared to SnO_2 nanoparticles, SnO_2 with a HoMSs configuration demonstrated enhanced

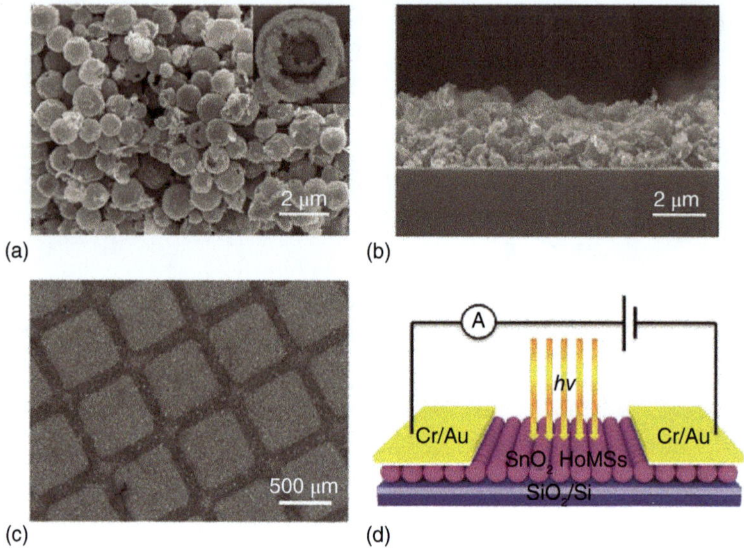

Figure 7.19 (a) A SEM image of the membrane of the SnO_2 HoMSs with an inset showing a broken sphere of SnO_2 HoMSs. (b) A cross-sectional SEM image of the SnO_2 HoMSs membrane. (c) An SEM image showing the surface condition of the whole photodetector device, where the brighter patterns are Cr/Au electrodes. (d) A schematic diagram of the SnO_2-HoMS-based UV light photodetector on a SiO_2/Si substrate with Cr/Au electrodes. Source: Reprinted with permission from Ref. [77]. Copyright 2019 Royal Society of Chemistry.

UV response and detection ability, owing to the inner cavity that facilitated light trapping and enhanced light absorption. Increasing the number of shells in the HoMSs structure further increased the photodetector's performance. As a result, photodetectors based on SnO_2 HoMSs showed a responsivity of 1.01×10^4 A W^{-1}, an EQE of 5.23×10^6%, and a detectivity of 3.83×10^{12} Jones.

7.4.6 Other Metal-Oxide UV Photodetectors

NiO_x is nonconductive at room temperature, with a resistivity of about 10^{13} Ω cm^{-1} [78]. However, the introduction of impurities and defects during the preparation of NiO_x can lead to the generation of Ni vacancies, leading to *p*-type conductivity in NiO_x. In addition, the conductivity of NiO can be improved through doping.

Wu et al. prepared a high-performance UV photodetector based on *p*-GaN/NiO_x nanostructure/*n*-GaN sandwich structure, which consisted of NiO_x nanostructures grown on the *n*-GaN and *p*-GaN layers, as shown in Figure 7.20a [79]. Photodetectors based on *p*-GaN/NiO nanostructure/*n*-GaN sandwich structure showed high responsivity and fast response. While devices with direct contact between *p*-GaN and *n*-GaN exhibited a photocurrent on/off ratio of only 50, those with a *p*-GaN/NiO nanosheets/*n*-GaN structure showed a significantly higher value of 2.3×10^2. Besides, such devices also performed rise and fall times of 50 ms, according to Figure 7.20b, suggesting a rapid response.

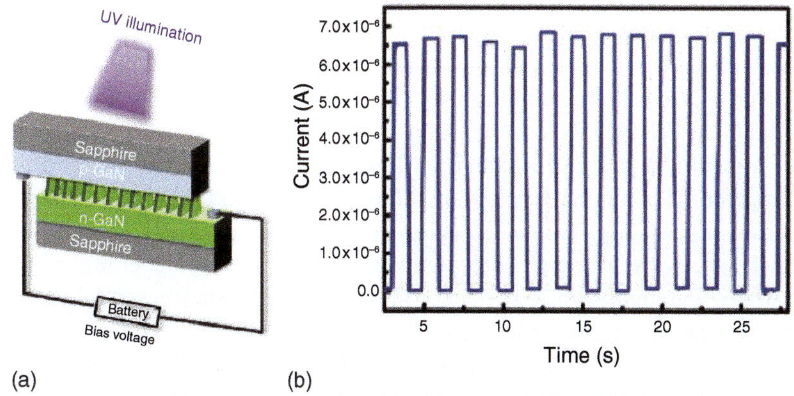

Figure 7.20 (a) The schematic diagram, (b) the current–time curve and enlarged view of a single on/off cycle of the p-GaN/NiO nanosheets/n-GaN direct-contacting UV photodetector. Source: Reprinted with permission from Ref. [79]. Copyright 2020 American Chemical Society.

Ao et al. prepared a self-powered UV detector based on a thermal oxidation process to create a heterostructure of *n*-GaN and *p*-NiO [80]. Compared to conventional Ni/GaN photodetectors, the heterostructure photodetectors had a higher effective barrier height and thus a larger turn-on voltage and a smaller dark current. At an external bias voltage of 0 V, the NiO/GaN photodetectors showed an improved photoresponsivity and UV–Vis rejection ratio from 0.08 A W^{-1} and 18–0.15 A W^{-1} and 406, respectively. Such an enhanced UV detector performance is attributed to the depletion region in the NiO/GaN heterostructure, which effectively eliminates trapped charge carriers at the metal/semiconductor interface.

Lin et al. utilized lithium-doped amorphous ZnSnO (a-ZTO) thin films for UV photodetectors by employing lithium-containing precursors and a sol-gel process [81]. The introduction of lithium doping (3 at%) significantly reduced the content of oxygen vacancies (V_O) in the a-ZTO films from 32.1% to 14.4%. The doped ZTO-based photodetectors exhibited a decreased dark current and an increased photocurrent, resulting in a high photo-dark current ratio of 1185. The reduction in V_O and the formation of strong Li–O bonds contributed to the improved performance of Li-doped ZTO photodetectors. These devices achieved a responsivity of 9.62 A W^{-1}, which was associated with a decrease in V_O, as V_O acted as shallow donors and trapping or recombination centers.

Yalcin et al. fabricated photodetectors based on Sr-doped yttrium-nickel oxide and analyzed their electrical properties [82]. The ideality factor, barrier height, and series resistance of the diode were calculated at different irradiance intensities in visible light, indicating the suitability of these photodetectors for optoelectronic device applications. *P*-Si/$Y_{1-x}Sr_xNiO_3$ heterojunction diodes with different Sr doping levels ($x = 0.00$, 0.05, 0.15, and 0.20) were fabricated, and the photoresponse behavior of photodetectors was investigated at different irradiation intensities.

The highest responsivity and detectivity were calculated as 25.34 mA W^{-1} and 7.84×10^{11} Jones, respectively.

Because of the comparable radius of Mg^{2+} (0.57 Å) and Zn^{2+} (0.60 Å) ions, the alloyed $Zn_{1-x}Mg_xO$ exhibits ultrahigh stability and tunable bandgaps from 3.37 to 7.5 eV, enabling deep UV detections. It has been found that the $Zn_{1-x}Mg_xO$ with a small Mg content below 0.5 exhibits a mixed phase and undefined bandgaps [83]. However, with the increase of Mg content from 0.5 to 0.7, the enlargement of bandgaps in $Zn_{1-x}Mg_xO$ is confirmed by the blueshift of cut-off wavelength in absorption spectra. As a result, the response of $Zn_{1-x}Mg_xO$ with higher Mg content demonstrates a gradual shift to the deep UV range, suggesting the key role of Mg alloying. Wang et al. fabricated deep UV MSM photodetectors based on $Zn_{1-x}Mg_xO$ by depositing semicircular and triangular metal electrodes, which exhibited 20.5% and 100% gain in photocurrent, respectively [84]. By optimizing the Schottky barriers and asymmetric electrode areas, the dark current of deep UV $Zn_{1-x}Mg_xO$ photodetectors was reduced to 5% of that in untreated devices while maintaining an unchanged photocurrent, implying the importance of electrode design in MSM photodetectors. Furthermore, Chen et al. reported self-powered deep UV photodetectors with a *p–n* photodiode structure, as displayed in Figure 7.21a [85]. The active detective layer consists of an organic/inorganic hybrid *p–n* junction with a $Zn_{1-x}Mg_xO$ film and densely doped polyaniline (PANI). Figure 7.21b demonstrates the *I–t* curves

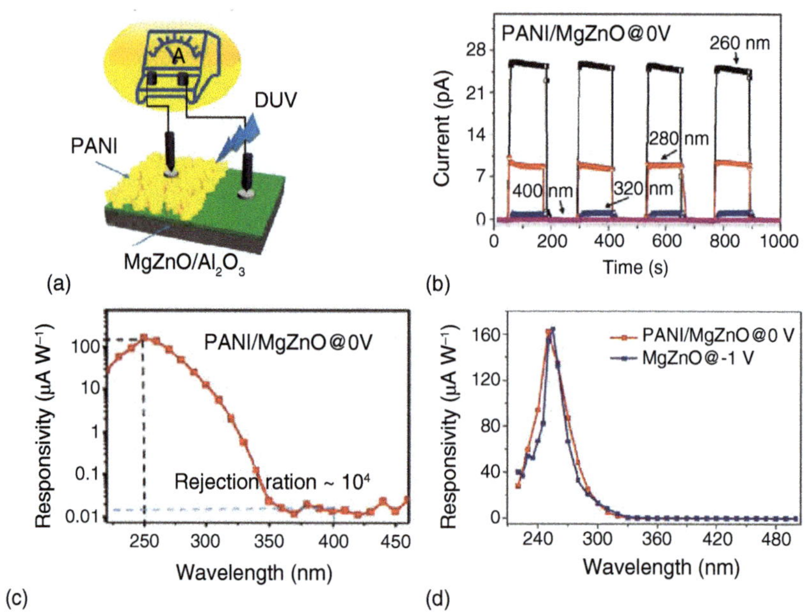

Figure 7.21 (a) A schematic diagram of a $Zn_{1-x}Mg_xO$/PANI heterojunction deep UV photodetector operated without a bias voltage. (b) Time-dependent photoresponse of the heterojunction photodetectors upon illuminations at different wavelengths. (c) Spectral response of $Zn_{1-x}Mg_xO$/PANI photodiodes on logarithmic scale. (d) Spectral response of $Zn_{1-x}Mg_xO$/PANI photodetectors at 0 V bias along with a pure $Zn_{1-x}Mg_xO$ detector at −1 V bias. Source: Reprinted with permission from Ref. [85]. Copyright 2016 John Wiley and Sons.

of the photodetectors without a bias voltage, in which a high I_{light}/I_{dark} ratio of ~10^4 is observed under 260 nm illumination with a light power of 130 mW cm^{-2}. However, under longer-wavelength UV illumination, the I_{light}/I_{dark} ratios reduce, showing a deep UV/visible rejection ratio (R_{250}/R_{400}) as high as ~10^4 (Figure 7.21c), suggesting a narrow deep UV detection window. The maximum responsivity of a photodetector based on $Zn_{1-x}Mg_xO$/PANI reaches 160 μA W^{-1}, as shown in Figure 7.21d. In comparison, photodetectors with pure $Zn_{1-x}Mg_xO$ films were found to work at a bias of −1 V, which exhibited a similar responsivity curve to hybrid *p–n* junction photodetectors without bias. This may indicate that the formation of hybrid *p–n* junction is a prerequisite for achieving self-powered function in UV photodetection.

7.5 Conclusion and Outlooks

In summary, this chapter first presents an introduction to UV photodetectors based on metal oxides with wide bandgaps and excellent stability, followed by a detailed description of device structures, including working mechanisms, advantages, and disadvantages. Next, important parameters for evaluating the detection performance of UV photodetectors based on metal oxides are summarized and discussed. Finally, the progress of typical UV photodetectors based on metal oxides, such as ZnO, Ga_2O_3, TiO_2, WO_3, SnO_2, $Zn_{1-x}Mg_xO$, and NiO, is proposed and examined. This chapter especially reviews and highlights the roles of bulk and nanostructured metal oxides in enhancing UV detectors.

Despite the significant achievements of UV photodetectors based on wide-bandgap metal oxides, there are still many challenges impeding their large-scale and commercial development. For example, the preparation of high-quality bulk metal oxides and the controllable growth of nanostructured metal oxides with homogeneous and uniform sizes, configurations, and orientations remain quite challenging. In addition, the effective *p*- or *n*-type doping of metal oxides is considered a major obstacle in constructing *p–n* and *p–i–n* devices due to the large activation energy at room temperature. Furthermore, to simultaneously achieve high responsivity, fast response speed, and low dark current in UV photodetectors, the scientific design of device structures should proceed via complex and multi-step manufacturing processes. As potential UV photodetectors, self-powered devices should be investigated in detail to improve their detection performance, reproducibility, and flexibility, but it poses a significant challenge.

There is still plenty of room for the development of wide-bandgap metal oxides and their applications in UV photodetectors. To solve the formidable issues mentioned above, new preparation methods, including compatible physical and chemical ones, need to be developed and promoted for synthesizing high-quality metal oxide materials. Besides, for nanostructure materials, the homogeneity and uniformity of sizes, configurations, and orientations can be improved with the assistance of templates and posttreatment. Furthermore, optimal doping technologies, such as precursor doping and doping-ion implantation, can be taken into consideration during the doping processes to realize controllable concentrations of *p*- and *n*-type doping at room temperature. To obtain high-performance UV photodetectors with high responsivity, fast response speed, and low dark current, compatible manufacturing methods and novel strategies involving plasmonic technologies and interface modification can be utilized and optimized to fabricate efficient devices. Finally, suitable metal oxides can be selected as absorption and active layers in accordance with the wavelength of the detective light to fabricate high-performance and self-powered photodetectors.

References

1 Chen, H., Liu, H., Zhang, Z. et al. (2016). Nanostructured photodetectors: from ultraviolet to terahertz. *Adv. Mater.* 28 (3): 403–433.

2 Razeghi, M. and Rogalski, A. (1996). Semiconductor ultraviolet detectors. *J. Appl. Phys.* 79 (10): 7433–7473.

3 Chen, H., Liu, K., Hu, L. et al. (2015). New concept ultraviolet photodetectors. *Mater. Today* 18 (9): 493–502.

4 Li, W. and Chou, S. (2010). Solar-blind deep-UV band-pass filter (250–350 nm) consisting of a metal nano-grid fabricated by anoimprint lithography. *Opt. Express* 18 (2): 931–937.

5 Pearton, S., Yang, J., Patrick, H. et al. (2018). A review of Ga_2O_3 materials, processing, and devices. *Appl. Phys. Rev.* 5: 011301.

6 Liao, M., Koide, Y., Alvarez, J. et al. (2008). Persistent positive and transient absolute negative photoconductivity observed in diamond photodetectors. *Phys. Rev. B* 78 (4): 045112.

7 Zhang, Z., Long, H., Tong, X. et al. (2018). Ultrafast, self-driven, and air-stable photodetectors based on multilayer $PtSe_{(2)}$/perovskite heterojunctions. *J. Phys. Chem. Lett.* 9 (6): 1185–1194.

8 Zhu, L., Wang, W., Yao, Z. et al. (2013). Structural spectral response tuning in organic deep ultraviolet photodetectors. *Solid-State Electron.* 80: 14–18.

9 Fujita, S. (2015). Wide-bandgap semiconductor materials: for their full bloom. *Jpn. J. Appl. Phys.* 54 (3): 030101.

10 Monroy, E. and Omnès F., and Calle F. (2003). Wide-bandgap semiconductor ultraviolet photodetectors. *Semicond. Sci. Technol.* 18: 33–51.

11 Lincke, R. and Wilkerson, T. (1962). New detector for the vacuum ultraviolet. *Rev. Sci. Instruments* 33: 911–913.

12 Takahashi, S., Abe, T., Nakagawa, A. et al. (2014). Applicability of nitrogen-doped ZnO single crystals for photoconductive type UV sensors. *Phys. Status Solidi. C* 11 (7–8): 1304–1307.

13 Han, S. (1997). New developments in photoconductive detectors. *Rev. Sci. Instruments* 68: 647–652.

14 Buscema, M., Island, J., Groenendijk, D. et al. (2015). Photocurrent generation with two-dimensional van der Waals semiconductors. *Chem. Soc. Rev* 44 (11): 3691–3718.

15 Xu, Y., Cheng, C., Du, S. et al. (2016). Contacts between two- and three-dimensional materials: Ohmic, Schottky, and p-n heterojunctions. *ACS Nano* 10 (5): 4895–4919.

16 Zahedi, F., Dariani, R., and Rozati, S. (2013). Ultraviolet photoresponse properties of ZnO N p-Si and ZnO p-Si heterojunctions. *Sens. Actuators, A* 199: 123–128.

17 Zhao, Y., Qi, J., Chandan, B. et al. (2015). Local irradiation effects of one-dimensional ZnO based self-powered. *Mater. Chem. Phys.* 166: 116–121.

18 Sumayya; I., Vinayak, G., Shahin, S. et al. (2015). Effect of the buffer layer on the metal semiconductor metal UV photodetector based on Al-doped and undoped ZnO thin films with different device structure. *Phys. Status Solidi A* 212 (8): 1704–1712.

19 Singh, S. (2016). Al doped ZnO based metal-semiconductor-metal andmetal-insulator-semiconductor-insulator-metal UV sensors. *Optik* 127: 3523–3526.

20 Menzel, A., Subannajui, K., Güder, F. et al. (2011). Multifunctional ZnO-nanowire-based sensor. *Adv. Funct. Mater.* 21: 4342–4348.

21 Xie, C., Wang, Y., Zhang, Z. et al. (2018). Graphene semiconductor hybrid heterostructures for optoelectronic device applications. *Nano Today* 19: 41–83.

22 Liu, K., Sakurai, M., and Aono, M. (2010). ZnO-based ultraviolet photodetectors. *Sensors* 10 (9): 8604–8634.

23 Liu, W., Yao, B., Li, B. et al. (2010). MgZnO/ZnO p-i-n junction UV photodetector fabricated on sapphire substrate by plasma-assisted molecular beam epitaxy. *Solid State Sci.* 12: 1567–1569.

24 Leung, Y., He, Z., Luo, L. et al. (2010). ZnO nanowires array homojunction and its application as a visible-blind ultraviolet photodetector. *Appl. Phys. Lett.* 96: 053102.

25 Bie, Y., Liao, Z., Zhang, H. et al. (2011). Self-powered, ultrafast, visible-blind UV detection and optical logical operation based on ZnO/GaN nanoscale p-n junctions. *Adv. Mater.* 23 (5): 649–653.

26 Shi, L., Wang, F., Li, B. et al. (2014). A highly efficient UV photodetector based on a ZnO microwire p–n homojunction. *J. Mater. Chem. C* 2: 5005.

27 Cai, J., Chen, X., Hong, R. et al. (2014). High-performance 4H-SiC-based p-i-n ultraviolet photodiode and investigation of its capacitance characteristics. *Opt. Commun.* 333: 182–186.

28 Bugallo, A.L., Tchernycheva, M., Jacopin, G. et al. (2010). Visible-blind photodetector based on p-i-n junction GaN nanowire ensembles. *Nanotechnology* 21 (31): 315201.

29 Xie, C., Lu, X., Tong, X. et al. (2019). Recent progress in solar-blind deep-ultraviolet photodetectors based on inorganic ultrawide bandgap semiconductors. *Adv. Funct. Mater.* 29: 1806006.

30 Konstantatos, G. and Sargent, E. (2010). Nanostructured materials for photon detection. *Nat. Nanotechnol.* 5 (6): 391–400.

31 Xie, C., Mak, C., Tao, X. et al. (2017). Photodetectors based on two-dimensional layered materials beyond graphene. *Adv. Funct. Mater.* 27: 1603886.

32 Zhao, B., Wang, F., Chen, H. et al. (2015). Solar-blind avalanche photodetector based On single ZnO-$Ga_{(2)}O_{(3)}$ core-shell microwire. *Nano Lett.* 15 (6): 3988–3993.

33 Hsu, C., Hsu, D., Hsueh, T. et al. (2017). Transparent gas senor and photodetector based on Al doped ZnO nanowires synthesized on glass substrate. *Ceram. Int.* 43: 5434–5440.

34 Podder, S., Basumatary, B., Gogoi, D. et al. (2021). Pyro-phototronic application in the Au/ZnO interface for the fabrication of a highly responsive ultrafast UV photodetector. *Appl. Surf. Sci.* 537: 147893.

35 Shabannia, R. (2018). High-sensitivity UV photodetector based on oblique and vertical Co-doped ZnO nanorods. *Mater. Lett.* 214: 254–256.

36 Look, D. (2001). Recent advances in ZnO materials and devices. *Mater. Sci. Eng. B* 80: 383–387.

37 Sun, H., Zhang, Q., and Wu, J. (2006). Electroluminescence from ZnO nanorods with an n-ZnO/p-Si heterojunction structure. *Nanotechnology* 17: 2271–2274.

38 Leonard, J. and Lu, Y. (2011). ZnO Schottky barriers and Ohmic contacts. *J. Appl. Phys.* 109: 121301.

39 Takeuchi, M., Kashimura, S., and Ozawa, S. (1990). Photoconductive and gas sensitive properties of ultrafine ZnO particle layers prepared by a gas evaporation technique. *Vaccum* 41: 1636–1637.

40 Harish, K. and Gupta, V. (2012). A comparative study of ultraviolet photoconductivity relaxation in zinc oxide (ZnO) thin films deposited by different techniques. *J. Appl. Phys.* 111: 102809.

41 Xu, Q., Zhang, J., Ju, K. et al. (2006). ZnO thin film photoconductive ultraviolet detector with fast photoresponse. *J. Crys. Growth* 289: 44–47.

42 Liang, S., Sheng, H., Liu, Y. et al. (2001). ZnO Schottky ultraviolet photodetectors. *J. Crys. Growth* 225: 110–113.

43 Jiang, D., Zhang, J., Lu, Y. et al. (2008). Ultraviolet Schottky detector based on epitaxial ZnO thin film. *Solid State Electron.* 52: 679–682.

44 Jin, Y., Wang, J., Sun, B. et al. (2008). Solution-processed ultraviolet photodetectors based on colloidal ZnO nanoparticles. *Nano Lett.* 8: 1649–1653.

45 Nie, B., Hu, J., Luo, L. et al. (2013). Monolayer graphene film on ZnO nanorod array for high-performance Schottky junction ultraviolet photodetectors. *Small* 9 (17): 2872–2879.

46 Zhang, T., Wu, G., Wang, J. et al. (2017). A sensitive ultraviolet light photodiode based on graphene-on-zinc oxide Schottky junction. *Nanophoton.* 6 (5): 1073–1081.

47 Tian, C., Jiang, D., Li, B. et al. (2014). Performance enhancement of ZnO UV photodetectors by surface plasmons. *ACS Appl. Mater. Interfaces* 6 (3): 2162–2166.

48 Rasool, A., Kumar, M., Mamat, M. et al. (2020). Analysis on different detection mechanisms involved in ZnO-based photodetector and photodiodes. *J. Mater. Sci. Mater. Electron.* 31: 7100–7113.

49 Doni, P., Joseph, W., Hariprasad, K. et al. (2021). Enhancement of optoelectronic properties of ZnO thin films by Al doping for photodetector applications. *Superlattice Microstruct.* 151: 106790.

50 Young, S., Liu, Y., Shiblee, N. et al. (2020). Flexible ultraviolet photodetectors based on one-dimensional gallium-doped zinc oxide nanostructures. *ACS Appl. Mater. Interfaces* 2: 3522–3529.

51 Damanpreet, K. and Kumar, M. (2021). A strategic review on gallium oxide based deep-ultraviolet photodetectors: Recent progress and future prospects. *Adv. Opt. Mater.* 9: 2002160.

52 Varshney, U., Aggarwal, N., and Gupta, G. (2022). Current advances in solar-blind photodetection technology: Using Ga_2O_3 and AlGaN. *J. Mater. Chem. C* 10: 1573.

53 Chen, X., Xu, Y., Zhou, D. et al. (2017). Solar-blind photodetector with high avalanche gains and bias-tunable detecting functionality based on metastable phase alpha-$Ga_{(2)}O_{(3)}$/ZnO isotype heterostructures. *ACS Appl. Mater. Interfaces* 9 (42): 36997–37005.

54 Zhang, D., Zheng, W., Lin, R. et al. (2018). High quality β-Ga_2O_3 film grown with N_2O for high sensitivity solar-blind-ultraviolet photodetector with fast response speed. *J. Alloys Compound.* 735: 150–154.

55 Li, M., Yang, N., Wang, G. et al. (2019). Highly preferred orientation of Ga_2O_3 films sputtered on SiC substrates for deep UV photodetector application. *Appl. Surf. Sci.* 471: 694–702.

56 Zhou, H., Cong, L., Ma, J. et al. (2020). High-performance high-temperature solar-blind photodetector based on polycrystalline Ga_2O_3 film. *J. Alloys Compound.* 847: 156536.

57 Mu, W., Jia, Z., Yin, Y. et al. (2017). One-step exfoliation of ultra-smooth β-Ga_2O_3 wafers from bulk crystal for photodetectors. *Cryst. Eng. Comm.* 19: 5122–5127.

58 Wang, S., Sun, H., Wang, Z. et al. (2019). In situ synthesis of monoclinic β-Ga_2O_3 nanowires on flexible substrate and solar-blind photodetector. *J. Alloys Compound.* 787: 133–139.

59 Zhang, M., Kang, S., Wang, L. et al. (2021). Facile synthesis of β-Ga_2O_3 nanowires network for solar-blind ultraviolet photodetector. *J. Phys. D Appl. Phys.* 54 (17): 175106.

60 Ashish, K. and Bag, A. (2019). High responsivity of quasi-2D electrospun β-Ga_2O_3-based deep-UV photodetectors. *IEEE Photon. Technol. Lett.* 31 (8): 619–622.

61 Oh, S., Kim, C., and Kim, J. (2017). High responsivity β-Ga_2O_3 metal-semiconductor-metal solar-blind photodetectors with ultraviolet transparent graphene electrodes. *ACS Photon.* 5 (3): 1123–1128.

62 Xue, H., Kong, X., Liu, Z. et al. (2007). TiO_2 based metal-semiconductor-metal ultraviolet photodetectors. *Appl. Phys. Lett.* 90: 201118.

63 Xing, J., Wei, H., Guo, E.J. et al. (2011). Highly sensitive fast-response UV photodetectors based on epitaxial TiO_2 films. *J. Phys. D Appl. Phys.* 44 (37): 375104.

64 Moudgil, A., Member, S., Sharma, K. et al. (2020). In_2O_3/TiO_2 heterostructure for highly responsive low-noise ultraviolet photodetector. *IEEE T. Electron Dev.* 67 (1): 166–172.

65 Zhang, D., Liu, C., Xu, R. et al. (2017). The effect of self-depleting in UV photodetector based on simultaneously fabricated TiO(2)/NiO *p-n* heterojunction and Ni/Au composite electrode. *Nanotechnology* 28 (36): 365505.

66 Sun, L., Wang, C., Ji, T. et al. (2017). Self-powered UV-visible photodetector with fast response and high photosensitivity employing an Fe:TiO_2/n-Si heterojunction. *RSC Adv.* 7 (81): 51744–51749.

67 Patel, D., Chauhan, K., Park, W. et al. (2017). Tunable TiO_2 films for high-performing transparent Schottky photodetector. *Mater. Sci. Semicond. Processing* 61: 45–49.

68 Zhang, H., Abdiryim, T., Jamal, R. et al. (2022). Self-powered TiO_2 NRs UV photodetectors: heterojunction with PTTh and enhanced responsivity by Au nanoparticles. *J. Alloys Compound.* 899: 163279.

69 Yan, T., Cai, S., Hu, Z. et al. (2021). Ultrafast speed, dark current suppression, and self-powered enhancement in $TiO_{(2)}$-based ultraviolet photodetectors by organic layers and Ag nanowires regulation. *J. Phys. Chem. Lett.* 12 (40): 9912–9918.

70 Cao, R., Xu, J., Shi, S. et al. (2020). High-performance self-powered ultraviolet photodetectors based on mixed-dimensional heterostructure arrays formed from NiO nanosheets and TiO_2 nanorods. *J. Mater. Chem. C* 8 (28): 9646–9654.

71 Li, M., Xu, J., Zhu, K. et al. (2021). The fabrication of a self-powered $CuInS_2/TiO_2$ heterojunction photodetector and its application in visible light communication with ultraviolet light encryption. *J. Mater. Chem. C* 9 (41): 14613–14622.

72 Chen, K., Chang, S., and Lin, C. (2018). Effect of oxygen vacancy concentration on indium tungsten oxide UV-Λ photodetector. *RSC Adv.* 9 (1): 87–90.

73 Hai, Z., Akbari, M., Xu, H. et al. (2017). Photodetector with superior functional capabilities based on monolayer WO_3 developed by atomic layer deposition. *Sensor Actuators B* 245: 954–962.

74 Kim, Y., Lee, S., Jeong, S. et al. (2020). Conversion of $WO_{(3)}$ thin films into self-crosslinked nanorods for large-scale ultraviolet detection. *RSC Adv.* 10 (24): 14147–14153.

75 Cheng, W. and Niederberger, M. (2016). Evaporation-induced self-assembly of ultrathin tungsten oxide nanowires over a large scale for ultraviolet photodetector. *Langmuir* 32 (10): 2474–2481.

76 Yan, J., Chen, Y., Wang, X. et al. (2019). High-performance solar-blind $SnO_{(2)}$ nanowire photodetectors assembled using optical tweezers. *Nanoscale* 11 (5): 2162–2169.

77 Li, M., Mao, D., Wan, J. et al. (2019). Hollow multi-shell structured SnO_2 with enhanced performance for ultraviolet photodetectors. *Inorg. Chem. Front.* 6: 1968.

78 Reddy, K., Sahatiya, P., Santos-Sauceda, I. et al. (2020). One-step fabrication of 1D p-NiO nanowire/n-Si heterojunction: development of self-powered ultraviolet photodetector. *Appl. Surf. Sci.* 513: 145804.

79 Yu, N., Li, H., and Wu, Y. (2020). A high-sensitivity, fast-response, rapid-recovery UV photodetector based on p-GaN/NiO nanostructures/n-GaN sandwich structure. *Solid State Sci.* 104: 106206.

80 Li, L., Liu, Z., Wang, L. et al. (2018). Self-powered GaN ultraviolet photodetectors with p-NiO electrode grown by thermal oxidation. *Mater. Sci. Semicond. Processing* 76: 61–64.

81 Lin, P., Huang, W., Lou, Y. et al. (2021). Solution-processed Li-doped ZnSnO metal-semiconductor-metal UV photodetectors. *J. Phys. D Appl. Phys.* 54: 345107.

82 Yalcin, M., Ozmen, D., and Yakuphanoglu, F. (2020). Sr-doped yttrium nickel oxide-based photodetectors. *J. Mater. Sci. Mater. Electron.* 31 (4): 3441–3455.

83 Mia, M., Pervez, M., Hossain, K. et al. (2017). Influence of Mg content on tailoring optical bandgap of Mg-doped ZnO thin film prepared by sol-gel method. *Results Phys.* 7: 2683–2691.

84 Wang, P., Zhen, Q., Tang, Q. et al. (2013). Steady-state characteristics and transient response of MgZnO-based metal-semiconductor-metal solar-blind ultraviolet photodetector with three types of electrode structures. *Opt. Express* 21 (15): 18387–18397.

85 Chen, H., Yu, P., Zhang, Z. et al. (2016). Ultrasensitive self-powered solar-blind deep-ultraviolet photodetector based on all-solid-state polyaniline/MgZnO Bilayer. *Small* 12 (42): 5809–5816.

86 Nasiri, N., Bo, R., Wang, F. et al. (2015). Ultraporous Electron-Depleted ZnO Nanoparticle Networks for Highly Sensitive Portable Visible-Blind UV Photodetectors. *Adv. Mater.* 27 (29): 4336–4343.

8

Metal Oxide Semiconductors for Memory Applications

8.1 Introduction

Nowadays, high-speed/high-density memories and processors have shown crucial applications in computers and mobile electronic devices, providing powerful processing capabilities. Semiconductor memories can be separated as random-access memories (RAMs) and read-only memories (ROMs), where in RAMs writing and reading times are almost the same, and in ROMs writing takes more time than reading. While ROMs are nonvolatile devices, RAMs could be either volatile or nonvolatile memories, according to the retention of the stored information. For example, dynamic random-access memories (DRAMs) and static random-access memories (SRAMs) are volatile memories with the stored information being lost once the power is turned off. Typical known resistive random-access memories (RRAMs) and flash memories are nonvolatile devices, which have the information stored in them is able to remain even after the power supply is turned off. Figure 8.1 shows the categories of semiconductor memories.

As promising alternatives to Si-based technologies, metal oxide semiconductors have shown great potential in emerging computing applications, including but not limited to nonvolatile RAMs, normally-off processors, pseudo-complementary metal oxide semiconductor (CMOS) circuits, memory devices with floating gates, memristors, etc. Besides, they could offer advantages of high transparency and flexibility and are hence being considered as star materials for next-generation foldable and transparent electronic devices.

8.2 Resistive Random-Access Memory

The resistive switching behavior of several materials has attracted considerable attention, particularly for their applications in non-volatile memory (NVM) devices. One typical type of NVM device is RRAMs, which use materials having a change of resistance with applied voltages. Up to now, plenty of metal oxides have been reported capable of showing resistive switching behaviors. Most of them are transition metal oxides, and some of them are lanthanide series metal oxides, as summarized in Table 8.1. Those materials include oxide semiconductors, i.e.

Metal Oxide Semiconductors: Synthesis, Properties, and Devices,
First Edition. Zhigang Zang, Wensi Cai, and Yong Zhou.

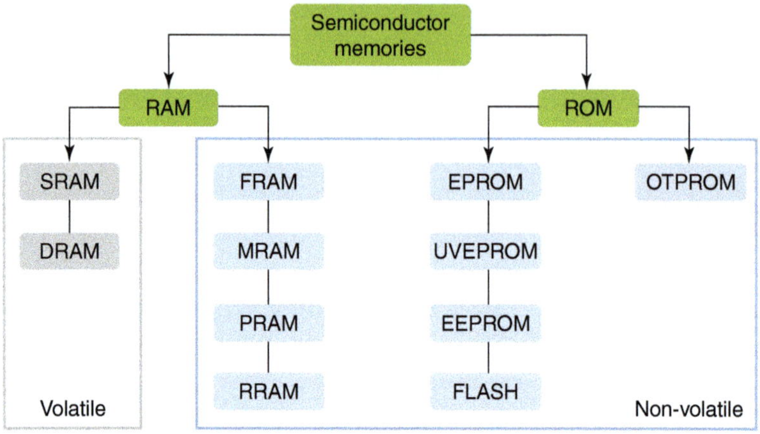

Figure 8.1 Catalogues of standard semiconductor memories.

Table 8.1 The periodic table of the elements, summarizing the elements that can be used in binary metal-oxide RRAMs.

Corresponding binary oxide that exhibits resistance switching

Metal that is used for electrode

^{1}H																	^{2}He
^{3}Li	^{4}Be											^{5}B	^{6}C	^{7}N	^{8}O	^{9}F	^{10}Ne
^{11}Na	^{12}Mg											^{13}Al	^{14}Si	^{15}P	^{16}S	^{17}Cl	^{18}Ar
^{19}K	^{20}Ca	^{21}Sc	^{22}Ti	^{23}V	^{24}Cr	^{25}Mn	^{26}Fe	^{27}Co	^{28}Ni	^{29}Cu	^{30}Zn	^{31}Ga	^{32}Ge	^{33}As	^{34}Se	^{35}Br	^{36}Kr
^{37}Rb	^{38}Sr	^{39}Y	^{40}Zr	^{41}Nb	^{42}Mo	^{43}Tc	^{44}Ru	^{45}Rh	^{46}Pd	^{47}Ag	^{48}Cd	^{49}In	^{50}Sn	^{51}Sb	^{52}Tb	^{53}I	^{54}Xe
^{55}Cs	^{56}Ba	^{57}La	^{72}Hf	^{73}Ta	^{74}W	^{75}Re	^{76}Os	^{77}Ir	^{78}Pt	^{79}Au	^{80}Hg	^{81}Ti	^{82}Pb	^{83}Bi	^{84}Po	^{85}At	^{86}Rn
				^{58}Ce	^{59}Pr	^{60}Nd	^{61}Pm	^{62}Sm	^{63}Eu	^{64}Gd	^{65}Tb	^{66}Dy	^{67}Ho	^{68}Er	^{69}Tm	^{70}Yb	^{71}Lu

ZnO, NiO_x, CuO_x, and oxide insulators with a large bandgap, i.e. AlO_x, HfO_x, and TaO_x [1, 2]. Compared with other traditional memories, RRAMs show advantages, particularly in terms of operating speed, scalability potential, and CMOS architectural compatibility. Considering the aim of this book, here in this subchapter, we systematically discuss RRAMs based on oxide semiconductors, focusing on their working mechanisms, performance characterizations, and representative works.

8.2.1 Resistive Switching Mechanisms

RRAMs were reported to have various structures, including but not limited to sandwich, planar, and laterally bridged [3–5]. Nevertheless, a basic RRAM should consist of two opposite electrodes and a storage material, as shown in Figure 8.2a. The resistance state of RRAMs can be varied from high resistance state (HRS) to low

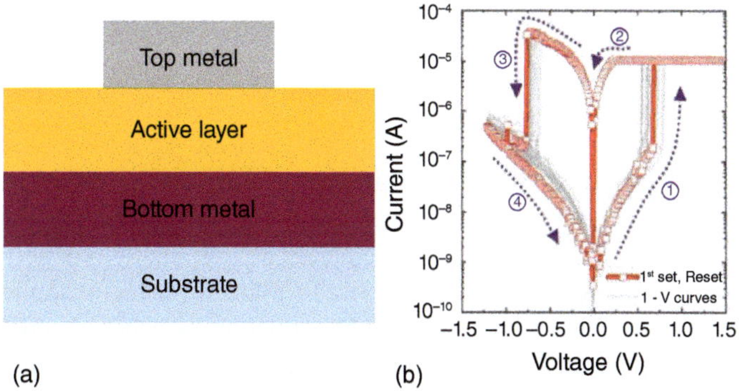

Figure 8.2 (a) A schematic diagram of RRAMs. (b) Representative *I–V* characteristics of RRAMs.

resistance state (LRS) or vice versa according to the applied voltage. The transition from HRS to LRS is generally called SET progress, while the change from LRS to HRS is RESET. For an as-prepared switching cell, the defects at initial states are generally quite low, making it behave as highly resistive. Therefore, an applied voltage larger than its forming voltage is required for the coming resistive switching. Such a process is called electroforming or forming. When the voltage is increased to forming voltage, the cell will change to LRS. During such a progress, conductive defects (either metal cations or oxygen anions) are generated within the active layer, forming conductive filaments with repeatable resistance effects. The device is then scanned negatively, and when the voltage is increased to the RESET transition voltage, the conductive filaments inside the active layer break, forcing the device to change from LRS to HRS, as shown in the procedures ② and ③ of Figure 8.2b.

In RRAMs, resistive switching can be separated into three different types, which are unipolar switching, bipolar switching, and nonpolar switching, depending on the polarity of the applied bias. In unipolar switching, the SET and RESET occur in the same direction, offering the advantages of high resistance ratios. However, it also has drawbacks such as high RESET current, poor uniformity, and low reliability. Bipolar switching, on the other hand, has SET and RESET occurring at an opposite bias polarity with the switching process, which could be either filamentary or non-filamentary. Besides, a few RRAMs also exhibit nonpolar switching, where the SET and RESET occur under any polarity. For the resistive switching behavior, a compliance current (CC) needs to be set through the semiconductor analyzer during the test to prevent damage to the devices from overcurrent. Alternatively, one can use a transistor, a diode, or a series of resistors to achieve such an aim. In order to read the data from the cell, a low read voltage is required, which should not affect the detection of cell to HRS or LRS.

Based on the switching mechanisms, RRAMs can be categorized into several groups, including electrochemical metallization types, valance change memory types, and thermochemical reaction types. In electrochemical metallization, one electrode must be active material and another must be inert or a counter electrode.

The active electrode materials here are mainly Cu and Ag, owing to the small electrical potential of Ag and Cu and the low standard Gibbs free energy for the formation of Ag and Cu oxides, making Ag^+ and Cu^{2+} easily and electrochemically dissolved. Other metals, including Ni, Al, Ti, Zn, Nb, and Au were also previously reported as active electrodes, despite not being as frequent as Ag and Cu. The switching principle of electrochemical metallization devices is the formation and rupture of the conductive filament. The conductive filament forms via the electrochemical dissolution process on the active electrode and finally redeposits on the counter electrode. The growth process of metallic filament under an applied bias can be separated into three steps, namely the oxidation of active metal to metal ions at the anode side, the electromigration of mobile ions from the anode to cathode, and the reduction of mobile ions to metal atoms at the top of counter electrode. After the growth process, the complete formation of metallic filament can conduct resistive switching.

In valance-change memory-type devices, the defects of oxygen vacancies are the main cause of them. In such cells, transition metal oxides, such as TiO_x, NiO_x, and WO_x are generally used as the storage layer with inert metals, such as Pt and Au, being used as electrodes [1, 2, 6]. Valance-change-memory type devices could be either filamentary switching or interface switching. For filamentary devices, resistive switching occurs due to the formation of a vacancy-based filament. By positively biasing the top electrode, the oxygen ions are drifted toward the top interface and accumulated due to the use of an inert metal, making the interface of the top electrode/active layer behave like an oxygen reservoir. The process starts with an initial soft breakdown due to the high electric field, which is known as the forming process. The voltage required for achieving resistive switching is generally lower than the forming voltage, and the size and thickness of the filament can be modified through current compliance. For non-filamentary-type valance-change memory-type devices, the modulation of the tunneling barrier thickness is crucial for controlling the switching process. While, in general, the on/off ratio in non-filamentary devices is lower than in filamentary ones, the switching stability is better in such devices.

As for thermochemical reaction-type memories, the active layer could be transition metal oxides, i.e. NiO_x and TiO_x. The resistive switching behavior in such devices is attributed to the temperature-induced change of valence state of transition metal ions.

8.2.2 Performance Characterization

Apart from the previously mentioned RESET and SET voltage, parameters including current on/off ratio, endurance, data retention, and switching time are also crucial parameters in detecting the performance of a RRAM.

Current on/off ratio: The current ratio between HRS and LRS of the switching cell. For RRAMs, since a CC is always applied to avoid the breakdown, the current at LRS is generally limited by the CC. In reality, the current on/off ratio needs to be as large as possible to reduce the load on amplifiers and simply the circuits.

Endurance: The number of cycles that a RRAM can be switched between two or more resistive states and simultaneously maintain a sufficient resistance ratio (>5) between them.

Retention time: A value that is used to evaluate the nonvolatility of various resistance states of the device, which is also called long-term stability. A long data retention time of more than 10 years is desired for RRAM devices. One common extrapolation method reported in the literature involves measuring the devices at a high temperature (≥85 °C), monitoring the resistance of the devices by applying pulses at certain time intervals (≥10^5 seconds), and extrapolating the resistance evolution line to the ten-year point.

Switching time: The time needed for RRAMs to switch between HRS and LRS. Generally, short switching times are required in real applications to enable quick read/write operations.

8.2.3 Representative Devices

As early as 1960s, the phenomena of resistive switching and negative differential resistance were reported in TiO_2 [7]. However, it was not until the 2000s that their applications in RRAMs were activated and reported. Waser's work in 2007 demonstrated both bipolar and unipolar resistive switching behaviors in Pt/TiO_2/Pt stacks, depending on the current compliance during the electroforming process [8]. Bipolar resistive switching behaviors were observed under a low current compliance (<0.1 mA), while unipolar ones were observed with a higher current compliance (1–10 mA). Furthermore, a permanent transition from bipolar to unipolar could be triggered by applying a voltage with a higher current compliance of approximately 3 mA. This work was interesting since the transition metal oxides reported before this work were considered to show only unipolar resistive switching behaviors.

Later in 2010, Hwang et al. directly observed the conduction filaments in TiO_2 during resistive switching using a transmission electron microscope (TEM) [9]. As shown in Figure 8.3a, the existence of Magnéli phase (Ti_nO_{2n-1}, $n = 4$) was confirmed using high-resolution transmission electron microscope (HRTEM) electron diffraction analyses. *In situ I–V* and low-temperature conductivity measurements further demonstrated that the overall resistive switching behaviors were related to the formation and disruption of the Magnéli phase filaments. Especially, they found that the SET operation was governed by the Magnéli nanofilaments, while the RESET operation corresponded to the phase transformation of nanofilaments from Magnéli to other stoichiometric phases. Another Magnéli phase filament (Ti_nO_{2n-1}, $n = 5$) was later observed by the same group, as shown in Figure 8.3b [10].

Due to its advantageous properties, including low cost, ease of deposition, and environmentally friendly nature, ZnO is considered as a promising candidate in RRAMs. For example, unipolar ZnO RRAMs with a Pt/ZnO/Pt structure were reported by Tsai et al. in 2008, achieving a current on/off ratio of 10^3–10^4 within 100 cycles of testing as well as a low required forming electric field [11]. The restive switching behavior in such devices was found to be related to the formation of straight filaments along the flat grain boundaries in highly (002)-oriented and

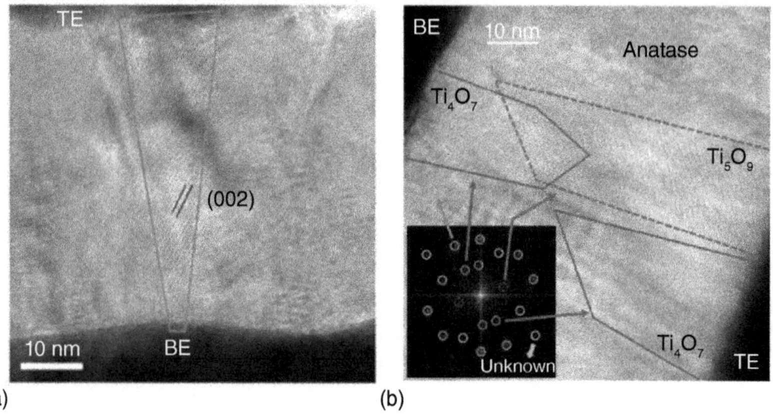

Figure 8.3 High-resolution TEM images of (a) Ti_4O_7 and (b) Ti_5O_9 filaments. Source: (a) Reproduced with permission from Kwon et al. [9]. © 2010, Spring Nature. (b) Reproduced with permission from Kim et al. [10]. 2011, AIP Publishing.

columnar-structured ZnO thin films. The conduction mechanism was demonstrated to be Ohmic behavior and Poole–Frenkel emission for LRS and HRS, respectively. This work opened the door for applying oxide semiconductors as the storage material in resistive switching devices.

However, compared to their amorphous counterparts, crystalline oxide semiconductors exhibit poor uniformity, especially over large areas, due to the existence of grain boundaries. To address this issue, Bao et al. used a low-temperature photochemical solution deposition method, which combined chemical solution deposition with UV irradiation treatment, to prepare amorphous indium gallium zinc oxide (IGZO), and reported RRAMs with a Pt/IGZO/Pt structure [12]. Several UV treatment times were examined, and the optimized one was found to be four hours. As shown in Figure 8.4, such RRAMs exhibited reproducible and highly uniform resistive switching properties in SET and RESET voltages, as well as a stable distribution of LRS and HRS. The excellent performance here might be attributed to the reduction of residual content, the formation of enhanced M–O and M–OH bonding, and the passivation of charge traps by UV-irradiated hydrogen bonding.

The previously mentioned works with a Pt/oxide semiconductor/Pt structure refer to valance memory type devices, where inert metals are used for both bottom and top electrodes. The use of a high-work-function metal as the cathode may create a high interface barrier, inducing resistive switching properties and resulting in a higher on/off ratio. However, inert metal anodes do not always provide better performance. For example, oxidizable metal anodes could form a thin interfacial metal oxide layer at the top electrode/active layer interface, controlling the outflow of oxygen to the environment during SET progress, which is beneficial for achieving long endurance [13]. The influence of metal electrodes on the device's performance was studied by Wang et al., who fabricated ZnO RRAMs using Ag, Ti, or Pt as the top electrodes [14]. As shown in Figure 8.5, applying a low positive voltage would form an Ag bridge from bottom to top in Ag/ZnO/Pt devices. Since the voltage is not large

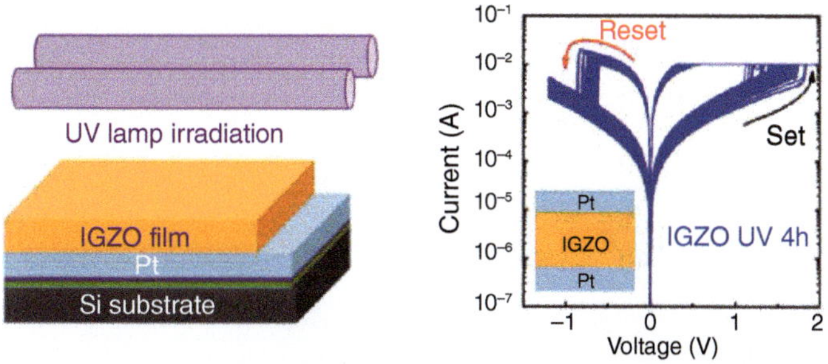

Figure 8.4 A schematic showing the deposition of amorphous IGZO and *I*–*V* characteristics of IGZO RRAMs. Source: Reproduced from Hu et al. [12]/with permission of American Chemical Society.

enough to induce the migration of oxygen vacancies and/or ions between electrodes, the conduction of SET and RESET processes can be successfully formed in HRS. In Ti/ZnO/Pt devices, a large amount of oxygen ions from ZnO are attracted to the ZnO/Ti interface due to the oxygen-getting ability of Ti. The interfacial ion migration can cause the redox reaction of oxygen vacancies in conducting filaments, leading to the formation and rupture of the conducting filaments, but with relatively poor endurance. In Pt/ZnO/Pt devices, once the filaments consisting of oxygen vacancies in the storage layer are formed, they are difficult to disrupt, resulting in permanent LRS after a forming-like process.

In addition to *n*-type oxide semiconductors, *p*-type ones, especially NiO_x, can also be used as the storage material in RRAM applications with various metals such as Pt, Au, and Ni being previously reported as the electrodes in NiO_x memories. Similar to most resistive switching materials, electroforming is necessary to trigger unipolar switching in NiO_x. The mechanisms involve the formation and rupture of conduction filaments, as reported by Kim et al., who used a current-bias method to observe the intermediate resistance states and anomalous resistance fluctuations between resistance states, demonstrating the resistive transition from HRS to LRS [15]. The formation and removal of filaments for HRS and LRS in NiO_x were further confirmed by Shin and Son, using a conducting atomic force microscope [16]. Using TEM analysis, Seo's group observed the Ni phase segregation at the grain boundaries of NiO_x film, suggesting that metallic Ni obtained from the phase separation of NiO_x was the main source of the conduction filaments [17].

Achieving a reasonably low reset current is important for promoting the practical use of RRAMs, which might be doable by reducing the parasitic capacitance, as demonstrated by Kinoshita et al. [18]. In their work, they studied the relationship between reset current, I_{reset}, and CC, I_{comp}, of a Pt/NiO_x/Pt structure in five different measurement configurations, as shown in Figure 8.6a. Figure 8.6b clearly shows that I_{reset} could saturate at a higher I_{comp} for structures with a larger parasitic capacitance. Devices with 1T1R structures showed more ideal behaviors, approaching $I_{reset} = I_{comp}$, even at a low CC. This also implies that I_{reset}–I_{comp} characteristics can approach asymptotically to $I_{reset} = I_{comp}$ with a reduction in CC, and the

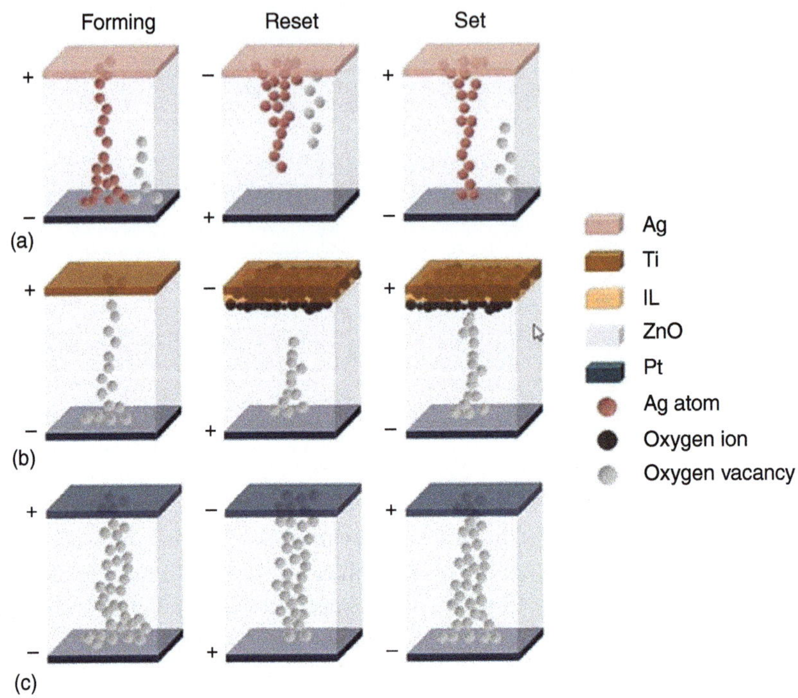

Figure 8.5 Schematic diagrams showing the resistive switching mechanisms of (a) Ag/ZnO/Pt, (b) Ti/ZnO/Pt, and (c) Pt/ZnO/Pt devices. Source: Reproduced from Wang et al. [14]/with permission of AIP Publishing.

achievement of $I_{reset} = I_{comp}$ is expected using an ideal current limiter with zero parasitic capacitance.

Further enhancing the resistive switching behaviors of NiO_x RRAMs might also be doable by introducing a bilayer structure, as demonstrated by Zhou et al., using an $Ag/NiO_x/TiO_2/FTO$ structure [19]. While a single NiO_x active layer shows typical electronic resistive switching memory behaviors and a single TiO_2 layer shows an obvious rectification effect at positive regions, an enhanced resistive switching memory phenomenon can be observed for bilayer NiO_x/TiO_2, showing a resistance ratio of ~10^3, a switching cycle endurance of 10^2, and a long retention time of 10^4 seconds. In general, the resistive switching in such RRAMs can be attributed to the formation and rupture of the redox-based Ag conduction filaments, construction and breaking of Ni paths, charge trapping and de-trapping from the trap sites near the electrodes, and the migration of oxygen vacancies. The analysis of conduction mechanisms and the variation of metal electrodes demonstrated that the migration of Ag^+ and Ni^{2+} filaments, and the diffusion of oxygen vacancies were the dominant ones for the reported RRAMs.

Other than NiO_x, copper oxides could also be used as the active layer in RRAMs. Cu_xO-based RRAM cells with an endurance cycle of 10^4 were first reported by Lin et al. [20]. The enhanced endurance of resistive switching is attributed to the reduction of damage during the forming process, facilitated by the formation of a thin CuO

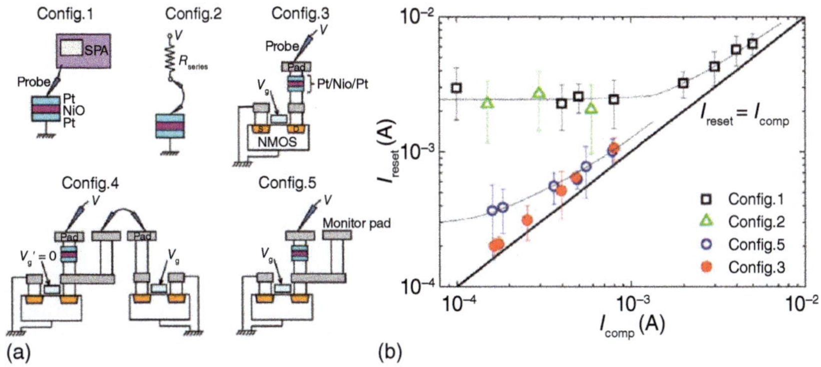

Figure 8.6 (a) Five configurations used to measure I_{reset}–I_{comp} characteristics. (b) Mean I_{reset} values of successive 30 switching cycles in different configurations. Source: Reproduced from Kinoshita et al. [18]/with permission of AIP Publishing.

upper layer through the plasma oxidation of Cu substrates. Hsu and Lin studied the electrode dependence of resistive switching characteristics in CuO memory devices [21]. The resistive switching behavior in CuO was caused by the copper ion filaments in the bulk CuO layer, resulting in a voltage-polarity-dependent SET process in Ag/CuO/ITO RRAMs since the irreversible resistance switching inferred the presence of abundant copper ion filaments. Alternatively, Al/CuO/ITO RRAMs showed stable resistive switching characteristics, which might be related to the existence of an Al_xO_y layer at the interface between Al and CuO, beneficial for the formation and rupture of the copper ion filaments.

Despite showing great progress, especially in the last two decades, challenges still exist in oxide semiconductor-based RRAMs, restricting their potential applications. At the device performance level, endurance cycles are also important, with various factors such as materials, processing techniques, device structures, and operation schemes affecting them. This includes two parts. The first is that during cycling, the HRS resistance tends to slowly decrease and eventually becomes stuck in LRS, unable to reset back to HRS due to the accumulation of too many defects. Those defects are mainly related to oxygen vacancies generated at or near the electrode/oxide interface, in or near the filament, and in the oxide matrix. As for LRS states, it is also noticed that during cycling, the LRS tends to increase due to the formation of an interfacial oxide layer between the electrode and the oxide [22].

Besides, a long data retention at a thermal stress of up to 85 °C as well as a small electrical stress, ideally longer than ten years, is desirable for NVM. However, until now only a few reported devices showed a retention of 10^6 seconds. Moreover, devices that maintain the resistance window for over 10^4 or 10^5 seconds cannot guarantee that the resistance window will still be adequate after 10^7 or 10^8 seconds, as the devices might collapse abruptly rather than gradually [23].

As for commercialization, there is a scaling trend toward the nanometer regime. In principle, RRAMs can be potentially scaled to sub-10 nm dimensions. Actually, the resistance of HRS increases as the inverse of the cell area, roughly following

Ohm's law, while LRS resistance shows less dependence on the cell area, suggesting that decreasing the device area would be beneficial for pursuing a high HRS/LRS resistance ratio. The difficulty of scaling is more related to the patterning process, especially for those RRAMs deposited by a solution-based method.

Another important feature is related to multilevel cell storage, which is highly required for information processing technologies. The LRS resistance can be modulated by setting current compliance possibly due to the modulation of the diameter or number of conduction filaments, while HRS resistance can be controlled by the reset voltage owing to the modulation of the ruptured conduction filament length [24]. Oxide RRAMs with up to eight level states have already been demonstrated, showing potential for multilevel cell storage [25]. However, further research is still required to investigate more potential active materials and device configurations, as well as achieve even more level states with good cycling endurance and retention.

Actually, different applications can be targeted based on device characteristics. For example, retention loss that mimics the dynamics of biological memory is beneficial for artificial neural network applications [26]. In neuromorphic applications, a large on/off ratio is not always a major concern [27]. This combined with the quite promising performance achieved so far, makes oxide RRAMs highly promising for future IoT applications, although they are not yet a matured technology.

8.3 Transistor-Structured Memory Devices

Besides the previously mentioned RRAMs with a diode-like structure, NVM devices could also be made with a transistor-like structure. A typical type of this is flash memory, which shows critical applications, especially as storage in computer systems. Unlike conventional hard-disk data storage, flash memory devices with a solid-state disk follow a purely electronic mechanism that does not require moving parts during the writing and reading. Besides, high-density integration could be achieved with flash memory technologies, making it the most compelling nonvolatile storage in today's computing.

8.3.1 Working Mechanisms

The operation of a flash memory is based on the change of transistors' threshold voltage, modulated depending on the degree of charge trapping in the gate dielectric layer. Specifically, when holes or electrons are trapped, threshold voltage decreases or increases, resulting in the erase or program state in a flash cell, respectively. Flash memory has two types of integration architectures: NOR and NAND, as shown in Figure 8.7. Clearly, in NOR architectures, each cell can be accessed separately, meaning that the erasing and programming could be done independently from other cell states. In contrast, flash memories with a NAND structure comprise serially connected memory cells with the source and drain of each transistor shared with the neighboring ones.

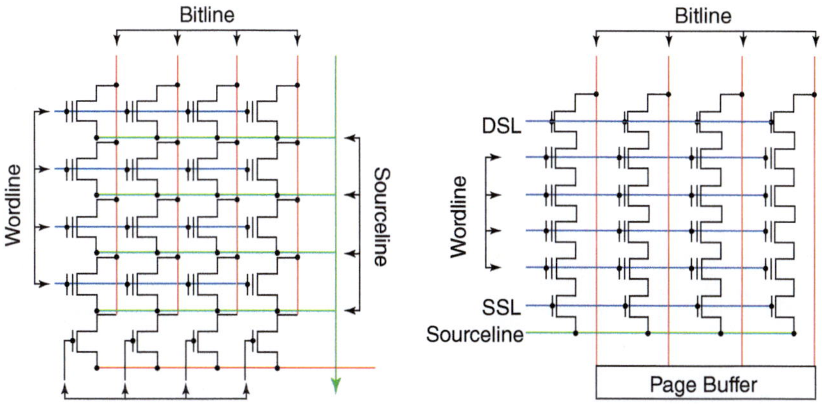

Figure 8.7 Schematic diagrams of flash memory devices with (a) NOR and (b) NAND architectures.

Now, we take a deep look at the single transistor used inside the memory cells. Traditional forms of transistor-structured memory devices can be classified into three categories: floating gate transistor memory (FGTM), charge trap transistor memory (CTTM), and ferroelectric field-effect transistor memory (FeFETM), as shown in Figure 8.8b–d. In normal transistors (Figure 8.8a), the flow of charges through the channel is controlled by adjusting the electric field between gate and source voltage, as detailed in Chapter 4. Transistor-structured memory devices have a similar structure to normal transistors, except for the inclusion of an extra charge-capture layer in/as the dielectric layer. During programming, the carriers are forced to move from the channel layer to the charge-capture layer due to the applied electric field. By applying a reverse gate voltage to the gate electrode, the prestored carriers are released back into the semiconducting channel, behaving as erasing. The presence of carriers in the charge-capture layer affects the conductivity of the channel, resulting in a change in the threshold voltage, as shown in Figure 8.8e. By analyzing the transfer curves under the initial and programming states, it is possible to determine whether the stored carriers are holes or electrons. Besides, by applying a suitable positive or negative gate bias, the transistor can repeatedly switch between erasing and programming, achieving binary "0" and "1" states. In transistor-structured memory devices, apart from the well-known characteristics such as device lifetime and retention time, the memory window is also an important parameter. The memory window refers to the threshold voltage difference between the erasing and programming states (Figure 8.8e).

To ensure programming and erasing operations, different device structures are required for each type of transistor-structured memory. In FGTM, low-dimensional materials such as metal nanoparticles or low-dimensional halide perovskites are commonly used as the floating gate with a relatively thick insulator layer being used to prevent charge carrier transfer from the floating gate to the gate electrode, and a thin tunneling layer to help the charge transfer into the floating gate. Notably, in

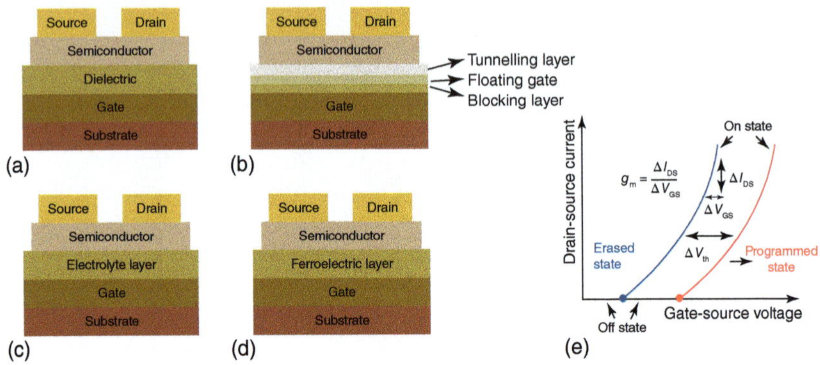

Figure 8.8 Schematic diagram of a (a) normal thin-film transistor, (b) floating-gated transistor, (c) electrolyte-gated transistor, and (d) ferroelectric-gated transistor. (e) Transfer characteristics of a transistor-structured memory.

order to achieve fast programming/erasing and low operating voltage, the tunneling layer needs to be very thin; however, this might also compromise stability.

CTTM has a similar structure to FGTM, but it utilizes a charge-trapping layer, normally an electrolyte, as the charge capture layer to store charges. Electrolytes are electrically insulating but ion-conducting materials that can be used to enable transistors with a low operating voltage thanks to the so-called electric-double-layer effects introduced at the dielectric/channel interface. Alternatively, electrolyte-gated transistors can operate based on electrochemical doping or redox mechanisms, depending on the applied electric field [28]. This is to say, for electrolyte-gated transistors, ions (i.e. Li^+ and H^+) in the electrolyte accumulate at the channel/electrolyte interface and modulate the channel conductance via electric-double-layer gating effects [29, 30], whereas the ions can penetrate into the channel under a relatively high gate voltage with the channel conductance being able to be modulated by electrochemical doping or redox [31, 32]. As a result, electrolyte-gated transistors based on electrochemical doping and redox exhibit a long retention time as the intercalated ions can remain trapped in the channel even after the gate pulse.

Memory behaviors can also be achieved using transistors with ferroelectric materials as the gate dielectrics, referred to as FeFETM. The ferroelectric effect refers to the spontaneous polarization of a material at a certain temperature. When applying a high voltage to the gate electrode of such thin-film transistors (TFTs), the ferroelectric polarization switches, causing the transistor characteristics to transition between the binary states, which can be used for memory functions.

8.3.2 Representative Devices

Due to their nonvolatile and easily erasing nature, flash memories have shown great potential, especially in computer-related applications. However, traditional silicon-oxygen-nitrogen-oxygen-silicon (SONOS) type structures are facing limitations in scaling down, which has led to the exploration of 3D architectures and novel memory or computing-in-memory approaches for next-generation NVMs. Among

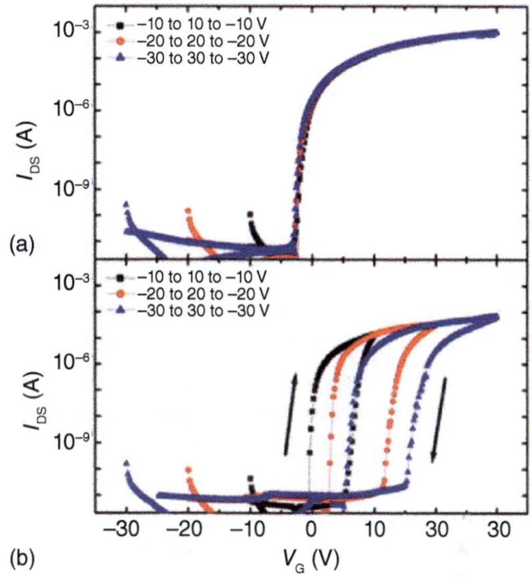

Figure 8.9 Transfer characteristics of IGZO transistors (a) without and (b) with Pt nanoparticles. Source: Reproduced from Suresh et al. [33]/with permission of AIP Publishing.

the potential alternatives, oxide semiconductors have gained attention. They can not only be combined with other trapping materials to achieve NVM behaviors but can also be used directly as charge-trapping layers. In this section, we review the recent progress of oxide semiconductor-based transistor-structured memories.

One of the most common types of NVM devices is FGTM, where the floating gate is typically made of low-dimensional metals. Muth et al. reported a transparent IGZO FGTM with platinum (Pt) nanoparticles added to the gate dielectric as the floating gate [33]. The presence of Pt nanoparticles was found to contribute to charge trapping, as evidenced by the hysteresis in the transfer characteristics of TFTs (Figure 8.9). In contrast, TFTs without Pt nanoparticles exhibited no hysteresis. Electron injection occurred under positive gate voltage stress, while holes could not be injected due to the characteristics of IGZO. The devices exhibited good retention, with only a 25% loss after 10^4 seconds, showcasing their potential for system-on-panel applications in transparent displays.

FGTMs have also been reported in various configurations, such as ZnO transistors with Al nanoparticles [34], IGZO transistors with Au nanoparticles [35], and IGZO transistors with Pt-Fe_2O_3 composite nanoparticles [36]. Alternatively, oxide semiconductors can serve as the floating gate to enable memory behaviors in a transistor. ZnO, in particular, is considered a promising charge-trapping material due to the abundance of oxygen vacancies in its thin film, which allows it to capture a large number of electrons and achieve a significant memory window. For example, Okyay et al. reported ZnO-based single-transistor memory cells using an Al_2O_3/ZnO/Al_2O_3 gate stack [37]. Yoon et al. demonstrated ZnO-based flexible memory devices with a wide memory window of 14.7 V and a large memory margin between on and off states [38]. Similarly, Ye et al. used ZnO as the charge-trapping layer and alumina as the tunneling layer and achieved transparent flash memory transistors with a maximum memory window of 15 V and excellent fatigue and retention properties [39].

In addition to ZnO, other oxide semiconductors, particularly IGZO, have been investigated as charge-trapping layers [40, 41]. Yin et al. first studied and demonstrated amorphous IGZO TFTs with a fully functional NVM structure [40]. The devices featured Al_2O_3/IGZO/Al_2O_3 stacked gate dielectrics as charge tunneling/storage/blocking layers, with IGZO serving as the channel layer. The typical transfer characteristics of such devices in programming and erasing modes suggested a distinct positive and negative shift in the transfer characteristics, attributing to channel hot electron injection in the programming mode and band-to-band tunneling-induced hot hole injection in the erasing mode, respectively.

In addition to serving as both the charge-trapping layer and channel layer, oxide semiconductors could also be combined with other stacked charge-trapping materials to behave as flash memory transistors. For example, Chin and coworkers reported nonvolatile IGZO channel-based memory transistors with a gate stack comprising double 2.5 nm SiO_2/12 nm HfLaO blocking oxides, double 5 nm Si_3N_4/5 nm HfON trapping layers, and double 4 nm HfLaO/2.5 nm SiO_2 tunneling oxides [43]. Such devices showed an initial memory window of 1.6 V and an extrapolated 10-year window of 1.2 V, which occurred at a programming/erasing voltage of 10 V/−12 V, with a fast response speed of 1 ms/100 μs.

In 2018, Hwang et al. reported IGZO-based flash memory devices, utilizing a 45 nm thick IGZO prepared by MOCVD or sputtering as the channel layer, 3 nm thick ALD-deposited Al_2O_3 as the tunneling layer, and 5 nm thick low-pressure CVD Si_3N_4 as the charge-trapping layer, as shown in Figure 8.10a [42]. Unlike most cases where sputtered IGZO showed better performance, these memories achieved better programming characteristics with amorphous IGZO deposited by MOCVD, possibly due to the 0.3 eV higher Fermi level position of MOCVD IGZO compared to sputtered IGZO (Figures 8.10b,c). The higher Fermi level position is a result of hydrogen atom incorporation in MOCVD IGZO films, which induces a higher carrier concentration and leads to a lower tunneling barrier between IGZO and Al_2O_3. MOCVD IGZO TFTs showed a programming threshold voltage shift of ∼1.8 and −3 V by applying

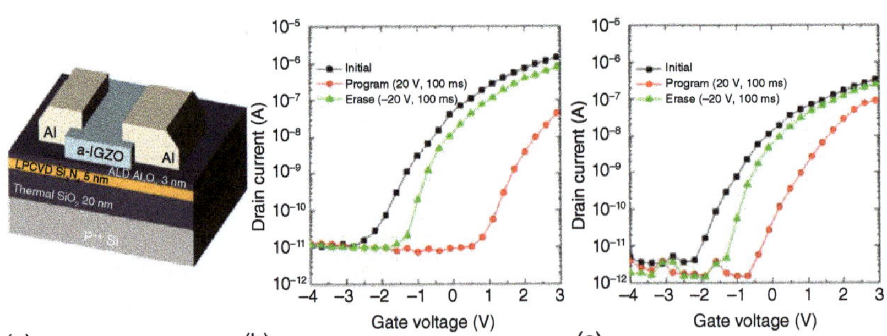

Figure 8.10 (a) A schematic of the device structure. Transfer characteristics of (b) MOCVD IGZO and (c) sputtered IGZO TFTs under programming and erasing states. Source: Reproduced from Hwang et al. [42]/with permission of IOP Publishing.

a 20 V gate voltage for 1 and 10 ms, respectively. Besides, applying an erase voltage of −20 V for 1 ms resulted in the recovery of more than 70% of the threshold voltage shift. The results demonstrate the crucial role of the band structure in determining the programming performance of flash memory devices, as it is closely related to the electron tunneling process from the channel to the charge-trapping layer.

A similar work was recently reported by Choi et al., who achieved flash memory devices using amorphous IGTO as a channel, Al_2O_3 as blocking and tunneling layers, and HfO_2 as the charge-trapping layer [45]. They observed that IGTO flash memories exhibited improved memory characteristics compared to those using IGZO, including faster programming/erasing speeds, wider memory window, and longer retention behaviors. This improvement can be attributed to the lower electron affinity and higher oxygen vacancies in IGTO thin films, which provide more holes. Furthermore, the researchers employed a light-assisted negative gate bias erasing method to enhance the erasing efficiency, as photo transitions are closely related to oxygen vacancies.

Apart from FGTMs, another well-known transistor-structured memory is CTTMs, which utilize electrolytes as the charge capture layer. Such a type of memory has the advantages of easy deposition since only one layer of gate dielectric is required. As early as 2010, Iwasa et al. demonstrated that by selectively driving protons or hydroxyl ions to the ZnO channel in protic ionic liquid-gated transistors, it was possible to extend the functionality of electrolyte-gated transistors to include proton memory and surface atom recognition [44]. As shown in Figure 8.11, the application of a 3 V gate voltage induces the generation of protons from water dissociation, which then move to the ZnO surface and interact with the surface atoms to behave as a "write" operation. When switched off, the trapped protons remain chemically bonded to the ZnO surface due to large bonding energy, acting as a "read" operation. When negatively charged, surface hydroxylation occurs, and the protons are driven away from the surface, equivalent to an "erase" operation. The authors also studied the response time of the proton transport and found that the current response to a stepwise gate potential was faster than 0.5 seconds (the measurement interval) even without any device optimization.

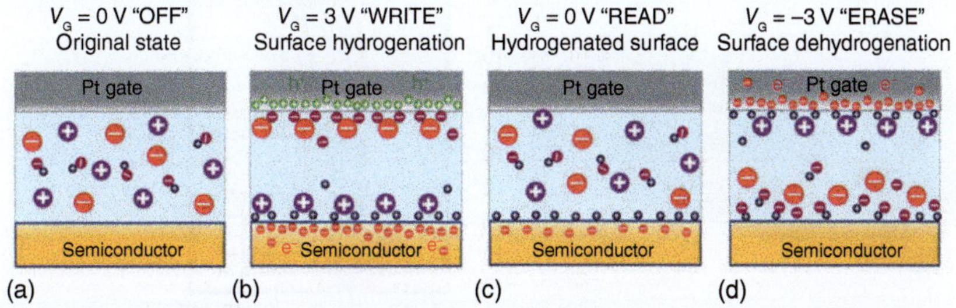

Figure 8.11 Schematic diagrams showing the different operating mechanisms of (a) off state, (b) write state, (c) read state and (d) erase state of the electrolyte-gated memory device. Source: Reproduced from Yuan et al. [44]/with permission of American Chemical Society.

The enabling of electrolyte-gated transistors as memory devices was further demonstrated by Kang's group, who reported NVM devices by inserting Au nanoparticles between ion gels and the IGZO channel layer [47]. The Au nanoparticles served as trapping sites for induced charges, storing electrical input signals. Two types of nanoparticles were used, including one prepared using direct thermal evaporation and the other through a colloidal process. Here, both cases showed an enlarged coverage of Au nanoparticles on IGZO, improving the programmed/erased memory signal ratio (I_{prog}/I_{erase}). Also, the insulating organic ligands attached to the nanoparticles prevented the escape of the trapped charges, further enhancing the retention characteristics of the programmed/erased signals. As a result, the reported IGZO-based non-volatile transistors showed an I_{prog}/I_{erase} larger than 10^3 at programming and erasing voltages of lower than 10 V, which could be maintained for over 10^5 seconds.

In computer storage cells, thermal stability is a crucial requirement for each memory cell. To achieve this, ferroelectric materials have been combined with oxide semiconductors in the realization of FeFETMs. Fully transparent NVM transistors were reported by Yoon et al., using a device structure shown in Figure 8.12a [46], at a temperature below 200 °C. The incorporation of poly(vinylidene fluoride-trifluoroethylene) [P(VDF-TrFE)] enabled memory behaviors in such TFTs, achieving a memory window of 7.5 V during a gate voltage sweep of −10 to 10 V, as shown in Figure 8.12b. A memory window of 1.8 V could still be maintained even with a lower gate voltage sweep of −6 to 6 V, providing novel functionalities and design schemes for various system applications. Similarly, Park et al., reported a ferroelectric memory array, consisting of two IGZO TFTs and a P(VDF-TrFE)-gated ferroelectric transistor [48]. The fabricated memory array successfully demonstrated disturb-free write/readout operation with a current ratio of ~10^4 between "1" and "0" states.

Inorganic ferroelectric materials were also reported as the dielectric layer in making oxide-transistor-structured memory devices by Lee ad Kim, who used a $HfZrO_x$ (HZO) gate dielectric and an IGZO channel [49]. Similar to P(VDF-TrFE),

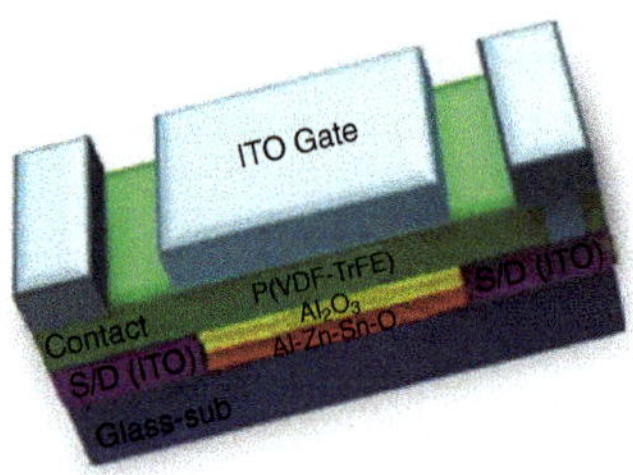

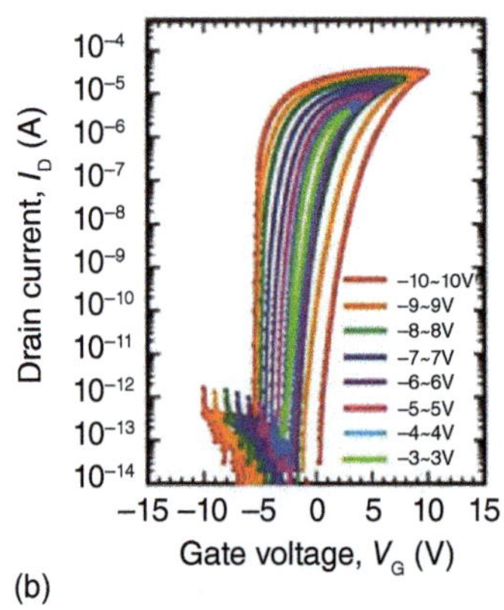

Figure 8.12 (a) A schematic showing the device structure of the ferroelectric TFTs. (b) Transfer characteristics of such TFTs under different gate voltage sweeps. Source: Reproduced from Yoon et al. [46]/with permission of John Wiley & Sons.

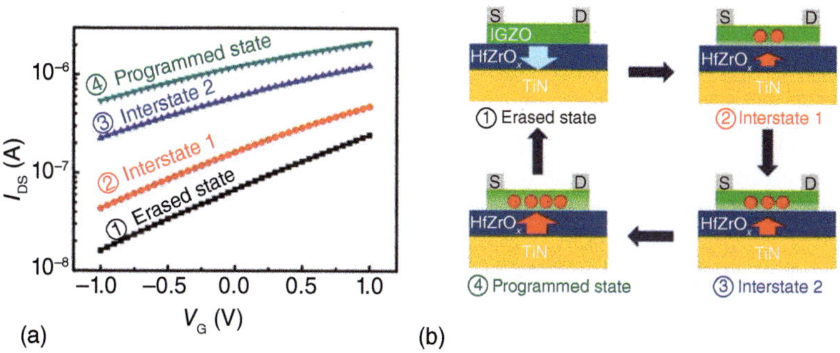

Figure 8.13 (a) The multilevel states of the ferroelectric TFTs. (b) Schematic diagrams of the suggested mechanism of multilevel data storage. Source: Reproduced from Kim and Lee [49]/with permission of American Chemical Society.

the polarization switching of HZO enables anticlockwise hysteresis in the transfer characteristics. The multilevel data storage of such TFTs is shown in Figure 8.13. First, the TFTs were set to an erase state by applying a negative bias pulse of −6 V for 30 ms to the gate electrode. At this time, the polarization of the HZO layer was pointed towards the gate electrode, and electrons inside the channel were depleted from the ferroelectric/channel interface, resulting in a low conductance of the channel. In contrast, the application of a positive bias pulse to the gate electrode switched the polarization towards the IGZO channel, enabling the accumulation of electrons at the interface and an increase in channel conductance with the increase in upward polarization [50]. Figure 8.13b clearly demonstrated that the channel conductance could be delicately modulated by changing the amplitude of applied bias pulses.

HZO has drawn increasing attention as a ferroelectric dielectric in oxide semiconductor-based transistor memories, particularly in the past decade, with related works being reported by Mo et al. [51], Lehninger et al. [52], Gong et al. [53], Jang et al. [54], and others. This is due to the good compatibility of HZO with existing complementary circuit technologies and the robust ferroelectricity of the material itself, even at an extremely thin thickness. These qualities make HZO a promising candidate for next-generation memory-related applications.

Clearly, all types of transistor-structured memory devices have garnered considerable attention owing to their potential integration with the existing CMOS-based technologies in computing-related fields. Oxide semiconductors, with their higher mobility, especially compared to organic semiconductors, and better uniformity and repeatability compared to polycrystalline silicon, hold promise as channel candidates in such devices. However, it should be noted that, currently, most of these memory transistors are fabricated on rigid substrates. Given the growing focus on flexible and wearable electronics, future directions for these devices may involve achieving high flexibility. Additionally, while current research primarily focuses on single transistors, the integration of these transistors to enable NAND or NOR architectures with multibit storage is of great importance for real commercialization.

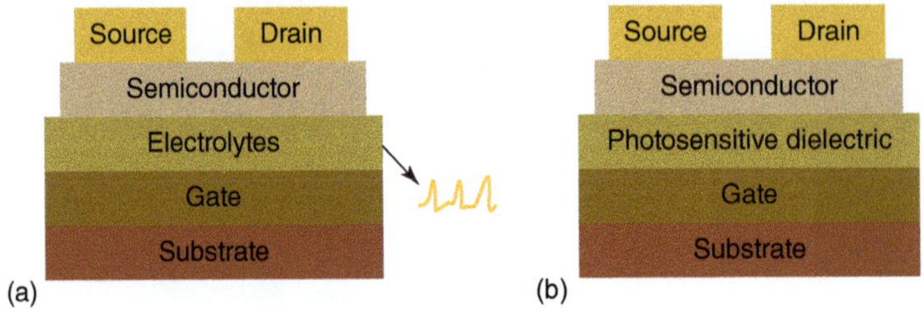

Figure 8.14 Schematic diagrams of (a) TSAS and (b) OMTs.

8.4 Other Memory Devices

Apart from the well-known RRAMs and transistor-structured memory devices, some novel types of memory devices have also been reported recently. A typical example is transistor-structured artificial synapse (TSAS), as shown in Figure 8.14a, which has a simultaneous signal transmission and learning, where a voltage spike applied to a gate electrode modulates the channel conductance.

Jin et al. reported ionic liquids (ILs, DEME-TFSI)-gated WO_3 transistors capable of achieving artificial synapses [55]. Such transistors successfully realized synaptic functions of both short-term and long-term plasticity (STP and LTP). One possible reason for this is that both electrostatic and electrochemical effects can be induced due to the electric double-layer effect and hydrolysis reaction, respectively, resulting in a modulation of channel conduction. As shown in Figure 8.15, at V_G less than the threshold value (V_T) of the hydrolysis reaction, $DEME^+$ ($TFSI^-$) ions in ILs accumulate at the IL/channel (IL/gate) interfaces to form an electric double layer, causing an effective decrease in channel resistance. The release of V_G relaxes the cations and anions in ILs, causing them to mix within seconds, resulting in a reversion of the channel resistance to its pristine value. At $V_G > V_T$, water molecules contaminated by ILs can dissociate into protons and hydroxyls, causing the insertion of the protons into the absent A-site of WO_3 via the strong electric field of electric double layer and a valence change from W^{6+} to W^{5+} to form a stable H_xWO_3 phase. This nonvolatile resistance state from the H_xWO_3 phase is maintained even after the removal of the applied voltage, and the ionic modulation of the channel resistance resembles the transmission process in biological synapses. This means that while STP can be mimicked with the help of electrolyte ion dynamics under low electrical bias, LTP can also be realized by the proton insertion in WO_3 under a high electrical bias. In addition to the demonstration of artificial synaptic activity, other essential synaptic behaviors, such as paired-pulse facilitation (PPF), the depression and potentiation of synaptic weight, and spike-timing-dependent plasticity, were also achieved in the reported transistors.

Electrolyte-gated transistors have shown great potential in TSAS. This is due to the ion-conducting nature of most electrolytes with ions inside them moving relatively slowly, restricting their potential applications in high-speed-required fields. However, on the other hand, this means that electrolyte-gated transistors can effectively

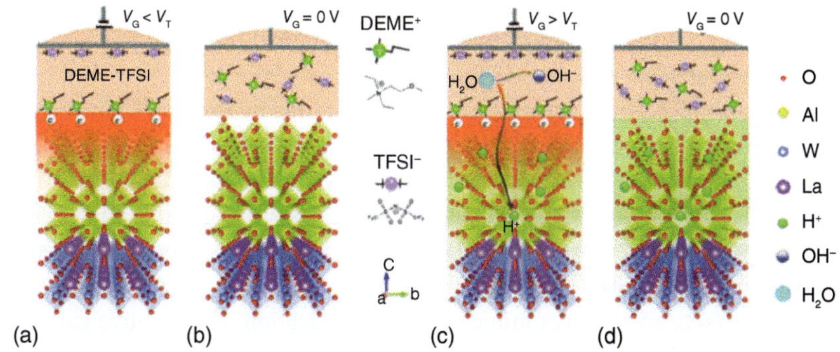

Figure 8.15 Schematic diagrams of (a) and (c) ion migration, and (b) and (d) relaxation for the gating dynamics. Source: Reproduced from Yang et al. [55]/with permission of John Wiley & Sons.

utilize ions in the electrolyte dielectric layer to modulate the channel conductivity of the devices. The migration of ions in the electrolytes under an applied electric field is very similar to the release of neurotransmitters in response to an action potential. Analogous to biological synapses, the gate electrode of such transistors can be regarded as the presynaptic membrane, while the semiconductor channel is considered as the postsynaptic membrane, and the channel conductance is regarded as the synaptic weight. When operating at electrostatic regimes, the gate-induced channel conductance change is reversible, providing the basis for mimicking STP. Operating in an electrochemical doping mode results in a partly irreversible change in the channel conductance owing to the doping of ions into the channel, providing the basis for emulating LTP. The ability to operate at a relatively low voltage also enables the possibility of achieving synaptic devices with ultralow power consumption. Thus, electrolytes, including but not limited to polymer electrolytes, ionic liquids, and porous SiO_2, have been previously reported as capable of enabling oxide transistors to show artificial synapse [56–58].

Besides, neuromorphic devices can be powered by light, with generated charges being stored in the electrolytes. In 2018, Yang et al. reported a photoelectric neuromorphic device based on resistor-load inverters using IGZO transistors [59]. Synaptic behaviors, including excitatory post-synaptic potential (EPSP), PPF, and LTP, can be mimicked in the form of photonic EPSPs. Under photon excitations, weakly bonded oxygen ions in the IGZO can easily hop, leaving a metastable donor state, which slows down the recombination of photocarriers. Hence, the protons at chitosan/IGZO interface help the light-induced electrons stay at the interface even longer, resulting in a decay of observed channel current after the light stimulation. Also, with gate modulations, the devices showed a transition from depression mode to potentiation mode, demonstrating their potential for use in neuromorphic computing systems.

Another new type of NVM is optical memory transistors (OMTs), which integrate an optically sensitive function with classic memory devices. OMTs utilize a change in the energy band of photosensitive materials under different laser pulses to implement charge storage. Generally, OMTs can be divided into two parts, including transistors using a photosensitive channel layer and those using a photosensitive gate dielectric layer (Figure 8.14b).

Light-erasable IGZO memory transistors were reported by Zan and Chen, using a simple structure with IGZO as channel and Si_3N_4 as gate dielectric layers [60]. The use of a low-pressure mercury lamp with an emission wavelength of 254 nm and a power density of 0.1 mW cm^{-2} successfully enabled the writing and erasing behaviors of the proposed memory. While the writing behavior was mainly attributed to the electron traps caused by the natural defects within the IGZO and/or at the dielectric interface, erasing was induced by the application of a negative gate bias accompanied by UV illuminations. Ceasing the UV illumination instantaneously recovered the writing mechanism of the transistors, making them also applicable as UV detectors. Similarly, UV-erasable IGZO OMTs were also reported by Ding et al. in 2013 [61]. While the devices themselves already showed UV-induced memory behavior, the sandwiching of self-assembled Pt nanocrystals between Al_2O_3 layers further enhanced the memory behaviors. During programming, electrons from the IGZO channel were easily trapped in the nanocrystals by F–N tunneling, causing an increase in threshold voltage shift. During erasing with UV light, the photogenerated electrons could jump to the CBM of IGZO, while the generated free holes might be trapped between the valence band maximums of IGZO and Al_2O_3, injected into the nanocrystals, and/or drifted towards the channel/insulator interface, becoming trapped in the preexisting interface traps [62]. As a result, the devices showed a positive threshold voltage shift of 2.25 V after 1 ms programming at 10 V and a negative threshold voltage shift of 3.48 V after five seconds of UV erasing.

Alternatively, the use of a photosensitive material as the gate dielectric layer could also enable the realization of OMTs. Among potential materials, halide perovskites, with a bandgap tunable by the halide components, have shown great potential in optoelectronics, such as solar cells, white light-emitting diodes, photodetectors, etc. [63–65], and are hence considered a promising material, especially as the tunneling layer in OMTs.

A typical example of using perovskites as the light absorber layer in realizing OMTs was reported by Ding et al., using a structure shown in Figure 8.16 [66]. In this case, $CsPbBr_3/CsPbI_3$ blend perovskite quantum dots (PQDs) were used as a floating gate, resulting in multilevel memory cells under programming and erasing modes, respectively, by imputing high-bandwidth optical signals. Three levels of programming (3.7, 1.9, and 0.8 V) and erasing windows (−1.9, −0.6, and −0.1 V) were obtainable under electrical pulses with different bandwidths of light inputs (500–550, 575–650, and 675–750 nm), respectively. A possible reason for this was that the blend PQDs had two different bandgaps, and different amounts of photo-generated carriers could be produced for optical inputs with different wavelengths. Four memory states were demonstrated in their devices, with gaps ranging from 1.12 to 5.61 V between each state. The memory transistors also showed good data retention and programming/erasing endurance, with the memory states easily switching by inputting different optical signals.

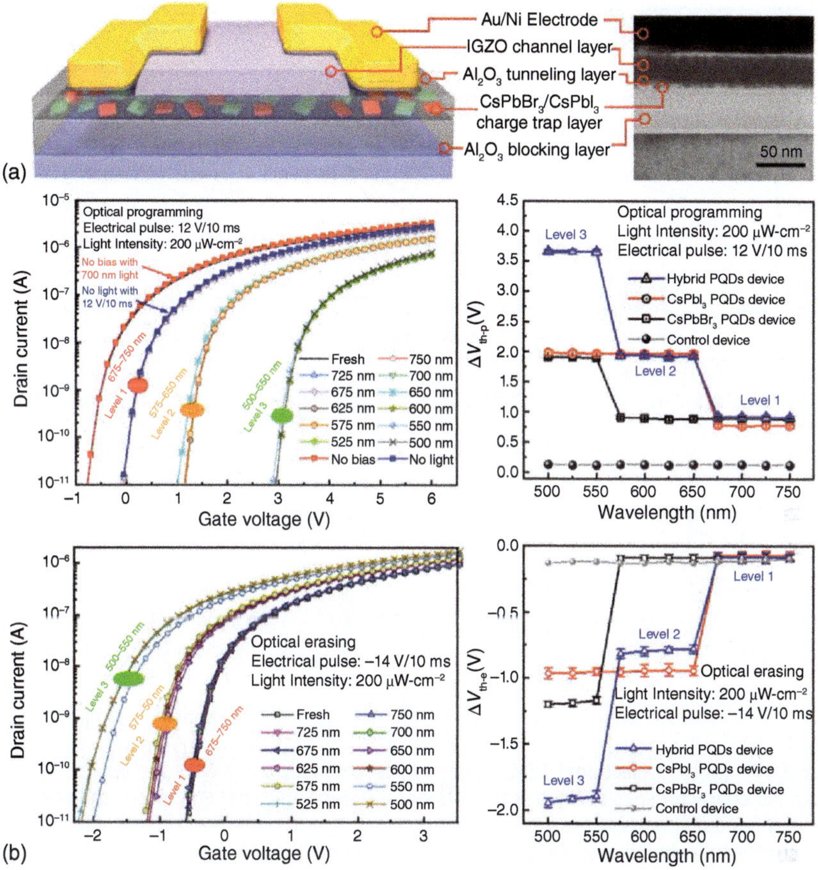

Figure 8.16 (a) Device structure and cross-sectional TEM image of IGZO memory transistors. (b) Transfer characteristics of the transistor under different light and electrical pulse conditions, showing three different levels of programming and erasing. Source: Reproduced with permission from Pei et al. [66]© 2022/IOP Publishing.

Recently, the same group also reported multifunctional transistors with stepped floating gates of PQDs possessing different bandgaps, as shown in Figure 8.17a [67]. $CsPbBr_3$ and $CsPbCl_2Br$ QDs were used as shallow and deep floating gates to trap charges and enable the memory behaviors by incident optical signals, while Al_2O_3 was used as blocking and tunneling layers, and IGZO was used as channel. As shown in Figure 8.17b, while no change of transfer characteristics was found under dark conditions with gate voltage increasing from −7 to −15 V, under light irradiations, the transfer characteristics showed an obvious change with three different transfer curves existed under a gate voltage of −13 V. The unchanged transfer characteristics under dark states might relate to the *n*-type nature of IGZO, making it difficult to generate holes under negative gate voltages (Figure 8.17c). When applying a light with a photon energy higher than 2.3 eV, photoexcitons were generated in $CsPbBr_3$ QDs, allowing the photoinduced electrons to tunnel through the 7 nm Al_2O_3 layer

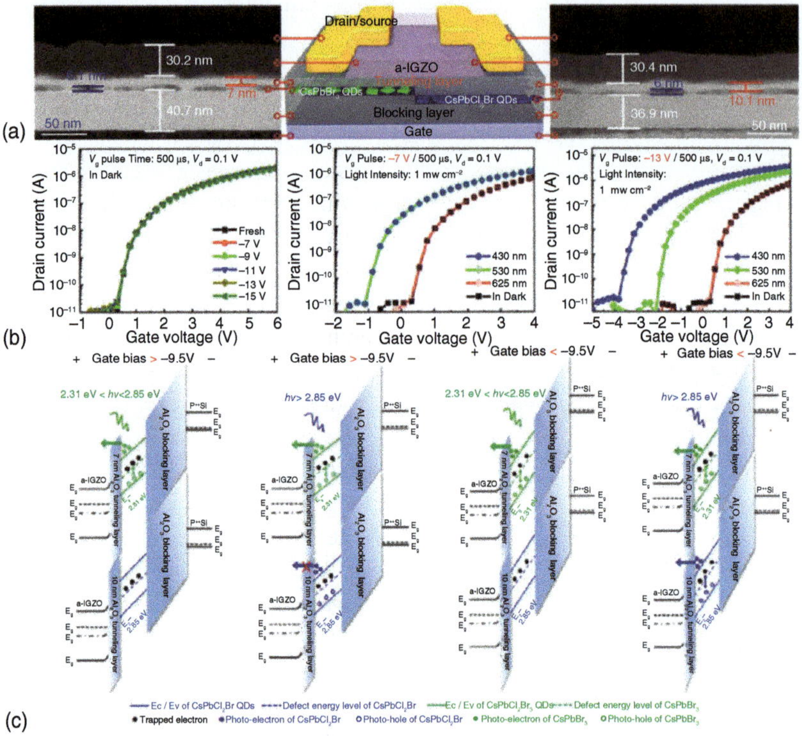

Figure 8.17 (a) Cross-sectional TEM images and a schematic diagram of the device. (b) Transfer characteristics of the transistor under different light and gate pulse conditions. (c) Energy band diagrams of the devices under different conditions. Source: Reproduced with permission from Pei et al. [67] © 2022 American Chemical Society.

to the channel under a relatively small gate voltage (i.e. −7 V), resulting in hole trapping in the floating gate and a left shift of the transfer curves. Further increasing the photon energy of light to higher than 2.8 eV generated excitons in both $CsPbBr_3$ and $CsPbCl_2Br$ QDs. In such cases, the small gate voltage only caused electron tunneling in the 7 nm Al_2O_3 tunneling layer because of the smaller tunneling energy barrier, and thus no further shift of transfer characteristics. Applying a higher gate voltage of −13 V led to electron tunneling in both 7 and 10.1 nm Al_2O_3 layers due to the stronger electric field, and thus a further left shift of transfer characteristics.

In addition to PQDs, OMTs can also be achieved by combining 2D perovskites with oxide semiconductors. Recently, Liao et al. reported flexible ambipolar SnO-based NVM transistors with multibit memory behaviors and an ultralong retention time using an Al_2O_3/2D $(PEA)_2PbI_4$ heterostructure dielectric layer [68]. The memory features were mainly attributed to the suppressed gate leakage by Al_2O_3 and hopping-like ionic transport in $(PEA)_2PbI_4$ with varying activation energy under different light intensities. These devices also showed good flexibility and long-term stability in atmospheric conditions and could detect light in a broadband spectrum ranging from 375 to 1064 nm with an ultrahigh specific detectivity of 2.7×10^{15} Jones.

To conclude, TSAS as an innovation in classic memory devices can implement fundamental synaptic functions with ultralow power consumption. Photonic TSAS, as a particular example, not only emulates retinal functions but also enables the development of conformal visual-perception systems. However, the realization of TSAS highly relies on the use of electrolytes, which are mostly in solution-based states, making thickness and shape control relatively difficult and hence less compatible with traditional all-solid-state electronics. Solid-state electrolytes, especially porous SiO_2, have shown potential in TSAS owing to their good compatibility with existing CMOS technologies, but further research is still required before pursuing commercial applications.

OMTs, on the other hand, validate a feasible route towards light-controlled cognitive nanodevices, with the irradiation history representing the memory process and nonvolatile state retention ensuring data preservation of data over time. Research on this novel type of memory device was not long but did attract significant attention until now. However, for OMTs with an *n*-type oxide semiconductor as channel layer, the relatively large bandgap of the material limits the choice of light sources. *P*-type oxide semiconductors with a smaller bandgap could be used to overcome this limitation, but the performance of *p*-type oxide TFTs is still inferior to that of *n*-type ones. For OMTs with photosensitive gate dielectric layers, perovskites are commonly used as the light absorbing layer, but they are highly sensitive to water and oxygen molecules in the air, regardless of their dimensions. Hence, further development in this field may require stability enhancement through the introduction of a passivation layer and/or the exploration of new types of materials.

References

1 Wong, H., Lee, H., Yu, S. et al. (2012). Metal–oxide RRAM. *Proc. IEEE* 100 (6): 1951–1970.

2 Carlos, E., Branquinho, R., Martins, R. et al. (2021). Recent progress in solution-based metal oxide resistive switching devices. *Adv. Mater.* 33 (7): 2004328.

3 Wang, Z., Xu, H., Zhang, L. et al. (2013). Performance improvement of resistive switching memory achieved by enhancing local-electric-field near electromigrated ag-nanoclusters. *Nanoscale* 5 (10): 4490–4494.

4 Zhuge, F., Peng, S., He, C. et al. (2011). Improvement of resistive switching in cu/zno/pt sandwiches by weakening the randomicity of the formation/rupture of cu filaments. *Nanotechnology* 22 (27): 275204.

5 Chuang, M., Chen, Y., Su, Y. et al. (2014). Negative differential resistance behavior and memory effect in laterally bridged ZnO nanorods grown by hydrothermal method. *ACS Appl. Mater. Interfaces* 6 (8): 5432–5438.

6 Jeong, D., Thomas, R., Katiyar, R. et al. (2012). Emerging memories: resistive switching mechanisms and current status. *Rep. Progr. Phys.* 75 (7): 076502.

7 Argall, F. (1968). Switching phenomena in titanium oxide thin films. *Solid-State Electron.* 11 (5): 535–541.

8 Jeong, D., Schroeder, H., and Waser, R. (2007). Coexistence of bipolar and unipolar resistive switching behaviors in a Pt/TiO_2/Pt stack. *Electrochem. Solid-State Lett.* 10 (8): G51.

9 Kwon, D., Kim, K., Jang, J. et al. (2010). Atomic structure of conducting nanofilaments in TiO_2 resistive switching memory. *Nat. Nanotechnol.* 5 (2): 148–153.

10 Kim, H., Lee, H., Seok, Y. et al. (2011). Improved endurance of resistive switching TiO_2 thin film by hourglass shaped magnéli filaments. *Appl. Phys. Lett.* 98 (26): 262901.

11 Chang, W., Lai, Y., Wu, T. et al. (2008). Unipolar resistive switching characteristics of zno thin films for nonvolatile memory applications. *Appl. Phys. Lett.* 92 (2): 022110.

12 Hu, W., Zou, L., Chen, X. et al. (2014). Highly uniform resistive switching properties of amorphous InGaZnO thin films prepared by a low temperature photochemical solution deposition method. *ACS Appl. Mater. Interfaces* 6 (7): 5012–5017.

13 Xu, N., Liu, L., Sun, X. et al. (2008). Characteristics and mechanism of conduction/set process in tin/zno/pt resistance switching random-access memories. *Appl. Phys. Lett.* 92 (23): 232112.

14 Wang, X., Qian, H., Guan, L. et al. (2017). Influence of metal electrode on the performance of zno based resistance switching memories. *J. Appl. Phys.* 122 (15): 154301.

15 Kim, D., Seo, S., Ahn, S. et al. (2006). Electrical observations of filamentary conductions for the resistive memory switching in nio films. *Appl. Phys. Lett.* 88 (20): 202102.

16 Son, J. and Shin, Y. (2008). Direct observation of conducting filaments on resistive switching of nio thin films. *Appl. Phys. Lett.* 92 (22): 222106.

17 Park, G., Li, X., Kim, D. et al. (2007). Observation of electric-field induced Ni filament channels in polycrystalline niox film. *Appl. Phys. Lett.* 91 (22): 222103.

18 Kinoshita, K., Tsunoda, K., Sato, Y. et al. (2008). Reduction in the reset current in a resistive random access memory consisting of niox brought about by reducing a parasitic capacitance. *Appl. Phys. Lett.* 93 (3): 033506.

19 Zhou, G., Xiao, L., Zhang, S. et al. (2017). Mechanism for an enhanced resistive switching effect of bilayer NiO_x/TiO_2 for resistive random access memory. *J. Alloys Compd.* 722: 753–759.

20 Yin, M., Zhou, P., Lv, H. et al. (2008). Enhancement of endurance for Cu_xO based rram cell. In: *2008 9th International Conference on Solid-State and Integrated-Circuit Technology*,IEEE pp. 917–920.

21 Hsu, C. and Lin, Y. (2019). Electrode dependence of resistive switching characteristics in copper(ii) oxide memory devices. *Semicond. Sci. Technol.* 34 (7): 075012.

22 Chen, B., Lu, Y., Gao, B. et al. (2011). Physical mechanisms of endurance degradation in TMO-rram. In: *2011 International Electron Devices Meeting*, IEEE pp. 12.13.11–12.13.14.

23 Gao, B., Kang, J., Zhang, H. et al. (2010). Oxide-based RRAM: physical based retention projection. In: *2010 Proceedings of the European Solid State Device Research Conference*,IEEE, pp. 392–395.

24 Lee, H., Chen, P., Wu, T. et al. (2008). Low power and high speed bipolar switching with a thin reactive Ti buffer layer in robust HfO_2 based RRAM. In: *2008 IEEE International Electron Devices Meeting*, IEEE, pp. 1–4.

25 Kim, T., Baek, I., and Cho, W. (2018). Resistive switching characteristics of solution-processed Al–Zn–Sn–O films annealed by microwave irradiation. *Solid-State Electron.* 140: 122–128.

26 Hansen, M., Zahari, F., Kohlstedt, H. et al. (2018). Unsupervised hebbian learning experimentally realized with analogue memristive crossbar arrays. *Sci. Rep.* 8 (1): 8914.

27 Pereira, M., Deuermeier, J., Nogueira, R. et al. (2020). Noble-metal-free memristive devices based on igzo for neuromorphic applications. *Adv. Electron. Mater.* 6 (10): 2000242.

28 Jiang, S., Nie, S., He, Y. et al. (2019). Emerging synaptic devices: from two-terminal memristors to multiterminal neuromorphic transistors. *Mater. Today Nano* 8: 100059.

29 Jiang, J., Guo, J., Wan, X. et al. (2017). 2D MoS_2 neuromorphic devices for brain-like computational systems. *Small* 13 (29): 1700933.

30 Liu, D., Shi, Q., Dai, S. et al. (2020). The design of 3D-interface architecture in an ultralow-power, electrospun single-fiber synaptic transistor for neuromorphic computing. *Small* 16 (13): 1907472.

31 Li, Y., Lu, J., Shang, D. et al. (2020). Oxide-based electrolyte-gated transistors for spatiotemporal information processing. *Adv. Mater.* 32 (47): 2003018.

32 Yang, J. and Xia, Q. (2017). Battery-like artificial synapses. *Nat. Mater.* 16 (4): 396–397.

33 Suresh, A., Novak, S., Wellenius, P. et al. (2009). Transparent indium gallium zinc oxide transistor based floating gate memory with platinum nanoparticles in the gate dielectric. *Appl. Phys. Lett.* 94 (12): 123501.

34 Park, B., Cho, K., Kim, S. et al. (2010). Nano-floating gate memory based on ZnO thin-film transistors and al nanoparticles. *Solid State Sci.* 12 (12): 1966–1969.

35 Naqi, M., Kwon, N., Jung, S. et al. High-performance non-volatile InGaZnO based flash memory device embedded with a monolayer Au nanoparticles. *Nanomaterials* 11 (5): 1101.

36 Hu, Q., Lee, S., Baek, Y. et al. (2013). Non-volatile nano-floating gate memory with Pt-Fe_2O_3 composite nanoparticles and indium gallium zinc oxide channel. *J. Nanoparticle Res.* 15 (2): 1435.

37 Oruc, F., Cimen, F., Rizk, A. et al. (2012). Thin-film ZnO charge-trapping memory cell grown in a single ALD step. *IEEE Electron Device Lett.* 33 (12): 1714–1716.

38 Kim, H., Kang, C., Kim, S. et al. (2019). Characterization on the operation stability of mechanically flexible memory thin-film transistors using engineered ZnO charge-trap layers. *J. Phys. D Appl. Phys.* 52 (32): 325106.

39 Zhang, N., Zhao, W., Zhang, X. et al. (2022). Transparent floating gate memory based on ZnO thin film transistor with controllable memory window. *IEEE J. Electron Devices Soc.* 10: 275–280.

40 Yin, H., Kim, S., Lim, H. et al. (2008). Program/erase characteristics of amorphous gallium indium zinc oxide nonvolatile memory. *IEEE Trans. Electron Devices* 55 (8): 2071–2077.

41 Yin, H., Kim, S., Kim, C.J. et al. (2008). Fully transparent nonvolatile memory employing amorphous oxides as charge trap and transistor's channel layer. *Appl. Phys. Lett.* 93 (17): 172109.

42 Hwang, E.S., Kim, J.S., Jeon, S.M. et al. (2018). $In_2Ga_2ZnO_7$ oxide semiconductor based charge trap device for nand flash memory. *Nanotechnology* 29 (15): 155203.

43 Su, N., Wang, S., and Chin, A. (2010). A nonvolatile InGaZnO charge-trapping-engineered flash memory with good retention characteristics. *IEEE Electron Device Lett.* 31 (3): 201–203.

44 Yuan, H., Shimotani, H., Tsukazaki, A. et al. (2010). Hydrogenation-induced surface polarity recognition and proton memory behavior at protic-ionic-liquid/oxide electric-double-layer interfaces. *J. Am. Chem. Soc.* 132 (19): 6672–6678.

45 Wang, Z., Xu, H., Zhang, Y. et al. (2022). Low temperature (<150 °c) annealed amorphous indium-gallium-tin oxide (IGTO) thin-film for flash memory application. *Appl. Surf. Sci.* 605: 154614.

46 Yoon, S., Yang, S., Byun, C. et al. (2010). Fully transparent non-volatile memory thin-film transistors using an organic ferroelectric and oxide semiconductor below 200 °C. *Adv. Funct. Mater.* 20 (6): 921–926.

47 Koo, J., Yang, J., Cho, B. et al. (2018). Nonvolatile electric double-layer transistor memory devices embedded with Au nanoparticles. *ACS Appl. Mater. Interfaces* 10 (11): 9563–9570.

48 Kim, B., Byun, C., Yoon, S. et al. (2011). Oxide-thin-film-transistor-based ferroelectric memory array. *IEEE Electron Device Lett.* 32 (3): 324–326.

49 Kim, M. and Lee, J. (2019). Ferroelectric analog synaptic transistors. *Nano Lett.* 19 (3): 2044–2050.

50 Kaneko, Y., Nishitani, Y., Tanaka, H. et al. (2011). Correlated motion dynamics of electron channels and domain walls in a ferroelectric-gate thin-film transistor consisting of a Zno/$Pb(Zr,Ti)O_3$ stacked structure. *J. Appl. Phys.* 110 (8): 084106.

51 Mo, F., Tagawa, Y., Jin, C. et al. (2020). Low-voltage operating ferroelectric FET with ultrathin IGZO channel for high-density memory application. *IEEE J. Electron Devices Soc.* 8: 717–723.

52 Lehninger, D., Ellinger, M., Ali, T. et al. (2021). A fully integrated ferroelectric thin-film-transistor – influence of device scaling on threshold voltage compensation in displays. *Adv. Electron. Mater.* 7 (6): 2100082.

53 Sun, C., Zheng, Z., Han, K. et al. (2021). Temperature-dependent operation of ingazno ferroelectric thin-film transistors with a metal-ferroelectric-metal-insulator- semiconductor structure. *IEEE Electron Device Lett.* 42 (12): 1786–1789.

54 Hasan, M., Mohit, I.M. et al. (2022). Improvement of amorphous ingazno thin-film transistor with ferroelectric ZrO_x/HfZrO gate insulator by 2 step sequential Ar/O_2 treatment. *IEEE Electron Device Lett.* 43 (5): 725–728.

55 Yang, J., Ge, C., Du, J. et al. (2018). Artificial synapses emulated by an electrolyte-gated tungsten-oxide transistor. *Adv. Mater.* 30 (34): 1801548.

56 Ling, H., Koutsouras, D.A., Kazemzadeh, S. et al. (2020). Electrolyte-gated transistors for synaptic electronics, neuromorphic computing, and adaptable biointerfacing. *Appl. Phys. Rev.* 7 (1): 011307.

57 Huang, H., Ge, C., Liu, Z. et al. (2021). Electrolyte-gated transistors for neuromorphic applications. *J. Semicond.* 42 (1): 013103.

58 He, Y., Yang, Y., Nie, S. et al. (2018). Electric-double-layer transistors for synaptic devices and neuromorphic systems. *J. Mater. Chem. C* 6 (20): 5336–5352.

59 Yang, Y., He, Y., Nie, S. et al. (2018). Light stimulated IGZO-based electric-double-layer transistors for photoelectric neuromorphic devices. *IEEE Electron Device Lett.* 39 (6): 897–900.

60 Chen, W. and Zan, H. (2012). High-performance light-erasable memory and real-time ultraviolet detector based on unannealed indium–gallium–zinc–oxide thin-film transistor. *IEEE Electron Device Lett.* 33 (1): 77–79.

61 Cui, X., Chen, S., Ding, S. et al. (2013). Unique UV-erasable In-Ga-Zn-O TFT memory with self-assembled Pt nanocrystals. *IEEE Electron Device Lett.* 34 (8): 1011–1013.

62 Oh, H., Yoon, S.-M., Ryu, M.K. et al. (2011). Transition of dominant instability mechanism depending on negative gate bias under illumination in amorphous In-Ga-Zn-O thin film transistor. *Appl. Phys. Lett.* 98 (3): 033504.

63 Wang, H., Li, H., Cai, W. et al. (2020). Challenges and strategies relating to device function layers and their integration toward high-performance inorganic perovskite solar cells. *Nanoscale* 12 (27): 14369–14404.

64 Cai, W., Li, H., Li, M. et al. (2021). Opportunities and challenges of inorganic perovskites in high-performance photodetectors. *J. Phys. D Appl. Phys.* 54 (29): 293002.

65 Mo, Q., Shi, Y., Cai, W. et al. (2022). Opportunities and challenges of low-dimensional hybrid metal halides in white light-emitting diodes. *J. Phys. D Appl. Phys* 55 (33): 333003.

66 Pei, J., Wu, X., Huo, J. et al. (2020). High-bandwidth light inputting multi-level photoelectric memory based on thin-film transistor with a floating gate of $CsPbBr_3$/$CsPbI_3$ blend quantum dots. *Nanotechnology* 32 (9): 095204.

67 Pei, J., Wu, X., Liu, W.-J. et al. (2022). Photoelectric logic and in situ memory transistors with stepped floating gates of perovskite quantum dots. *ACS Nano* 16 (2): 2442–2451.

68 Tian, Q., Hong, R., Liu, C. et al. (2022). Flexible SnO optoelectronic memory based on light-dependent ionic migration in ruddlesden–popper perovskite. *Nano Lett.* 22 (1): 494–500.

Index

Metal Oxide Semiconductors: Synthesis, Properties, and Devices,
First Edition. Zhigang Zang, Wensi Cai, and Yong Zhou.

r

s

t